Frontiers in Antimicrobial Agents

(Volume 2)

Current Trends in the Identification and Development of Antimicrobial Agents

Edited by

M. Amin-ul Mannan

Department of Molecular Biology and Genetic Engineering, School of Bioengineering and Biosciences, Lovely Professional University, Phagwara, Punjab-144011, India

Division of Infectious Disease, The Lundquist Institute for Biomedical Innovation at Harbor- University of California Los Angeles (UCLA) Medical Center, Los Angeles, California, USA

&

Gaurav Kumar

Department of Microbiology, School of Bioengineering and Biosciences, Lovely Professional University, Phagwara, Punjab-144011, India

Frontiers in Antimicrobial Agents

(Volume 2)

Current Trends in the Identification and Development of Antimicrobial Agents

Editors: M. Amin-ul Mannan and Gaurav Kumar

ISSN (Online): 2452-2589

ISSN (Print): 2452-2570

ISBN (Online): 978-981-5080-05-6

ISBN (Print): 978-981-5080-06-3

ISBN (Paperback): 978-981-5080-07-0

First published in 2023.

Bentham Science Publishers Pte. Ltd.
80 Robinson Road #02-00
Singapore 068898
Singapore
Email: subscriptions@benthamscience.net

BENTHAM SCIENCE

CONTENTS

FOREWORD

M. Amin-ul Mannan and Gaurav Kumar have done a wonderful job in compiling the book entitled Frontiers in Antimicrobial Agents Vol. 2, Current Trends in Identification and Development of Antimicrobial Agents. All chapters in the book are written by well-experienced researchers; however, keeping the convenience of readers in mind, simple language is used throughout the book. The book is also well supported with diagrams, photographs, tables, *etc.*, to make the content lucid and interesting.

I hope that this book will help the readers to update and upgrade their knowledge and skills in the field of anti-microbial drug discovery and development.

Further, I appreciate the editors, Dr. Mannan and Dr. Kumar, for taking the initiative to compile recent developments in the field of a very important global problem. Further, I also wish all the best to the contributors. I hope that this book will make its contribution to the field of anti-microbial drug discovery and development.

Manish Kumar
Associate Professor
Head of the Department of Biophysics
University of Delhi-South Campus
South Moti Bagh, New Delhi-110021
India

PREFACE

Microorganisms are ubiquitous organisms; we cannot imagine our existence without them. These are one of the world's greatest living populations. Some bacteria can harm people, others enable us to live on this planet, and a few have no effect on us. Microbial infections have long been one of the primary causes of morbidity and mortality around the world. Despite the fact that the discovery of a wide range of anti-microbial medicines has aided in the control of the impact of microbial diseases, the growth of drug resistance among bacteria has lowered the efficiency of anti-microbial agents, paving the need for novel anti-microbial treatments.

Keeping the recent developments in context, this book **"Frontiers in Antimicrobial Agents Vol. 2, Current Trends in the Identification and Development of Antimicrobial Agents"** is an effort to bring together the experience of eminent researchers to discuss various aspects of anti-microbial research. Different chapters of this book cover various aspects of anti-microbial drugs, drug resistance, emerging and re-emerging microbial diseases, drug discovery, sources of novel anti-microbial compounds, and future needs of anti-microbial drugs. The book will be useful for microbiologists, biotechnologists, pharmacologists, clinicians, researchers in biotechnology, or any scholar of the allied fields who have an interest in working on anti-microbial drug discovery and development.

The chapters incorporated have been written in simple and straightforward language, while significant emphasis has been given to providing updated information. Keeping the researchers in mind, each chapter is well supported with the latest references, which will assist them in following the information in more detail.

We would like to express our gratitude to all the authors of the chapters for contributing their knowledge and experience to make this book possible. We also like to thank Dr. Neeta Raj Sharma, Professor and Dean, School of Bioengineering and Biosciences, Lovely Professional University, for supporting us throughout. Lastly, we want to extend our sincere thanks for various favors extended by our family members at various stages of the publication. With this, we hope that the content of this book will be helpful for the readers to improve their knowledge in the field of anti-microbial drug discovery and development.

M. Amin ul Mannan
Department of Molecular Biology and Genetic Engineering
School of Bioengineering and Biosciences
Lovely Professional University
Phagwara, Punjab-144011
India

Division of Infectious Disease
The Lundquist Institute for Biomedical
Innovation at Harbor- University of California Los Angeles (UCLA) Medical Center
Los Angeles, California
USA

&

Gaurav Kumar
Department of Microbiology
School of Bioengineering and Biosciences
Lovely Professional University
Phagwara, Punjab-144011
India

List of Contributors

Abdel Rahman M. Al-Tawaha	Department of Biological Sciences, Al Hussein Bin Talal University, Ma'an, Jordan
Abhijit Dey	Department of Life Sciences, Presidency University, Kolkata, India
Amit Joshi	Department of Bioinformatics, School of Bio-Engineering and Bio-Sciences, Lovely Professional University, Punjab, India
Anwesha Das	Drug Design and Medicinal Chemistry Laboratory, Department of Pharmaceutical Chemistry, School of Pharmaceutical Education and Research, Jamia Hamdard, New Delhi 110062, India
Arijit Nandi	Dr. B.C. Roy College of Pharmacy and Allied Health Sciences, Durgapur-713206, West Bengal, India
Arun Karnwal	Department of Microbiology, School of Bio-engineering and Bio-sciences, Lovely Professional University, Phagwara-144411, Punjab, India
Arvind Sharma	Department of Biochemistry and Molecular Biology, Howard Medical Sciences, University of Nevada, Reno, Nevada, USA
Brijendra Pratap Mishra	Department of Biochemistry, Autonomous State Medical College Bahraich, KGM University, Lucknow, UP, India
Camil Rex	Department of Botany, St Joseph's University, Bengaluru, India
Devendra Kumar Pandey	School of Basic and Applied Sciences, Raffles University, Alwar, Rajasthan, India
Gaurav Kumar	Department of Microbiology, School of Bioengineering and Biosciences, Lovely Professional University, Phagwara, 144411, Punjab, India
Harishkumar Madhyastha	Department of Cardiovascular Physiology, Faculty of Medicine, University of Miyazaki, Miyazaki, Japan
Janmejay Pandey	Department of Biotechnology, School of Life Science, Central University of Rajasthan, Kishangarh, Ajmer, Rajasthan, India
Jay Kant Yadav	Department of Biotechnology, School of Life Science, Central University of Rajasthan, Kishangarh, Ajmer, Rajasthan, India
Kasi Gopinath	School of Materials and Energy, Southwest University in Chongqing, BeiBei, China
Kaushita Banerjee	Department of Biomedical Sciences, School of Biosciences and Technology, Vellore Institute of Technology, Vellore-632014, India
Lokesh Ravi	Department of Botany, St Joseph's University, Bengaluru, India
Manoj Kumar	Department of Research, Sidra Medicine, Doha, Qatar
Megha Kaushik	Department of Microbiology, Maharshi Dayanand University, Rohtak, Haryana, India
Neha Gogia	Department of Genetics and Neuroscience, Yale School of Medicine, New Haven, CT, United States

Neha Malhotra	Tuberculosis Research Section, Laboratory of Clinical Immunology and Microbiology, National Institute of Allergy and Infectious Diseases, National Institutes of Health, Bethesda, MD, 20892, United States
Nidhi Verma	Department of Biotechnology, School of Life Science, Central University of Rajasthan, Kishangarh, Ajmer, Rajasthan, India
Nitin Pal Kalia	Department of Biological Sciences (Pharmacology & Toxicology), National Institute of Pharmaceutical Education and Research (NIPER), Hyderabad, Telangana, India
Padmanabh Dwivedi	Laboratory of Plant Tissue Culture and Stress Physiology, Department of Plant Physiology, Institute of Agricultural Sciences, Banaras Hindu University, Varanasi, India
Payal Chakraborty	Department of Pharmaceutical Technology, JIS University, 81, Nilgunj Road, Kolkata – 700109, W.B. India
Pooja Gulati	Dept. of Microbiology, Maharshi Dayanand University, Rohtak, Haryana, India
Prabhjot Kaur	School of Basic and Applied Sciences, Raffles University, Alwar, Rajasthan, India
Pramod Kumar	Department of Molecular and Integrative Physiology, University of Illinois, 61801, United States
Pranjali Chole	Department of Botany, St Joseph's University, Bengaluru, India
Prasann Kumar	Department of Agronomy, School of Agriculture, Lovely Professional University, Punjab, India
R.M. Banik	School of Biochemical Engineering, Indian Institute of Technology (BHU), Varanasi, India
Ramakrishnan Sethu	Department of Microbiology, University of Illinois, Urbana, Illinois 61801, United States
Ritu Bala	Department of Microbiology, School of Bio-engineering and Bio-sciences, Lovely Professional University, Phagwara-144411, Punjab, India
Rohan Samir Kumar Sachan	Department of Microbiology, School of Bio-engineering and Bio-sciences, Lovely Professional University, Phagwara-144411, Punjab, India
Sabia Imran	Department of Botany, St Joseph's University, Bengaluru, India
Salim Faruk Bashir	Department of Microbiology, School of Bioengineering and Biosciences, Lovely Professional University, Phagwara, Punjab, India Department of Microbiology and Biotechnology, Faculty of Science, Federal University Dutse, Jigawa State, Nigeria
Samia Khanum	Department of Botany, University of the Punjab, Lahore, Pakistan
Sandeep Sharma	Department of Medical Laboratory Sciences, Lovely Professional University, Phagwara, Punjab, India
Sanjay K. Rohaun	Department of Microbiology, University of Illinois, Urbana, Illinois 61801, United States
Shefali Gupta	Department of Industrial Microbiology, Sam Higginbottom University of Agriculture, Technology and Sciences (SHUATS), Allahabad, India

Shipa Rani Dey	Department of Agronomy, School of Agriculture, Lovely Professional University, Punjab, India
Shiv Swaroop	Department of Biochemistry, School of Life Science, Central University of Rajasthan, Kishangarh, Ajmer, Rajasthan, India
Shivani Meena	Department of Microbiology, School of Bioengineering and Biosciences, Lovely Professional University, Phagwara, 144411, Punjab, India
Souhaila Al Khodor	Department of Research, Sidra Medicine, Doha, Qatar
Sunil Krishnan G.	Department of Bioinformatics, School of Bio-Engineering and Bio-Sciences, Lovely Professional University, Punjab, India
Tabarak Malik	Department of Biochemistry, College of Medicine and Health Sciences, University of Gondar, Gondar, Ethiopia
Vijay Kumar	School of Basic and Applied Sciences, Raffles University, Alwar, Rajasthan, India
Vikas Kaushik	Department of Bioinformatics, School of Bio-Engineering and Bio-Sciences, Lovely Professional University, Punjab, India
Wei-Yu Lin	Department of Medicinal and Applied Chemistry, Kaohsiung Medical University, Kaohsiung, Taiwan

CHAPTER 1

Anti-Microbial Drugs for Emerging and Re-emerging Microbial Diseases: Paradigm in the 21st Century

Shiv Swaroop[1], **Nidhi Verma**[2], **Jay Kant Yadav**[2,*] and **Janmejay Pandey**[2,*]

[1] *Department of Biochemistry, Central University of Rajasthan, Kishangarh, Ajmer, Rajasthan, India*

[2] *Department of Biotechnology, School of Life Science, Central University of Rajasthan, Kishangarh, Ajmer, Rajasthan, India*

Abstract: Infectious diseases are one of the leading causes of human deaths worldwide. They have devastated human life for a very long time; however, with advancements in diagnostics, prevention, and therapeutic approaches, they were controlled to a manageable extent during the second half of the 20th century. It was assumed that humankind has successfully defeated the threat of infectious diseases. However, many infectious diseases have undergone re-emergence and are now a major cause of concern. Besides, there is the emergence of several new infectious diseases. These diseases are termed re-emerging and emerging infectious diseases, and collectively account for more than 20 infectious diseases. World Health Organization has identified these diseases as the predominant health hazard faced by human beings. Owing to this situation, there is an urgent need to revisit infectious diseases and make efforts towards the development of anti-microbial drugs for emerging and reemerging microbial diseases. The present book chapter aims to provide a comprehensive account of re-emerging and newly emerging infectious diseases and the paradigm shift of anti-microbial drugs in the 21st century. It is expected to provide useful insight into this important research and development area.

Keywords: Anti-microbial drugs, Emerging and Re-emerging Microbial Diseases, Genomics guided discovery, Human pathogens, Infectious diseases.

* **Corresponding authors Jay Kant Yadav and Janmejay Pandey**: Department of Biotechnology, School of Life Science, Central University of Rajasthan, Kishangarh, Ajmer, Rajasthan, India; E-mails: jaykantyadav@curaj.ac.in and janmejay@curaj.ac.in

M. Amin-ul Mannan & Gaurav Kumar (Eds.)

INTRODUCTION

Evolution of Microbial Life & Co-evolution of the Human-human Microbiome

Prokaryotic microorganisms are the first form of life to evolve on earth several million years before the evolution of more complex forms of life [1]. The solar system, including earth and moon, is estimated to have originated ~ 4.54 billion years ago [2]. In the beginning, the Earth's environment was dominated by strong radiation, volcanic eruptions, and extremely high temperatures. These conditions were too harsh for the origin and evolution of life. With the passage of time and the alteration of Earth's environment, prokaryotic microorganisms originated, evolved, and migrated to acquire a ubiquitous presence on earth [3]. The evolution of prokaryotic microorganisms has continued ever since, and now they are acknowledged to be ubiquitously present on earth. Their evolution has resulted in microorganisms being endowed with the capability of thriving in a wide variety of microenvironments, including those characterized by extremes of physio-chemical conditions.

One of the most prominent ecological niches having the presence of prokaryotic microorganisms is on and within other living being, including human beings [4, 5]. It is now well-acknowledged scientific knowledge that human being consists of ~ 10^{13} human cells and ~ 10^{14} microbial cells. Collectively, the microflora associated with the human body is referred to as human microbiome and represents an enormous genetic diversity repertoire [6]. Recent studies have defined human microbiome as a vital determinant of human health and results obtained with these studies have suggested that well-being depends upon the complex interactions that occur between the environmental factors, human nuclear DNA, mitochondrial DNA, and the microbiome [7]. The Human microbiome is also argued to have a critical role in promoting human evolution [8].

Owing to the close interaction of human beings with microorganisms, human evolution is suggested to have progressed through a long and tough contest for survival between human beings and microorganisms. In this contest, human beings have been successful in attenuating or even domesticating potential pathogens. It is apparent from the observation that microorganisms cause diseases in human being only under the following conditions: (i) the immune system is weakened; (ii) microorganisms and viruses infect a sterile part of the body.

HUMAN PATHOGENS – PROKARYOTIC MICROORGANISMS & VIRUSES

A small fraction of total prokaryotic microorganisms and viruses cause diseases in human hosts even when they are not immunocompromised or injured [9]. Further, they can exit the host body to affect other non-infected hosts [5, 6]. Despite these common characteristics, each pathogen implements a unique mechanism for eliciting the disease, making it extremely challenging to understand the underlying mechanism of different infections [6]. Yet another challenge related to infectious microbial disease is the lack of a comprehensive list of all pathogens for a very long time. Recently, Taylor and co-workers generated a list of all known human microbial pathogens; it contained entries of each microbial species known to be causing disease in human beings [10]. There are ~ 1400 species of infectious microorganisms that cause diseases to humans. These include 217 viruses and prions, 538 bacteria and rickettsia, 307 fungi, 66 protozoa and 287 helminths [10, 11]. An alarming observation from this list is the fact that ~ 175 pathogenic species have been identified for the first time only during the last 2 decades. They cause diseases that are considered emerging infectious diseases. Additionally, the last 2-3 decades have also seen the re-emergence of infectious diseases that were earlier thought to be successfully treated with available countermeasures [12, 13].

EMERGING AND RE-EMERGING INFECTIOUS DISEASES

During the second half of the 20th century, it was proposed that the long war against microbial pathogens had been won with the advancement of countermeasures, including antibiotics, vaccines, *etc*. By the 1970s, effective treatments were available against the most prevalent infectious diseases. The situation was aptly complemented by improved methods for sanitation, food handling, pasteurization, *etc*., to cut down on the variety of food & water-borne infectious diseases [11]. As a consequence, exemplary success was achieved with the mitigation of many infectious diseases. One such example was 'smallpox'; in 1979, the World Health Organization (WHO) declared smallpox the first-ever disease to be completely eradicated by technological intervention. Inferring from this success, the scientific and medical community projected that "infectious diseases would be completely eliminated in the near future" [14]. However, microorganisms have exhibited remarkable adaptability and adapted to circumvent anti-microbial therapeutics. It has resulted in: (i) non-pathogenic microorganisms becoming pathogenic, and; (ii) susceptible pathogenic microorganisms becoming resistant towards anti-microbial therapeutics. The former is referred to as "Emerging pathogens", while the latter is referred to as "Re-emerging pathogens". According to the definitions by the National Institute of Allergy and Infectious Diseases (NIAID), "**Emerging Infectious Diseases**" are infections that have newly appeared in a population or have existed previously but are rapidly increasing in incidence [15]. The Human Immunodeficiency Virus (HIV) and Severe Acute Respiratory Syndrome (SARS)-coronavirus (SARS-CoV) are typical examples. These diseases were never faced by human beings before the 1980s and 2003, respectively [16]. "**Re-Emerging Infectious Diseases**" as diseases that reappear after being on a significant decline. Influenza-A virus pandemics of 1957, and 1968 are examples of re-emerging infectious diseases. Other examples include malaria, tuberculosis, cholera, pertussis, influenza, pneumococcal disease, gonorrhea, *etc.* [17]. Another category of infectious diseases observed in the recent past is "**Deliberately Emerging Infectious Diseases**". This category includes diseases that have emerged due to the deliberate release of the infectious agent. The bioterrorist attack of anthrax on the United States of America in the year 2001 falls into this category [12]. With strict global regulations on the deliberate release of infectious pathogens, incidences of deliberately emerging infectious diseases have been kept in a strong check.

Incidences of both emerging and re-emerging diseases have had a societal and economic impact at the global levels [13]. The recent COVID-19 pandemic is another example of an emerging infectious disease. On March 11, 2020 (*i.e.*, within 3 months of the first report of its outbreak), the World Health Organization declared the COVID-19 outbreak a global pandemic [18]. It has been predicted to adversely affect the global economy over the next 12-18 months [19]. It will cause massive losses in GDP in the year 2020 [20]. The secondary effects of the COVID-19 pandemic are also reasoned to be quite severe and long-lasting [21]. Therefore, such emerging and re-emerging infectious diseases must not be overlooked as they may cause public fear, economic loss, and other adverse outcomes.

Causative Agents of Emerging and Re-Emerging Infectious Disease

Amongst the 175 new infectious microorganisms identified during the last 3 decades, ~ 50 have been identified by the Center for Disease Control and Prevention for causing life-threatening diseases and affecting infections include millions of people across the globe [22]. Amongst these, the most threatening is SARS, MERS, Ebola, Chikungunya, Avian Flu, Swine Flu, and Zika Virus. In 2018, the list of major emerging infectious diseases identified during the last 3 to 4 decades are dominated by viruses the WHO committee stated, "given their potential to cause public health emergency; there is an urgent need for R&D on Crimean-Congo hemorrhagic fever, Ebola virus and Marburg virus disease, Lassa fever, Middle East Respiratory Syndrome (MERS) and Severe Acute Respiratory

Syndrome (SARS), Henipaviral Diseases, Rift Valley Fever, and Zika Virus Disease [23]. The list of major emerging infectious diseases identified during the last 3 to 4 decades are dominated by viruses [11]. A summary of the broad taxonomic distribution of emerging and re-emerging microbial human pathogens is presented in Fig. (**1**).

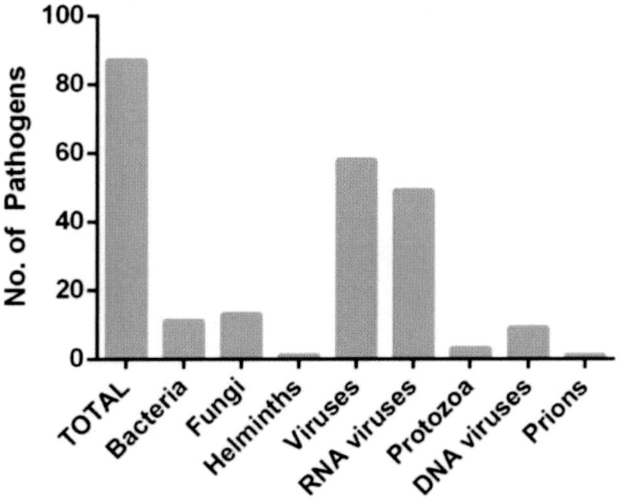

Fig. (1). Broad taxonomic distribution of emerging and re-emerging microbial human pathogens; dataset adapted from [11].

Most of the prominent emerging and re-emerging infectious pathogens have also been characterized up to their species level. A summary of species, the disease caused, and the year of identification of some of the major bacterial and emerging viral pathogens are presented in Table **1**.

Table 1. Major emerging bacterial and viral pathogens (since 1970); adapted from [22, 24].

Pathogenic Bacterial Species	Year Identified	Diseases Caused in Human Beings
Emerging Bacterial Pathogens		
Campylobacter spp.	1973	Diarrhea
Clostridium difficile	1974	Pseudo-membrane colitis & Toxic megacolon
Streptococcus bovis	1974	Endocarditis
Legionella pneumophilia	1976	Lung infection
Staphylococcus aureus	1981	Toxic shock syndrome
Escherichia coli O157:H7	1982	Hemorrhagic colitis
Chlamydia pneumoniae	1983	Lung infection

(Table 1) cont.....

Pathogenic Bacterial Species	Year Identified	Diseases Caused in Human Beings
Helicobacter pylori	1983	Gastric ulcers
Rhodococcus equi	1986	Pneumonia in immunocompromised patient
Spotted fever group Rickettsia spp.	1990	Spotted fever rickettsiosis
Vibrio choleraeO139	1992	Diarrhoea
Aerococcus spp.	1992	Urinary Tract Infections & Endocarditis
Wolbachia spp.	1995	Lymphatic filariasis
Simkania negevensis	1997	Lung infection
Actinobaculum schaalii	1997	Urinary Tract Infections
Parachlamydia acanthamoebae	1997	Lung infection
Waddlia chondrophila	2007	Miscarriage
Neoehrlichia mikurensis	2010	Neoehrlichiosis: systemic inflammation
Escherichia coli (O104:H4)	2011	hemolytic uremic syndrome
Emerging Viral Pathogens		
Ebola virus	1976	Hemorrhagic fever
Human T cell leukemia virus (HTLV) 2	1982	Adult T-cell leukemia/lymphoma
Human immunodeficiency virus (HIV)	1983	Acquired immunodeficiency syndrome
Hepatitis E virus	1988	Hepatitis
Human Herpes virus6 (HHV6)	1988	Roseola
Hepatitis C virus	1989	Hepatitis
Guanarito virus	1991	Hemorrhagic fever
Sabia virus	1994	Hemorrhagic fever
Hendra virus	1994	Hemorrhagic fever, Encephalitis, Pneumonia
Human Herpes virus 8 (HHV8)	1995	Kaposi sarcoma
Avian Influenza virus	1997	Pneumonia
Nipah virus	1999	Encephalitis
Human metapneumovirus	2001	Bronchiolitis, Pneumonia

The geographical distribution of incidences of emerging infectious diseases is yet another critical point for deliberation. Characteristically such incidences have been reported from diverse regions of the world; for example, *E.coli* strain O157:H7; and Hendra Virus were identified from Japan & Australia, while H3N2v influenza and Human Monkeypox virus were identified from the United States. Some of the most dreaded emerging infectious agents, *e.g.*, HIV, Ebola Virus, Zika Virus, and drug-resistant Tuberculosis, were identified in Central and Sub-Saharan Africa. A comprehensive account of the global geographical pattern for emerging and re-emerging infectious diseases has been provided by the

National Institute of Allergy and Infectious Diseases (NIAID), NIH - USA [25]. A graphical representation of data accessible from this report is presented in Fig. (**2**).

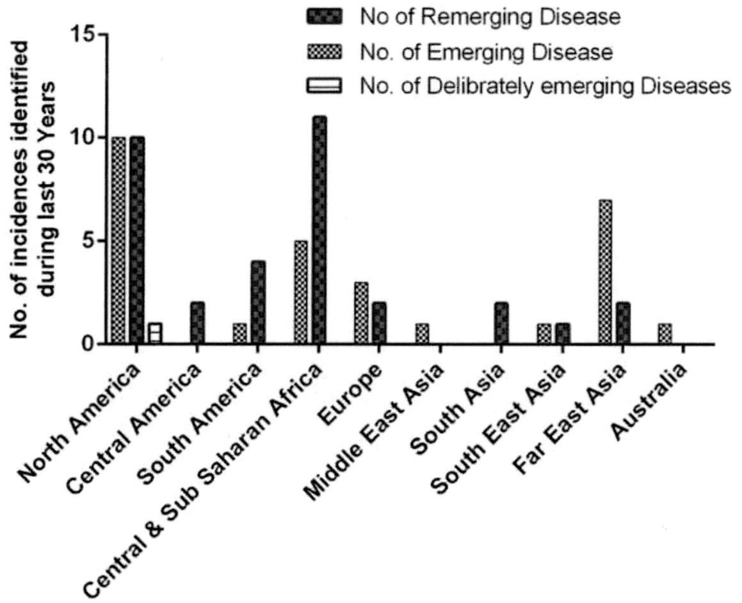

Fig. (2). Geographical distribution of first reported human cases of disease caused by major emerging, re-emerging, and deliberately emerging infectious diseases. Dataset adapted from [25].

North America and Central & Sub Saharan Africa have been the major sites for the identification of both emerging and re-emerging infectious diseases, whereas almost negligible cases of new infectious diseases have been reported from Australia & Oceania and Polar Regions. It could be argued that successful identification of emerging and re-emerging infectious may be biased and prominently influenced by the availability of adequate diagnostic tools [24].

Another important observation with regard to emerging and re-emerging infectious diseases is the fact that ~ 75% are zoonotic in nature [12, 22]. Many of the recent infections have probably emerged due to human interactions with animal hosts [12]. This observation corroborates well with increased human-wildlife interactions through activities related to domestic animals, companion animals, animal exhibits, animal-centric tourism, and above all, exotic animal cuisine. Once any pathogen finds its way into the human body, it could infect other human beings leading to incidences of disease pandemics.

Although, the increased human interaction with wildlife can explain the emergence of new infectious diseases; however, it does not explain the re-emergence of infectious diseases that were previously controlled with countermeasures. Also, it does not provide an appropriate explanation for the emergence of virulent strains of non-virulent species (*e.g.*, hemorrhagic *E.coli* strain O104:H4). This suggests that the phenomenon of emerging and re-emerging infectious diseases is much more complex than previously envisaged. A theory in this regard suggests that emerging diseases have been present in the Earth's environment; however, they were not detected in the past due to a lack of infection incidences and the non-availability of diagnostic tools [24]. The other theory is based on 'antigenic drift' and 'acquisition of anti-microbial resistance due to random mutations' [16, 24]. A more elaborate deliberation of the major determinants of emerging and re-emerging infectious diseases is presented below.

Determinants of Emerging and Re-Emerging Infectious Diseases

Infectious diseases have been emerging and re-emerging for millions of years, however, the rate of incidences has increased significantly during the past few decades [13]. This increase has been brought about by the alteration of environmental factors, anthropological activities, and increased microorganism and human interaction [26]. Present-day societies find it difficult to handle emerging and re-emerging infectious diseases [12]. The possible explanations rest in present-day infectious diseases being multifaceted and influenced by factors including **(i)** global climate changes; **(ii)** alterations in population composition **(iii)** acquisition of antibiotic resistance and antigenic drift in microorganisms; **(iv)** acquisition of pesticides resistance amongst vector pests; **(v)** change in human demographics and behavior; lack of public health services and infrastructure; **(vi)** international trade and commerce; **(vii)** increased global interdependence and business-related global travel; and **(vii)** expansion of agricultural land, *etc.* [27]. These determinants are broadly classified into 3 categories as listed in Table **2**.

Table 2. Classification of major determinants of emerging and re-emerging infectious diseases.

Major Determinants of Emerging and Re-emerging Infectious Diseases		
Environment Specific	**Host Specific**	**Pathogen Specific**
Global Climate Change Environmental pollution Alteration of ecological niches Expansion of agriculture land use Expansion of industrial activities Mass scale deforestation Altered demography of reservoir host Contamination of food or water supplies	Non-judicious antibiotics use Wildlife encroachment Increased susceptibility Altered human demographic Global trade and travel Occupational health hazard Lack of health services Poverty and social inequality	Genetic mutations Pesticide resistant

Delineation of each factor is technologically challenging; however, some of the factors are now possible with the advent of technologies such as 'pathogen genomics' [28]. Results from studies on pathogen genomics indicate that most emerging and re-emerging infectious microorganisms are genetically adapted for heightened transmission and pathogenic characteristics. Examples of such pathogens include multidrug-resistant and extensively drug-resistant (MDR, XDR) *Mycobacterium tuberculosis*, methicillin-resistant *Staphylococcus aureus* (MRSA), *etc.* [29]. Ecological changes in agricultural, industrial, and clinical practices are other major determinants for incidences of emerging and re-emerging infectious diseases. Short-term change in weather conditions has impacted emerging and re-emerging infectious diseases through alteration of vector behavior or changed behavior of the human being [30]. Medical treatments and clinical practices, like blood transfusion and organ transplantation and xenotransplantation, have created new ways for infectious microorganisms to spread [31]. The spread of HIV through blood transfusions during the mid-1980s to 1990s is a prime example of the same [32].

Altered human demographics, including recreational activities, wildlife poaching, bush-meat harvesting, and sexual contact with wildlife animals, have emerged as major risk factors for animal-derived emerging and re-emerging infections [33]. Changes in approaches for food processing and packaging, animal farming, and rapid dissemination of food supplies have also had a profound effect on incidences of emerging and re-emerging diseases [34]. Global trade and commerce-related international travel have emerged as yet another major demographic change-related determinant for emerging and re-emerging diseases [35]. The ease with which human beings and commercial goods, including animals, are transported around the world has significantly increased the incidences and spread of infectious agents. Poor public health measures, breakdown of public health initiatives, poor hygiene, and lack of drinkable water have also accelerated the incidences of infectious diseases [36]. Incomplete immunization programs have also contributed to the emergence of diseases that were previously not experienced [37, 38]. Last but not least, changes in human lifestyle and the prevalence of lifestyle diseases have increased human susceptibility to infectious agents.

Adaptive Microbial Physiology for Resistance to Anti-Microbials

It is widely acknowledged that microorganisms have acquired resistance to anti-microbials through spontaneous mutations, however, in the recent past, adaptive microbial physiology has also been recognized as a vital mechanism for resistance towards conventional anti-microbials [39]. It defines resistance on the basis of **(i)** Monoclonal heterogeneity; **(ii)** Microbial persistence; **(iii)** Microbial dormancy &

(iv) Microbial biofilm formation [40]. A small minority of otherwise genetically homogeneous microbial populations shows resistance to inhibitory concentrations of anti-microbial agents; and competence transformation with extracellular DNA [41]. Clonal heterogeneity has attained a very important place in the recent past in both basic as well as applied research. It is now being considered the fundamental question to address for tackling the issues pertaining to microbial resistance. According to the present understanding, the microbial population may become resistant to conventional anti-microbials through one of the four possible routes, as depicted in Fig. (**3**). It is critical to address them to evade the problems of emerging and re-emerging infectious diseases in general and bacterial infections in particular.

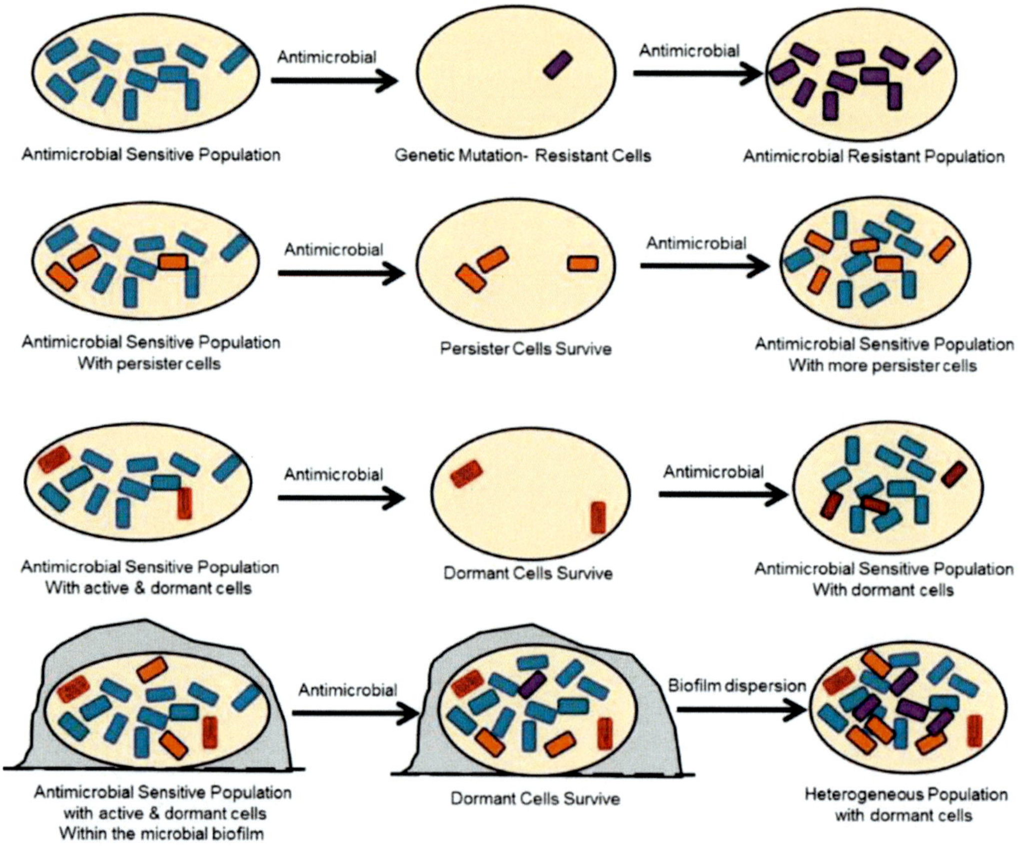

Fig. (3). Graphical representation of possible routes for microbial population becoming resistant to conventional anti-microbials. *i.e.*, due to (i) event of genetic mutation; (ii) occurrence of persister cells in the population; (iii) occurrence of dormant cells in the population; (iv) formation of microbial biofilms.

The theory of adaptive physiology is critical in explaining 'multiple drug resistance'; A phenomenon that is difficult to explain only on the basis of spontaneous genetic mutation. The discovery that resistance could be conferred by cytoplasmic factors in addition to the chromosomal mutation provided one of the first examples for further discoveries being made subsequently in this direction. Noticeably, the cytoplasmic resistance factors were later identified for conferring resistance toward multiple antibiotics.

Countermeasures for Emerging and Re-Emerging Diseases

The conventional countermeasures have suffered the loss of efficiency and efficacy against emerging and re-emerging diseases. Antimicrobial-resistant phenotype against most of the classes of antibiotics has happened within a few years of their deployment. A graphical representation showing the timeline for the deployment of antibiotics and identification of the resistant phenotype is shown in Fig. (**4**).

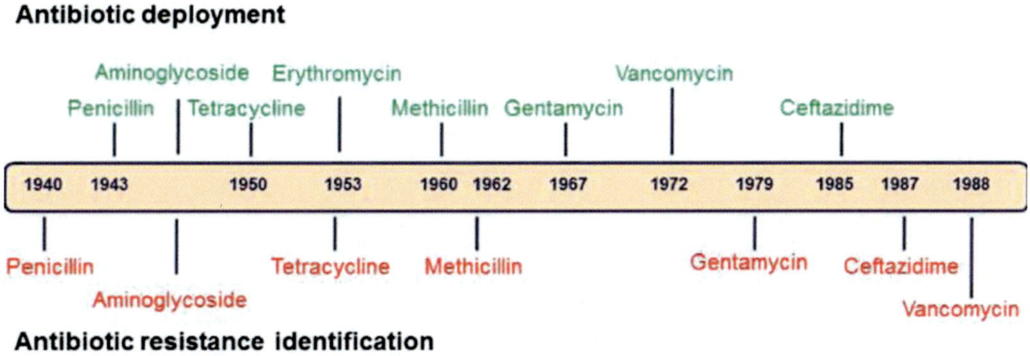

Fig. (4). Graphical representation showing the timeline of anti-microbial resistance development relative to the time of deployment of anti-microbial agents. (Dataset adapted from CDC's "Antibiotic resistance threats in the United States, 2013 [42].

One of the noticeable observations from the above graphical representation is the discovery of a resistant phenotype against a particular class of antibiotic within a few years of its deployment; for example, the identification of 'penicillin-resistant phenotype' in 1943 following its deployment in 1940. This observation suggests that microorganisms may be inherently bestowed with resistance capabilities against antibiotics. Mutations may be an accelerating force for escalating adaptation to natural and modified antibiotics derived from chemical modification of the naturally occurring antibiotics [43]. The phenomenon has also rapidly transmuted into an alarming situation due to the non-discovery of any new class of antibiotics for almost 3 and half decades now. There is an emergent need for alteration of the treatment regime for effectively managing and mitigating the emerging and re-emerging infectious diseases in the 21st Century.

INFECTIOUS DISEASES: TREATMENT PARADIGM IN THE 21ST CENTURY

The emerging and re-emerging infectious diseases are complex multi-facet problems. Consequently, the treatment regime for them also needs to be multi-facet and integrated to address different categories of major determinants of emerging and re-emerging diseases. The treatment regime must have a program consisting of **(i)** legislative regulation for improvement of global environmental conditions, sanitation, and public health conditions; food and water availability; strict regulation on trans-border transportation of plant and animal material; reduction of human-animal interactions for poaching, recreation or commercial purposes; and **(ii)** scientific advancement for characterization of mechanisms underlying the emergence and re-emergence of infectious diseases; and **(iii)** development of innovative treatment procedures [44]. The latter will be accomplished by scientific disciplines such as genomics, proteomics, structure biology, bioinformatics, cheminformatics, nanotechnology, *etc.*

Anti-Microbials for Infectious Diseases in 21st Century

While other countermeasures, including legislative regulations, are still being defined, the progress concerning the development of anti-microbials has been quite substantial. Many scientific discoveries have been made recently with regard to emerging and re-emerging infectious diseases. A notable example has been the design and development of ~ 20 anti-retroviral drugs that have significantly reduced HIV -associated morbidity and mortality [45]. Many other anti-microbials against emerging and re-emerging infections are at various stages of development. At this point, it is pertinent to highlight that the process of 'Anti-microbial Therapeutics Development for Human Application' is extremely rigorous and involves several levels of laboratory, pre-clinical and clinical trials prior to deployment for clinical application; it takes 10 – 15 years and costs about 0.2 – 1.3 Billion US$ [46]. A graphical representation showing the timeline for successful drug development is shown in Fig. (**5**).

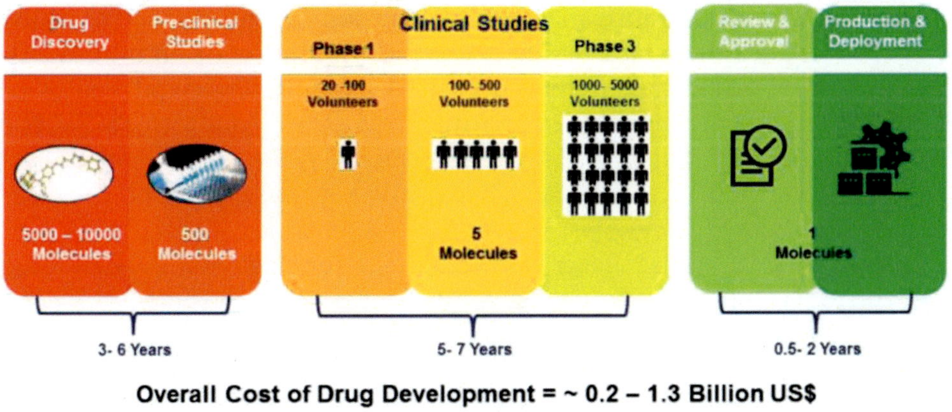

Fig. (5). Graphical representation of the typical timeline for drug discovery, development, and deployment. Adapted from [46].

The timeline shown in this figure also stands true for vaccines and anti-microbial drugs. However, with recent advancements in the fields of *in silico* drug discovery and development, the time duration for drug deployment is expected to reduce significantly. Still, the time required for a new drug to be deployed in the market would range from ~ 5 years [46, 47]. This timeline may not be acceptable for the mitigation of emerging and re-emerging diseases since they usually acquire pandemic status and require immediate therapeutic interventions [48]. The approach suitable for the discovery and development of 'anti-microbial therapeutics against emerging and re-emerging infectious diseases' should be wholistic and compatible with the expedited approval process. Many therapeutic regimes are suited to these characteristics. A brief description of progress made concerning some of these approaches is presented below:

Vaccines for Emerging and Re-Emerging Diseases

Scientific understanding of the human immune system has grown substantially during the past 4-5 decades; consequently, it is now possible to rationally design and develop vaccines against most infectious pathogens. Several neo-classical vaccine platforms based on the use of recombinant proteins, short peptide antigens, viral vector-based vaccines, genes encoding for immunogenic chimeric proteins, and pseudo-virions-based vaccines have been developed [49, 50]. The critical feature of most modern-day vaccine design approaches is the use of these novel vaccine candidates [15]. Vaccines being developed on these new approaches are being referred to as rationally designed vaccines and include **(i)** Viral vectors-based vaccines; **(ii)** Virus-like particles; **(iii)** Nucleic acid vaccines; and **(iv)** Subunit vaccines. A brief description of these platforms is presented here:

Viral vectors-based vaccines: Some viral families could be effectively used as vectors for expressing the antigen in the background of an active heterologous virus. This stimulates immune responses for the generation of both humoral and cellular immunity [51]. One of the successful examples of viral vector-based vaccines is based on the vesicular stomatitis virus (VSV) platform. VSV is non-pathogenic to humans; therefore, a recombinant VSV virus harboring any antigen-expressing gene can induce an immune response without causing any disease to the human being [52]. Another successful example is Simian Adenoviruses (SAV); its E1 gene is deleted and replaced with a heterologous antigen gene [53].

Virus-like particles: These are assemblies of viral proteins that mimic the native virus. They do not contain viral genetic materials and provide a safe and noninfectious alternative to traditional live virus-based vaccines [54]. A unique property of VLPs is their particulate nature which is extremely useful in the development of effective and affordable recombinant vaccines [54]. They elicit a strong immunogenic response like those elicited by the whole virus vaccines. The structures of VLP are unique and allow for the insertion of foreign antigenic sequences [55, 56]. VLPs have been used to develop the FDA-approved vaccines for Hepatitis B and Human Papillomavirus; these vaccines are now commercially available. Despite the advantages associated with VLP, this platform faces significant challenges in translation from laboratory models to clinical deployment. Therefore, to date, only a very few VLP vaccines have been adopted into the vaccination schedule [55].

Nucleic acid vaccines: Nucleic acid vaccines have been proposed as a safer alternative to live and inactivated vaccines. They contain only DNA or mRNA which is taken up by cells and converted to proteins by using host cells' transcription and translational machinery. Despite their relatively nascent nature, these vaccines are being applied to a wide range of infectious and malignant diseases [57]. DNA vaccines have been successfully evaluated against many emerging viruses like Dengue [58], CHIKV [59], and EBOV [60] using different animal models. More recently, mRNA or vaccine development has been getting great attention due to its high potency, low manufacturing cost, and safe administration mode [61]. In several preclinical and clinical studies, mRNA-based vaccines have shown a safe and durable immune response in animal models and humans [62].

Subunit vaccines: Subunit vaccines consist of synthetic peptides or purified recombinant proteins, which include well-defined epitopes of a viral protein. This approach presents antigens to the immune system without introducing viral particles. The major advantages of subunit vaccines include ease of production, upscaling at the time of high demand, stability, high shelf life, and no risk associated with the use of the whole virus, *etc*. An alternative method of making a subunit vaccine is based on putting the antigen coding gene into a viral vector or attenuated bacterium that serves as a recombinant vaccine (called a recombinant subunit vaccine) [63]. There are few subunit vaccines presently in use, while others are being developed against difficult-to-vaccinate viruses, *e.g*., Ebola virus [64, 65] and HIV [66].

Genomics Guided Discovery of New Classes of Antibiotics

Microbial bioactive secondary metabolites are an important source of anti-microbial drugs. To date, ~ 20,000 microbial metabolites have been isolated and characterized for diverse biological activities. However, the past 3 decades have experienced an abrupt decline in the discovery of new classes of antibiotics. This phenomenon is attributed to several factors, including technical challenges related to isolation and purification of antibiotics; extremely low concentrations of antibiotics produced under standard laboratory conditions; lack of scientific understanding of parameters for robust expression of genes encoding the antibiotics; and, above all, the redundant discovery of previously known natural products [67].

To harness the repertoire of natural anti-microbials, several efforts have been made for culture-independent and genomics-guided discovery of novel bioactive natural products. More than 30,000 complete bacterial genomes were sequenced and annotated by 2015 [68], and > 1,50,000 bacterial genomes are expected to be sequenced and annotated by 2022 [69]. The total number of *Actinobacterial* genomes completely sequenced to date is ~ 3074. Computational analyses of these genomes have led to the development of 'pipelines' that can be used for the prediction of non-ribosomal peptides and type I and II polyketides [70]. Several tools based on gene cluster prediction algorithms, *e.g*., SMBP, and MIBiG database [71], have been developed for predicting the genomic potential for synthesis of novel bioactive secondary metabolites. A graphical representation highlighting the process of genomics-guided discovery of novel anti-microbial therapeutics is presented in Fig. (**6**).

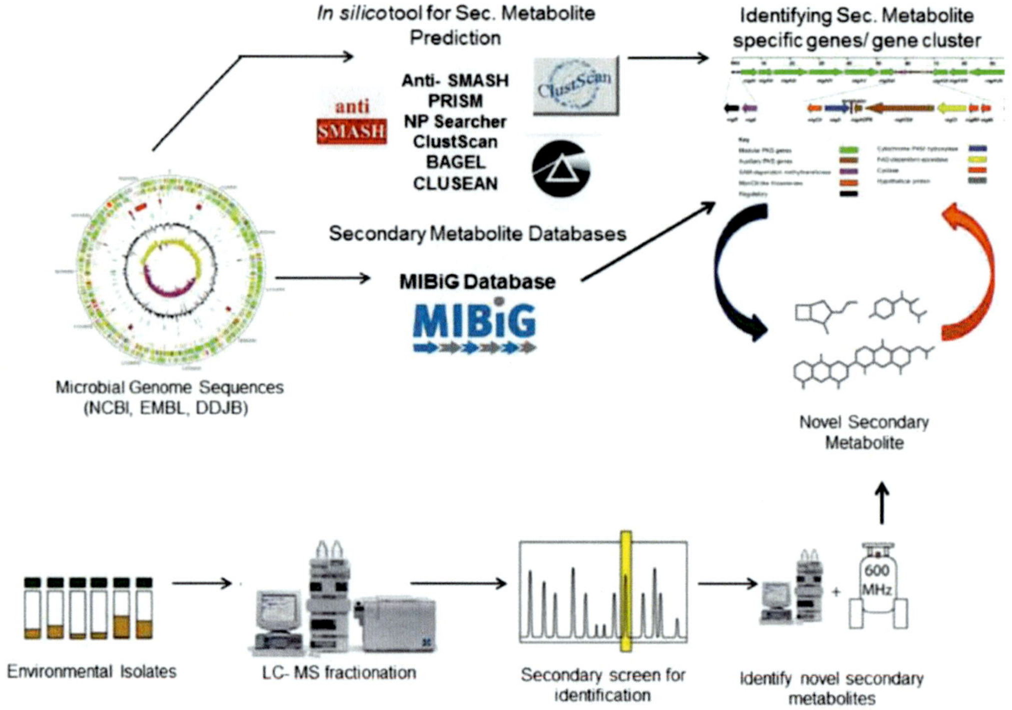

Fig. (6). Graphical representation of the genomics-guided and conventional approaches for identification novel secondary metabolite. The genomics guided approach maximizes the prospective of identifying genes/ gene clusters with the potential to synthesize novel secondary metabolites.

This approach has been successfully used for the identification of a few novel secondary metabolites during the past few years. A novel antibiotic substance was recently identified from *Allosalinactinospora lopnorensis* strain CA15-2T using this approach [72]. Noticeably, the genome analyses of *A. lopnorensis* CA15-2T showed that it contains 17 gene clusters that could potentially encode known or novel secondary metabolites [73]. Genomics-guided mining has also been implemented for the identification of secondary metabolites in fungal strains. As many as 28 novel molecules have been identified in crude extracts of *Aspergillus pachycristatus* [74].

Structure-Function Guided Discovery of Inhibitors of Drug Targets

The discovery and development of anti-microbial therapeutics have undergone a remarkable paradigm shift in the recent past. From times of serendipitous discovery of penicillin by Sir Alexander Fleming in the 1930s to the late 1990s, the discovery of anti-microbial therapeutics had remained overly dependent upon screening-based methods. This approach was quite successful in identifying a number of new anti-microbials. However, there has been a marked decline discovery, and development of new antibiotics since the 1990s. It has been realized that the use of a screening-based approach may not be sufficient for the mitigation of infectious diseases in the 21st century and beyond [75]. Fortunately, the 1990s also experienced the advent of computational approaches for explaining the physico-chemical characteristics of biological macromolecules. Since its advent has substantiated and expedited the discovery and development process of countermeasures against 'emerging and re-emerging infectious microorganisms'. Collectively, the computational approaches have resulted in the inception of a new paradigm for the search for new anti-microbial therapeutics. It integrates the basic understanding of microbial cell physiology and biochemistry with the principles of mathematic modeling, virtual screening, and molecular dynamic simulations. It has resulted in the development of several innovative platforms for carrying out hypothesis-driven screening of drugs instead of empirical labor-intensive and non-economical screening for bioactive metabolites [76]. In addition, these approaches have also paved the way for the rational designing of novel drugs, selective modification of existing drugs and, thereby, minimizing the efforts and investments required for development and eventual deployment. In contrast to the conventional approach, the 'structure-function guided' non-conventional approach focuses on target identification, determination of structural characteristics of identified targets, and virtual screening of the most potent drug(s). The only major limitation of this approach is the identification of a very large number of potentially active modulator compounds (referred to as hits). The large number of hits makes establishing prioritization for experimental validation difficult. Another interesting notion with regards to the comparison of conventional and non-conventional approaches for drug discovery is: "either of the approaches alone may not be sufficient to provide a sustainable system for discovery and development of novel antibacterial therapeutics" [75].

The key factors for the use of the neo – conventional approach is the identification of druggable microbial targets that are: (i) previously characterized as essential for the growth and survival of microorganisms; and (ii) not homologous to any essential gene within the human genome. From the point of view of implementation, this approach heavily depends upon X-ray crystallography and/ or NMR resolved structure determination of target proteins. However, crystallography can be extremely challenging, especially for structural proteins, membrane-bound proteins, and protein-protein complexes. Additionally, it requires extensive upstream experimentation that is tedious, time-consuming, and non-economical [77]. These limitations are addressed by computational biology approaches of homology modeling and *de novo* modeling as alternative routes for predicting the structure of target proteins for structure-guided drug design, discovery, and development [78].

Once a high-quality protein structure or model is available, the structure-based drug design can be applied in a variety of methods. The most common path passes through the identification of chemical starting points through virtual screening and docking of potential drug molecules. Molecules with the best docking characteristics are used for the generation of a virtual protein-drug complex, which is subsequently subjected to molecular dynamics (MD) simulation to improve the quality and confidence of *in silico* scoring. The drugs found to have stable protein-drug complexes are eventually tested in *in vitro* validation experiments [79]. Despite the challenges, this approach is quite impressive and has been successful in the identification of a number of drug molecules, including anti-microbial therapeutics. It is worth highlighting that while the conventions screening approach has a hit rate of ~ 0.30%, whereas, the structure-guided drug discovery may have a hit rate of >10% [80]. A comprehensive review of the same is presented by Staker and a co-worker [81].

Innovative Novel Alternative Anti-Microbial Therapeutics

Apart from the advancements in the above-mentioned anti-microbial therapeutic regimes, there is also a need for the development of innovative novel alternative anti-microbial therapeutics. It is essential because microorganisms can rapidly adapt and disseminate antibiotic resistance phenotypes [82]. Novel anti-microbial therapeutic approaches would be based on the convergence of material science, nanotechnology, synthetic chemistry, *etc.* Some examples of such alternative therapeutics are at various stages of development, ranging from laboratory studies to pre-clinical and clinical studies.

Nanomaterials as Potent Alternative Anti-microbial Therapeutics

Nanomaterials (including nano-fibers, nano-particles) have been reported for anti-microbial activities against a wide variety of microorganisms ranging from non-pathogenic model organisms, *e.g.*, *E.*coli and *Bacillus subtilis* [83] to pathogens, *e.g.*, Methicillin-resistant *Staphylococcus aureus* (MRSA) and *Pseudomonas aeruginosa* [84]. Conventionally, nanomaterials are categorized into 2 broad categories (*i.e.*, metallic nanomaterial and non-metallic nanomaterial). Nanomaterials of both metallic and non-metallic types in diverse shapes and forms have been proposed as potent anti-microbials [85]. Initial discoveries in this regard established silver nanoparticles (AgNPs) as important antibacterial, even against antibiotic-resistant strains [86]. AgNPs are of special interest due to their relatively easy synthesis, chemical stability, catalytic activity, high conductivity, and surface plasma resonance characteristics that make them ideal for their use as anti-microbial therapeutics. The anti-microbial characteristic of AgNPs is determined by their higher surface-to-volume ratio, providing better contact with microorganisms. Even the AgNPs synthesized with the use of green synthetic processes show noticeable anti-microbial activity against wound-infecting pathogens, including *Bacillus subtilis, Enterococcus faecalis, Staphylococcus epidermidis,* multidrug-resistant *Staphylococcus aureus* [87]. The efficacy and stability of metallic nanomaterial, including AgNPs in biomedical applications, were major concerns for a long time; the same has been addressed with the use of biopolymers based on noninflammatory capping agents like collagen and peptides. Similarly, AgNPs have been combined with other nanomaterials, such as graphene oxide (GO), having antibacterial activity for enhanced anti-microbial properties [88]. The mechanism for this synergistic effect was defined based on GO wrapping around bacterial cells, while the Ag component of the NPs killed the bacteria with its toxicity.

Nanoparticles made up of non-silver metals, *e.g*., gold-based nanomaterials or titanium-based nanomaterials, are also reported for anti-microbial activity. They exhibit strong anti-microbial effects against bacterial strains that cause wound infections [89]. Photocatalytically active TiO_2 is reported for anti-microbial activity against several microorganisms, including microbial strains that are resistant again multiple drugs [90]. Yet another development in the field of nanomaterial-based anti-microbials has been the development of ferric oxide (Fe_3O_4) NPs that exhibit anti-microbial activity and also act synergistically with other anti-microbial substances. These novel nanomaterials are adequate for biomedical applications, such as hospital tools, but also food preservation or wastewater treatment.

Just like the metallic nanomaterial, the non-metallic polymer-based nanomaterials have also been reported as potent anti-microbial therapeutics. Particularly, positively charged biomolecules, *e.g*., chitosan nanomaterials. They are positively charged and interact efficiently with negatively charged microbial cell membranes to induce cell damage or cell lyses. A few studies have shown enhanced anti-microbial activities for chitosan nanoparticles modified with the integration of antibacterial substances. In one such study, chitosan NPs containing protamine, a natural anti-microbial peptide showed significantly enhanced anti-microbial activity towards pathogenic *E. coli* [91]. These examples establish that nanomaterials are suitable candidates for future development of alternative anti-microbial to be used for existing as well as emerging, re-emerging infectious pathogens. In hindsight, it is important to emphasize that further research would be essential to determine the physiological and mechanistic features of nanomaterial-mediated anti-microbial activities. It would also be important to address how combined approaches that employ both nanomaterials and small molecule-based drugs would perform in developing sustainable anti-microbial nanomaterials.

The Anti-Microbial Peptide as Potent Alternative Anti-Microbial Therapeutics

Anti-microbial peptides (AMPs) are short, positively charged peptides with potent antibacterial, antiviral, and antifungal activity. They are also known as host defense peptides. They are found in widespread forms of life, ranging from microorganisms to humans [92]. AMPs constitute an important component of the first line of defense against foreign attacks, and they are indispensable components of the innate immune system in different life forms [93, 94]. AMPs exert their anti-microbial activities through disruption of the bacterial cell membranes, modulation of the immune response, and regulation of inflammatory activities. They are also indicated to be able to render synergistic activities along with conventional antibiotics. Hence, AMPs are projected as potent alternative anti-microbial therapeutics of the future. They are gaining priority in the development of anti-infective drugs, particularly against drug-resistant pathogens [95] and also against emerging and re-emerging infectious pathogens [96]. AMPs were first discovered in the 1980s; thereafter, they have been a subject of curiosity and research for enhancing their efficacy and formation as anti-microbial drugs. According to the anti-microbial peptide database, 3234 anti-microbial peptides from six kingdoms (358 bacteriocins/peptide antibiotics from bacteria, 5 from archaea, 8 from protists, 20 from fungi, 360 from plants, and 2400 from animals, including some synthetic peptides) have been identified to date (http://aps.unmc.edu/AP/main.php). A few of them have been approved for clinical application by the U.S. Food and Drug Administration (FDA).

As stated earlier, most AMPs are short peptides with a net positive charge responsible for their attraction toward negatively charged membranes of bacteria [97]. The hydrophilic and hydrophobic amino acids of AMPs are precisely segregated in such a manner that their solubility is ensured in both aqueous and lipid-rich environments. The shortest AMPs are as small as 10 to 20 amino acids in length, and the largest ones are in the range of ~100-150 amino acid residues. Structurally, the shorter AMPs possess simple secondary structures, including alpha-helix, beta-sheet, or unstructured, however, certain AMPs, particularly of the defensins family, may have elaborate three-dimensional structures. For most of the past 25 years, AMPs were regarded to be nonspecific and functionally redundant. Furthermore, it was believed that the biochemical simplicity of AMPs reflected fundamentally irresistible modes of action, including permeabilization of the cell envelope through the formation of open pores, which was assumed to largely prevent bacterial evolution of resistance.

While looking for alternative anti-microbial therapeutics to be used against drug-resistant microorganisms and emerging & re-emerging infectious microorganisms; AMPs are projected as a promising solution, due to their wide spectrum of activity against several microorganisms [96]. AMPs hold potential for translational applications, however, rigorous research would be required to establish a fundamental understanding towards maximizing their potential. Optimization of their biochemical properties and pharmacodynamics will render them, ideal candidates, against which evolution of resistance might be rare or substantially less frequent.

Translation development of AMP-based anti-microbial preparations is presently faced with several limitations; one such limitation is their proteolytic degradation. The advent of AMPs-based nanoparticles is projected to circumvent this limitation. Currently, AMPs-conjugated nanoparticles have attracted the attention of the scientific community due to their enormous advantages in terms of effective delivery and stability [98]. The main goals of AMPs- conjugated nanoparticles are to increase therapeutic efficacy, minimize side effects, offer a controlled pharmacokinetic profile, protect the encapsulated peptide from degradation, and reduce toxicity. Various types of nanomaterials, such as polymers, hydrogels, microgels, liposomes, self-assembly systems, *etc*., are being explored to enhance their (AMPs) therapeutic efficacies.

CONCLUSION

While Humanity continues to face the threatening crisis posed by incidences of emerging and re-emerging infectious diseases in the 21st century, there is also a tremendous opportunity for scientific advancements for developing innovative countermeasures and therapeutic paradigms with the development of novel anti-microbial drugs. Given the inherent uncertainty associated with emerging and re-emerging infectious diseases, the global community needs to shift from the '**Wait & Watch**' approach to a more '**Proactive Prediction**' approach for predicting future incidences of emerging and re-emerging diseases. Questions such as why a certain disease or certain types of diseases emerge at a particular location and a particular time may provide valuable information and facilitate a reasonable disease prediction system. Address these questions and be technologically prepared to tackle the possible threats posed to the very survival of human beings.

CONSENT FOR PUBLICATION

Not applicable.

CONFLICT OF INTEREST

The author declares no conflict of interest, financial or otherwise.

ACKNOWLEDGEMENT

Declared none.

REFERENCES

[1] Oró J, Miller SL, Lazcano A. The origin and early evolution of life on Earth. Annu Rev Earth Planet Sci 1990; 18(1): 317-56. [http://dx.doi.org/10.1146/annurev.ea.18.050190.001533] [PMID: 11538678]

[2] Hayashi C, Nakazawa K, Nakagawa Y. Formation of the solar system. prpl 1985; 1100-53.

[3] Liebes S, Sahtouris E, Swimme B. A walk through time: From stardust to us: The evolution of life on Earth. 1998.

[4] Whitman WB, Coleman DC, Wiebe WJ. Prokaryotes: The unseen majority. Proc Natl Acad Sci USA 1998; 95(12): 6578-83. [http://dx.doi.org/10.1073/pnas.95.12.6578] [PMID: 9618454]

[5] Madigan MT, Martinko JM, Parker J. Brock biology of microorganisms: Prentice hall Upper Saddle River, NJ 1997.

[6] Alberts B, Johnson A, Lewis J. Molecular Biology of the Cell in Cell 4th. 2002.

[7] Garagnani P, Pirazzini C, Giuliani C, Candela M, Brigidi P, Sevini F, *et al.* The three genetics (nuclear DNA, mitochondrial DNA, and gut microbiome) of longevity in humans considered as metaorganisms. BioMed Research International. 2014.

[8] Moya A, Peretó J, Gil R, Latorre A. Learning how to live together: genomic insights into prokaryote–animal symbioses. Nat Rev Genet 2008; 9(3): 218-29. [http://dx.doi.org/10.1038/nrg2319] [PMID: 18268509]

[9] Hugon P, Dufour JC, Colson P, Fournier PE, Sallah K, Raoult D. A comprehensive repertoire of prokaryotic species identified in human beings. Lancet Infect Dis 2015; 15(10): 1211-9. [http://dx.doi.org/10.1016/S1473-3099(15)00293-5] [PMID: 26311042]

[10] Taylor LH, Latham SM, woolhouse MEJ. Risk factors for human disease emergence. Philos Trans R Soc Lond B Biol Sci 2001; 356(1411): 983-9. [http://dx.doi.org/10.1098/rstb.2001.0888] [PMID: 11516376]

[11] Woolhouse M, Gaunt E. Ecological origins of novel human pathogens. Crit Rev Microbiol 2007; 33(4): 231-42. [http://dx.doi.org/10.1080/10408410701647560] [PMID: 18033594]

[12] Fauci AS. Emerging and reemerging infectious diseases: the perpetual challenge. Acad Med 2005; 80(12): 1079-85. [http://dx.doi.org/10.1097/00001888-200512000-00002] [PMID: 16306276]

[13] Morens DM, Fauci AS. Emerging infectious diseases: threats to human health and global stability. PLoS Pathog 2013; 9(7): e1003467. [http://dx.doi.org/10.1371/journal.ppat.1003467] [PMID: 23853589]

[14] Snowden FM. Emerging and reemerging diseases: a historical perspective. Immunol Rev 2008; 225(1): 9-26. [http://dx.doi.org/10.1111/j.1600-065X.2008.00677.x] [PMID: 18837773]

[15] Morens DM, Folkers GK, Fauci AS. The challenge of emerging and re-emerging infectious diseases.

Nature 2004; 430(6996): 242-9.
[http://dx.doi.org/10.1038/nature02759] [PMID: 15241422]

[16] Ka-Wai Hui E. Reasons for the increase in emerging and re-emerging viral infectious diseases. Microbes Infect 2006; 8(3): 905-16.
[http://dx.doi.org/10.1016/j.micinf.2005.06.032] [PMID: 16448839]

[17] Ravin KA. What's Old is New: Re-emerging Infectious Diseases 2019.

[18] Cucinotta D, Vanelli M. WHO declares COVID-19 a pandemic. Acta Biomed 2020; 91(1): 157-60.
[PMID: 32191675]

[19] Atkeson A. Atkeson A. What will be the economic impact of covid-19 in the us? rough estimates of disease scenarios. National Bureau of Economic Research Report No.: 0898-2937 2020.

[20] Fernandes N. Economic effects of coronavirus outbreak (COVID-19) on the world economy. SSRN Electronic Journal 2020.
[http://dx.doi.org/10.2139/ssrn.3557504]

[21] Pritchard J, Collier A, Mundega M, Bartels SA. COVID in crisis: The impact of COVID-19 in complex humanitarian emergencies. Journal of Military, Veteran and Family Health. 2020.

[22] Feldmann H, Czub M, Jones S, *et al.* Emerging and re-emerging infectious diseases. Med Microbiol Immunol (Berl) 2002; 191(2): 63-74.
[http://dx.doi.org/10.1007/s00430-002-0122-5] [PMID: 12410344]

[23] Mehand MS, Al-Shorbaji F, Millett P, Murgue B. The WHO R&D Blueprint: 2018 review of emerging infectious diseases requiring urgent research and development efforts. Antiviral Res 2018; 159: 63-7.
[http://dx.doi.org/10.1016/j.antiviral.2018.09.009] [PMID: 30261226]

[24] Vouga M, Greub G. Emerging bacterial pathogens: the past and beyond. Clin Microbiol Infect 2016; 22(1): 12-21.
[http://dx.doi.org/10.1016/j.cmi.2015.10.010] [PMID: 26493844]

[25] Paules CI, Eisinger RW, Marston HD, Fauci AS. What recent history has taught us about responding to emerging infectious disease threats. Ann Intern Med 2017; 167(11): 805-11.
[http://dx.doi.org/10.7326/M17-2496] [PMID: 29132162]

[26] Alley C, Sommerfeld J. Infectious disease in times of social and ecological change. Med Anthropol 2014; 33(2): 85-91.
[http://dx.doi.org/10.1080/01459740.2013.850590] [PMID: 24512379]

[27] Mukherjee S. Emerging Infectious Diseases: Epidemiological Perspective. Indian J Dermatol 2017; 62(5): 459-67.
[PMID: 28979007]

[28] Spyrou MA, Bos KI, Herbig A, Krause J. Ancient pathogen genomics as an emerging tool for infectious disease research. Nat Rev Genet 2019; 20(6): 323-40.
[http://dx.doi.org/10.1038/s41576-019-0119-1] [PMID: 30953039]

[29] Jones KE, Patel NG, Levy MA, *et al.* Global trends in emerging infectious diseases. Nature 2008; 451(7181): 990-3.
[http://dx.doi.org/10.1038/nature06536] [PMID: 18288193]

[30] Ushie EF, Andefiki U. Climate Change and the Risk of Emerging and Re-Emerging Diseases. International Journal of Public Health and Health Systems 2018; 3(3): 33.

[31] Zumla A, Hui DSC. Emerging and re-emerging infectious diseases: global overview. Infectious Disease Clinics 2019; 33(4): xiii-x.
[PMID: 31668202]

[32] Scott SR, Wu Z. Risks and challenges of HIV infection transmitted *via* blood transfusion. Biosafety and Health 2019; 1(3): 124-8.

[http://dx.doi.org/10.1016/j.bsheal.2019.12.001]

[33] Yale G, Bhanurekha V, Ganesan P. Anthropogenic factors responsible for emerging and re-emerging infectious diseases. Curr Sci 2013; 940-6.

[34] Rohr JR, Barrett CB, Civitello DJ, *et al.* Emerging human infectious diseases and the links to global food production. Nat Sustain 2019; 2(6): 445-56.
[http://dx.doi.org/10.1038/s41893-019-0293-3] [PMID: 32219187]

[35] Rothe C, Jong EC. Emerging infectious diseases and the international traveler. 2017; 27.
[http://dx.doi.org/10.1016/B978-0-323-37506-1.00003-9]

[36] Abdallah M. Review on emerging and re-emerging infectious diseases and their origins. Microbiol Res J Int 2018; 1-5.

[37] Piot P, Larson HJ, O'Brien KL, *et al.* Immunization: vital progress, unfinished agenda. Nature 2019; 575(7781): 119-29.
[http://dx.doi.org/10.1038/s41586-019-1656-7] [PMID: 31695203]

[38] Osterholm MT. Global health security—an unfinished journey. Emerg Infect Dis 2017; 23(13) (Suppl. 1): S225.
[http://dx.doi.org/10.3201/eid2313.171528]

[39] Creager ANH. Adaptation or selection? Old issues and new stakes in the postwar debates over bacterial drug resistance. Stud Hist Philos Sci Part Stud Hist Philos Biol Biomed Sci 2007; 38(1): 159-90.
[http://dx.doi.org/10.1016/j.shpsc.2006.06.016] [PMID: 17324813]

[40] Dhar N, McKinney JD. Microbial phenotypic heterogeneity and antibiotic tolerance. Curr Opin Microbiol 2007; 10(1): 30-8.
[http://dx.doi.org/10.1016/j.mib.2006.12.007] [PMID: 17215163]

[41] Avery SV. Microbial cell individuality and the underlying sources of heterogeneity. Nat Rev Microbiol 2006; 4(8): 577-87.
[http://dx.doi.org/10.1038/nrmicro1460] [PMID: 16845428]

[42] Health UDo. Services H, Control CfD Prevention Antibiotic Resistance Threats in the United States 2019.

[43] Hiltunen T, Virta M, Laine A-L. Antibiotic resistance in the wild: an eco-evolutionary perspective. 2017.
[http://dx.doi.org/10.1098/rstb.2016.0039]

[44] Fauci AS. Biodefence on the research agenda. Nature 2003; 421(6925): 787.
[http://dx.doi.org/10.1038/nature01480] [PMID: 12594484]

[45] Dybul M, Fauci AS, Bartlett JG, Kaplan JE, Pau AK. Guidelines for using antiretroviral agents among HIV-infected adults and adolescents. Ann Intern Med 2002; 137(5_Part_2): 381-433.
[http://dx.doi.org/10.7326/0003-4819-137-5_Part_2-200209031-00001] [PMID: 12617573]

[46] Wong CH, Siah KW, Lo AW. Estimation of clinical trial success rates and related parameters. Biostatistics 2019; 20(2): 273-86.
[http://dx.doi.org/10.1093/biostatistics/kxx069] [PMID: 29394327]

[47] Pors K. Drug discovery into the 21st century. Chemistry 2011; 69-96.

[48] Liu K. Improving surveillance and prediction of emerging and re-emerging infectious diseases 2019.

[49] Charlton Hume HK, Lua LHL. Platform technologies for modern vaccine manufacturing. Vaccine 2017; 35(35): 4480-5.
[http://dx.doi.org/10.1016/j.vaccine.2017.02.069] [PMID: 28347504]

[50] Hayman B, Pagliusi S. Emerging vaccine manufacturers are innovating for the next decade. Vaccine X 2020; 5: 100066.

[http://dx.doi.org/10.1016/j.jvacx.2020.100066] [PMID: 32462140]

[51] Liu MA. Immunologic basis of vaccine vectors. Immunity 2010; 33(4): 504-15. [http://dx.doi.org/10.1016/j.immuni.2010.10.004] [PMID: 21029961]

[52] Regules JA, Beigel JH, Paolino KM, *et al.* A recombinant vesicular stomatitis virus Ebola vaccine. N Engl J Med 2017; 376(4): 330-41. [http://dx.doi.org/10.1056/NEJMoa1414216] [PMID: 25830322]

[53] Morris SJ, Sebastian S, Spencer AJ, Gilbert SC. Simian adenoviruses as vaccine vectors. Future Virol 2016; 11(9): 649-59. [http://dx.doi.org/10.2217/fvl-2016-0070] [PMID: 29527232]

[54] Buonaguro L, Tornesello ML, Buonaguro FM. Virus-like particles as particulate vaccines. Curr HIV Res 2010; 8(4): 299-309. [http://dx.doi.org/10.2174/157016210791208659] [PMID: 20353398]

[55] Donaldson B, Lateef Z, Walker GF, Young SL, Ward VK. Virus-like particle vaccines: immunology and formulation for clinical translation. Expert Rev Vaccines 2018; 17(9): 833-49. [http://dx.doi.org/10.1080/14760584.2018.1516552] [PMID: 30173619]

[56] Jain NK, Sahni N, Kumru OS, Joshi SB, Volkin DB, Russell Middaugh C. Formulation and stabilization of recombinant protein based virus-like particle vaccines. Adv Drug Deliv Rev 2015; 93: 42-55. [http://dx.doi.org/10.1016/j.addr.2014.10.023] [PMID: 25451136]

[57] Leitner WW, Ying H, Restifo NP. DNA and RNA-based vaccines: principles, progress and prospects. Vaccine 1999; 18(9-10): 765-77. [http://dx.doi.org/10.1016/S0264-410X(99)00271-6] [PMID: 10580187]

[58] Kochel TJ, Porter KR, Raviprakash K, Hoffman SL, Hayes CG. Dengue nucleic acid vaccines that induce neutralizing antibodies. 2002.

[59] Gao S, Song S, Zhang L. Recent progress in vaccine development against chikungunya virus. Front Microbiol 2019; 10: 2881. [http://dx.doi.org/10.3389/fmicb.2019.02881] [PMID: 31921059]

[60] Cashman KA, Wilkinson ER, Wollen SE, *et al.* DNA vaccines elicit durable protective immunity against individual or simultaneous infections with Lassa and Ebola viruses in guinea pigs. Hum Vaccin Immunother 2017; 13(12): 3010-9. [http://dx.doi.org/10.1080/21645515.2017.1382780] [PMID: 29135337]

[61] Zhang C, Maruggi G, Shan H, Li J. Advances in mRNA Vaccines for Infectious Diseases. 2019. [http://dx.doi.org/10.3389/fimmu.2019.00594]

[62] Ding S, Greenberg HB. Zika mRNA vaccine induces long-term protective immunity. AME Med J 2017; 2: 86. [http://dx.doi.org/10.21037/amj.2017.05.22]

[63] Vartak A, Sucheck S. Recent advances in subunit vaccine carriers. Vaccines (Basel) 2016; 4(2): 12. [http://dx.doi.org/10.3390/vaccines4020012] [PMID: 27104575]

[64] Chappell K, Watterson D, Muller D, Depelsenaire A, Young A, Marsh G, *et al.* A stabilized subunit vaccine for ebola virus. 2016.

[65] Lehrer AT, Wong TAS, Lieberman MM, *et al.* Recombinant subunit vaccines protect guinea pigs from lethal Ebola virus challenge. Vaccine 2019; 37(47): 6942-50. [http://dx.doi.org/10.1016/j.vaccine.2019.06.035] [PMID: 31324500]

[66] Armas M, Sahai J, Zhanel GG. Current Status of the HIV Vaccine. Can J Hosp Pharm 2018; 50(1).

[67] Pandey J, Nimesh S. Genomics-Guided Discovery of Novel Therapeutics of Actinobacterial Origin. Post-Genomic Approaches in Drug and Vaccine Development 2015; 5: 71.

[68] Land M, Hauser L, Jun SR, *et al.* Insights from 20 years of bacterial genome sequencing. Funct Integr Genomics 2015; 15(2): 141-61.
[http://dx.doi.org/10.1007/s10142-015-0433-4] [PMID: 25722247]

[69] Arriola LA, Cooper A, Weyrich LS. Palaeomicrobiology: Application of ancient DNA sequencing to better understand bacterial genome evolution and adaptation. Front Ecol Evol 2020; 8: 40.
[http://dx.doi.org/10.3389/fevo.2020.00040]

[70] Skinnider MA, Merwin NJ, Johnston CW, Magarvey NA. PRISM 3: expanded prediction of natural product chemical structures from microbial genomes. Nucleic Acids Res 2017; 45(W1): W49-54.
[http://dx.doi.org/10.1093/nar/gkx320] [PMID: 28460067]

[71] Kautsar SA, Blin K, Shaw S, *et al.* MIBiG 2.0: a repository for biosynthetic gene clusters of known function. Nucleic Acids Res 2020; 48(D1): D454-8.
[PMID: 31612915]

[72] Xu BH, Lu YQ, Ye ZW, *et al.* Genomics-guided discovery and structure identification of cyclic lipopeptides from the Bacillus siamensis JFL15. PLoS One 2018; 13(8): e0202893.
[http://dx.doi.org/10.1371/journal.pone.0202893] [PMID: 30169540]

[73] Huang C, Leung RKK, Guo M, *et al.* Genome-guided investigation of antibiotic substances produced by Allosalinactinospora lopnorensis CA15-2 T from Lop Nor region, China. Sci Rep 2016; 6(1): 20667.
[http://dx.doi.org/10.1038/srep20667] [PMID: 26864220]

[74] Perlatti B, Lan N, Jiang Y, An Z, Bills G. Identification of Secondary Metabolites from *Aspergillus pachycristatus* by Untargeted UPLC-ESI-HRMS/MS and Genome Mining. Molecules 2020; 25(4): 913.
[http://dx.doi.org/10.3390/molecules25040913] [PMID: 32085602]

[75] Farha MA, Brown ED. Unconventional screening approaches for antibiotic discovery. Ann N Y Acad Sci 2015; 1354(1): 54-66.
[http://dx.doi.org/10.1111/nyas.12803] [PMID: 26100135]

[76] Schneider MV. Defining systems biology: a brief overview of the term and field. *in silico* systems biology. Springer 2013; 1-11.

[77] DePristo MA, de Bakker PIW, Blundell TL. Heterogeneity and inaccuracy in protein structures solved by X-ray crystallography. Structure 2004; 12(5): 831-8.
[http://dx.doi.org/10.1016/j.str.2004.02.031] [PMID: 15130475]

[78] Wishart D. NMR spectroscopy and protein structure determination: applications to drug discovery and development. Curr Pharm Biotechnol 2005; 6(2): 105-20.
[http://dx.doi.org/10.2174/1389201053642367] [PMID: 15853690]

[79] Kuhn B, Gerber P, Schulz-Gasch T, Stahl M. Validation and use of the MM-PBSA approach for drug discovery. J Med Chem 2005; 48(12): 4040-8.
[http://dx.doi.org/10.1021/jm049081q] [PMID: 15943477]

[80] Liu B, He H, Luo H, Zhang T, Jiang J. Artificial intelligence and big data facilitated targeted drug discovery. Stroke Vasc Neurol 2019; 4(4): 206-13.
[http://dx.doi.org/10.1136/svn-2019-000290] [PMID: 32030204]

[81] Staker BL, Buchko GW, Myler PJ. Recent contributions of structure-based drug design to the development of antibacterial compounds. Curr Opin Microbiol 2015; 27: 133-8.
[http://dx.doi.org/10.1016/j.mib.2015.09.003] [PMID: 26458180]

[82] Malik B, Bhattacharyya S. Antibiotic drug-resistance as a complex system driven by socio-economic growth and antibiotic misuse. Sci Rep 2019; 9(1): 9788.
[http://dx.doi.org/10.1038/s41598-019-46078-y] [PMID: 31278344]

[83] Díez-Pascual AM. Antibacterial activity of nanomaterials 2018.

[http://dx.doi.org/10.3390/nano8060359]

[84] Aswathanarayan JB, Vittal RR. R Vittal R. Antimicrobial, biofilm inhibitory and anti-infective activity of metallic nanoparticles against pathogens MRSA and Pseudomonas aeruginosa PA01. Pharm Nanotechnol 2017; 5(2): 148-53.
[http://dx.doi.org/10.2174/2211738505666170424121944] [PMID: 28440203]

[85] Hemeg H. Nanomaterials for alternative antibacterial therapy. Int J Nanomedicine 2017; 12: 8211-25.
[http://dx.doi.org/10.2147/IJN.S132163] [PMID: 29184409]

[86] Rai MK, Deshmukh SD, Ingle AP, Gade AK. Silver nanoparticles: the powerful nanoweapon against multidrug-resistant bacteria. J Appl Microbiol 2012; 112(5): 841-52.
[http://dx.doi.org/10.1111/j.1365-2672.2012.05253.x] [PMID: 22324439]

[87] Al-Dhabi N, Mohammed Ghilan AK, Arasu M. Characterization of silver nanomaterials derived from marine Streptomyces sp. al-dhabi-87 and its *in vitro* application against multidrug resistant and extended-spectrum beta-lactamase clinical pathogens. Nanomaterials (Basel) 2018; 8(5): 279.
[http://dx.doi.org/10.3390/nano8050279] [PMID: 29701657]

[88] Vi T, Rajesh Kumar S, Rout B, *et al.* The preparation of graphene oxide-silver nanocomposites: the effect of silver loads on Gram-positive and Gram-negative antibacterial activities. Nanomaterials (Basel) 2018; 8(3): 163.
[http://dx.doi.org/10.3390/nano8030163] [PMID: 29538336]

[89] Elbagory A, Meyer M, Cupido C, Hussein A. Inhibition of bacteria associated with wound infection by biocompatible green synthesized gold nanoparticles from South African plant extracts. Nanomaterials (Basel) 2017; 7(12): 417.
[http://dx.doi.org/10.3390/nano7120417] [PMID: 29186826]

[90] Kubacka A, Ferrer M, Fernández-García M. Kinetics of photocatalytic disinfection in TiO2-containing polymer thin films: UV and visible light performances. Appl Catal B 2012; 121-122: 230-8.
[http://dx.doi.org/10.1016/j.apcatb.2012.03.016]

[91] Tamara F, Lin C, Mi FL, Ho YC. Antibacterial Effects of Chitosan/Cationic Peptide Nanoparticles. Nanomaterials (Basel) 2018; 8(2): 88.
[http://dx.doi.org/10.3390/nano8020088] [PMID: 29401728]

[92] Bahar A, Ren D. Antimicrobial Peptides. Pharmaceuticals (Basel) 2013; 6(12): 1543-75.
[http://dx.doi.org/10.3390/ph6121543] [PMID: 24287494]

[93] Zasloff M. Antimicrobial peptides of multicellular organisms. Nature 2002; 415(6870): 389-95.
[http://dx.doi.org/10.1038/415389a] [PMID: 11807545]

[94] Broekaert WF, Cammue BPA, De Bolle MFC, *et al.* Antimicrobial Peptides from Plants. Crit Rev Plant Sci 1997; 16(3): 297-323.
[http://dx.doi.org/10.1080/07352689709701952]

[95] Parai D, Dey P, Mukherjee SK. Anti-microbial Peptides: An Approach to Combat Resilient Infections. Curr Drug Discov Technol 2019.
[PMID: 31250760]

[96] Peters BM, Shirtliff ME, Jabra-Rizk MA. Antimicrobial peptides: primeval molecules or future drugs? PLoS Pathog 2010; 6(10): e1001067.
[http://dx.doi.org/10.1371/journal.ppat.1001067] [PMID: 21060861]

[97] Powers JPS, Hancock REW. The relationship between peptide structure and antibacterial activity. Peptides 2003; 24(11): 1681-91.
[http://dx.doi.org/10.1016/j.peptides.2003.08.023] [PMID: 15019199]

[98] Pal I, Brahmkhatri VP, Bera S, *et al.* Enhanced stability and activity of an antimicrobial peptide in conjugation with silver nanoparticle. J Colloid Interface Sci 2016; 483: 385-93.
[http://dx.doi.org/10.1016/j.jcis.2016.08.043] [PMID: 27585423]

CHAPTER 2

Emergence of Multidrug Resistance Microbes: Bacteria, Fungi, and Viruses

Neha Malhotra[1], **Pramod Kumar**[2], **Ramakrishnan Sethu**[3] and **Sanjay K. Rohaun**[3,*]

[1] *Tuberculosis Research Section, Laboratory of Clinical Immunology and Microbiology, National Institute of Allergy and Infectious Diseases, National Institutes of Health, Bethesda, MD, 20892, United States*

[2] *Department of Molecular and Integrative Physiology, University of Illinois, 61801, United States*

[3] *Department of Microbiology, University of Illinois, Urbana, Illinois, 61801, United States*

Abstract: Increased incidences of multidrug-resistant (MDR) microorganisms has become a global health concern for humans, animals, and agriculture. The advent of newer resistance mechanisms evolving in microorganisms at a high rate compared to the treatments available urges the need to understand its origin and reservoirs. The expanded use of antimicrobial drugs, inefficient diagnosis as well as broad use in agriculture and veterinary contributes to the emergence of resistance in microorganisms. Presently, almost all infectious agents (bacteria, fungi, and viruses) have developed MDR. About 7 lac people die of bacterial resistance to antibiotics every year, with an estimated ~10 million deaths by 2050. Similarly, MDR arising in pathogenic fungi like *Candida*, *Aspergillus,* or *Fusarium* to the limited therapeutic options is highly challenging. Bacteria and fungi develop resistance mainly due to biofilm formation, increased efflux pump activity, drug target mutations, drug binding alterations, chromosome abnormalities, and the ability to escape host immune defenses. The co-existence of MDR bacteria and fungi forming biofilms is even much of an alarm in medicinal applications. Apart from these, drug resistance to current antiviral therapeutics has imposed significant risk amidst life-threatening diseases caused due to viruses like HIV and influenza A. Owing to its severity and complexity, we aim to illustrate the detailed mechanism and evolution of MDR in bacteria, fungi, and viruses. We also review different approaches to deal with MDR, emphasizing alternatives, vaccine development, global surveillance programs and stewardship measures to combat resistance.

Keywords: Antibiotics, Bacteria, Candida, Copper-complex, Disease, Enzymes, Fungi, HIV, Influenza, MDR, MDR-MRSA, Membrane, Microbiota, Microbe, Oxidants, Phage, ROS, Virus, Vaccine, XDR.

* **Corresponding author Sanjay K. Rohaun**: Department of Microbiology, University of Illinois, Urbana, Illinois, 61801, United States; E-mails: sanjayihbt85@gmail.com and rohaun@illinois.edu

M. Amin-ul Mannan & Gaurav Kumar (Eds.)

INTRODUCTION

The presence of multidrug-resistant pathogens is one of the greatest challenges posing a serious health concern around the globe. Multidrug resistance (MDR), as the name suggests, arises when an organism is resistant to more than one class of drug. Several aspects, like high mutagenesis rate, overexpression of efflux pumps, and alterations in drug binding, are responsible for rendering drug resistance and making current clinical therapeutics ineffective. Moreover, selective pressure arising due to excessive use or misapplication of anti-infective agents in clinical setups, inefficient diagnosis, expanding usage in agriculture, animal feed and aquaculture, travels, and immigration has enabled pathogenic microbes to adapt and develop antimicrobial resistance. Additionally, the transmission of resistant genes through a horizontal gene transfer mechanism has led to the enhancement of resistant microbes. Multidrug resistance is a global threat as there exist limited treatment options. This condition worsens when there is resistance to second-line drugs or all therapeutics, as is the case for extremely drug-resistant (XDR) or drug-resistant (TDR) pathogens. Today, multidrug resistance has evolved in almost all bacterial, fungal and viral pathogens and has served to be an emerging crisis, especially for immunocompromised patients, chemotherapies, surgeries, tuberculosis, HIV-AIDS, *etc.* This has further resulted in prolonged illness, increased treatment costs, and high mortality rates.

Occurrence of multidrug resistance in nosocomial infecting ESKAPE pathogens, including *Enterococcus faecium*, *Staphylococcus aureus*, *Klebsiella pneumoniae*, *Acinetobacter baumannii*, *Pseudomonas aeruginosa*, and *Enterobacter cloacae* has resulted in high morbidities and mortalities. In addition, the emergence of New Delhi Metallo-β-lactamase-1 (NDM-1) and mobile colistin-resistant (MCR) strains has been another major concern for the clinical and agricultural society, where the food chain is a big reservoir for dissemination of resistance [1 - 4]. Moreover, drug-resistant *Mycobacterium tuberculosis* has created a state of emergency and needs urgent control to curb the spread. It has resulted in half a million new cases and 230 000 deaths reported globally in 2018, with about 8.5% of MDR-TB cases having XDR-TB [5].

Multidrug resistance in pathogenic fungal species like *Candida*, *Aspergillus*, *Cryptococcus,* and *Fusarium* spp. that causes life-threatening invasive diseases like Candidiasis, aspergillosis, fusariosis, meningitis, pneumonia, *etc*., is highly alarming. The treatment options for fungal infections are quite restricted; thus, the emergence of multidrug resistance becomes a great challenge. The recent catastrophic crisis from the multidrug-resistant *Candida Auris*, which is difficult to be identified with standard laboratory practices and is called a "mysterious bug" or "superbug" has created serious infectious and death outbreaks around the globe [6, 7].

The worldwide spread of life-threatening viral diseases has demanded strict adherence to anti-retroviral therapies but has also worsened the problem of multidrug resistance in viruses and has important economic implications.

High incidence of MDR pathogens necessitates the development of newer strategies, efficient diagnostics, contact tracing, the discovery of novel targets for microbial pathogenicity without affecting the normal microbial flora and the development of further resistance to other anti-infectives. Unlike the conventional approaches, which target affecting essential metabolic processes, several alternative methods of treatment are in demand to combat resistant microbes. These include the use of inhibitors for quorum sensing, biofilm formation, protein synthesis, cell wall synthesis, use of monoclonal antibodies, antimicrobial peptides, phage therapy, essential oils, nanotherapeutics, vaccine, and CRISPR-Cas9 system. Additionally, combinatorial therapy has profound applications to disseminate the wide majority of drug resistance pathogens [8]. Moreover, the advancement in whole-genome sequencing has led to a better understanding of specific diagnostic resistance markers, which would encourage the use of appropriate treatments and prevent the rapid disease transmission of MDR pathogens within the community. Therefore, it becomes important that these action plans should be globalized and implemented with appropriate financial support. In this book chapter, we bring together information on the evolutionary mechanisms of multidrug resistance in pathogenically significant bacteria, fungi, and viruses, focusing on their resistance markers to drugs. We also discuss alternative approaches and global action plans to circumvent the emergence and transmission of resistant microbes.

PROBLEM OF MULTIDRUG RESISTANCE

Treatment strategies to effectively control MDR infections faces many challenges. A few of these include detection at the right time, cost, toxicity of higher doses of different drug combinations, targeting the individual MDR microbe, and non-practical observational treatment. Globally, MDR is amongst the top ten threats to public health. Antimicrobial drugs are used to treat infections, exposed surgical procedures, and other medical advanced treatments such as transplants and chemotherapy. The presence of Certain bacteria inside the tumors inactivate the drug used to treat cancer, and thus, chemotherapy becomes ineffective. Moreover, chemotherapy-induced neutropenia may also trigger endogenous infections, as the case of bacteria in the GI tract may spread to the bloodstream by translocation and cause sepsis. Such procedures as transplantation, surgery, chemotherapy and HIV-AIDS suppress the immunity making the host more prone to other microbial infections. The condition even worsens if these microbes are MDR. In 2016, 490 000 people got MDR-TB infections globally and XDR-TB has also spread to 105 countries. In the case of HIV, some countries have up to 40% of people with resistant HIV to the first-line regimens. Additionally, co-infection of HIV with MDR bacteria (like *K. pneumoniae*, *Enterobacter aerogenes*, *Shigella*, and *Escherichia coli*), MDR fungus (like *Candida* spp., *Aspergillus* spp.) and other viruses (like influenza and paramyxoviruses) have created alarming situations around the globe. Therefore, there is an urgent need to understand the co-infection burden too. Apart from healthcare-associated MDR infections, practices like wastewater treatment, antimicrobial sprays and animal live stocks are also significant sources of MDR microbes in aquatic and agricultural environments (Fig. **1**).

Increasing resistance levels pose important economic and health implications. The second and third-line regimens are 3-18 times more expensive. It is expected that if no measures are taken, it would lead to an estimated death of about 300 million people in the coming 35 years and a loss of economic production too (~100 thousand billion dollars).

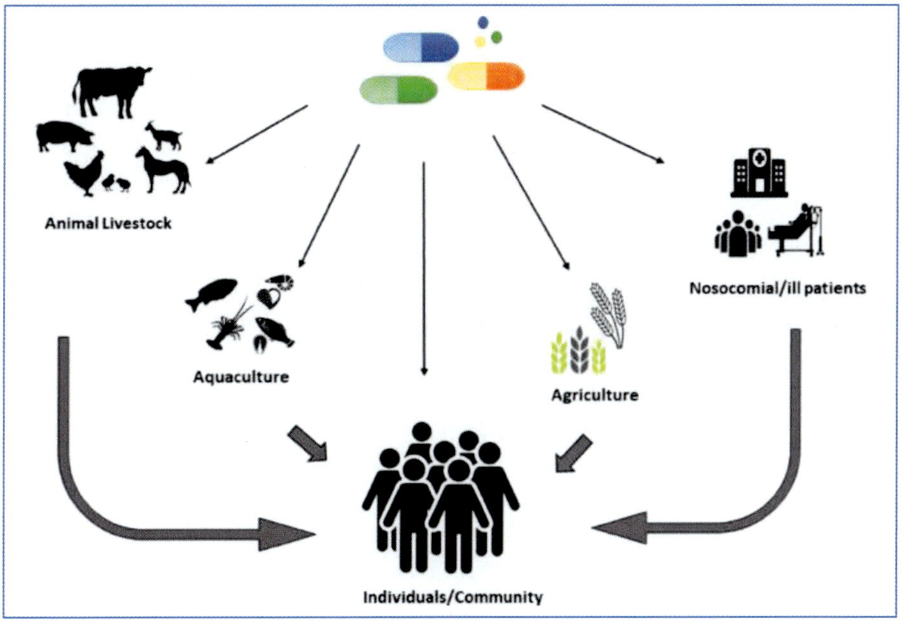

Fig. (1). Sources of MDR pathogens. MDR microbes are spread across all the sectors of life by the improper and/or overuse of antimicrobials in Agriculture, Aquaculture, wastewater treatment, livestock, and healthcare management, ultimately leading to the spread of resistant microbes in the community.

MECHANISM OF MULTIDRUG RESISTANCE

Resistance to anti-infectious agents is an ever-evolving, multifactorial process involving several molecular and biochemical key features like overexpression of efflux pumps, mutations, alterations in the drug target, drug sequestration, drug inactivation, *etc*. This can be a naturally occurring intrinsic/inherent property of an organism to be resistant to the action of the drug or acquired resistance in a previously susceptible organism. Intrinsic resistance is majorly contributed by the presence of efflux pump or membrane permeability or certain genes and loci. For example, multidrug resistance in Gram-negative (G-ve) bacteria arising due to impermeable outer membrane barrier and efflux pumps makes it tolerant to Gram-positive (G+ve) antibiotics as well as fluconazole resistance in *Fusarium* spp [9]. On the other hand, microbes may undergo selective pressure induced in the presence of excess anti-infectives leading to biochemical, mutational, and genetic adaptations, thus, Acquired Multidrug resistance. Acquired resistance arises mainly from the transmission of genetic determinants on plasmids, bacteriophages, and transposons. The spread of resistance genes is a dynamic phenomenon and occurs multi-directionally across different organisms. The resistance genes are present in humans, clinical patients, nosocomial setups, live stocks, agriculture, aquaculture, *etc*. that acts as reservoirs, and their dissemination happens rapidly, causing increased evidence for multidrug or pan/total-drug resistance situations for almost all microbes (Fig. **1**). Furthermore, the acquisition of resistance to multiple drugs may also arise due to either co-resistance or cross-resistance. Co-resistance occurs when there is a transfer of multiple genetic elements or plasmids in the same microbial cell or by the acquisition of mutations in different genetic elements conferring resistance to more than one class of antibiotics or antimicrobials. For example, methicillin-resistant *S. aureus* (MRSA) accumulates multidrug resistance to other drugs, including beta-lactams, macrolides, aminoglycosides, or fluoroquinolones [10]. In contrast, cross-resistance refers to the phenomenon when the single resistance determinant (like multidrug resistance efflux pumps or drug modifying enzymes) leads to resistance to an entire class of antimicrobials or even different classes of antimicrobials agents. For instance, aminoglycoside modifying enzymes results in resistance to several members of aminoglycosides; efflux pumps can extrude several different classes of antimicrobials and non-antimicrobial agents, leading to multi-drug resistance [11]. A few of the other examples of co- and cross-resistance contributing to MDR in fungus are mentioned in detail under next sections.

In the following section, we broadly define various intrinsic and acquired mechanisms of evolving drug resistance into two major categories: Genetic as well as Biochemical and mutational basis (Fig. **2**), while the role of specific resistance mechanisms is highlighted along with the respective MDR microbe in subsequent sections.

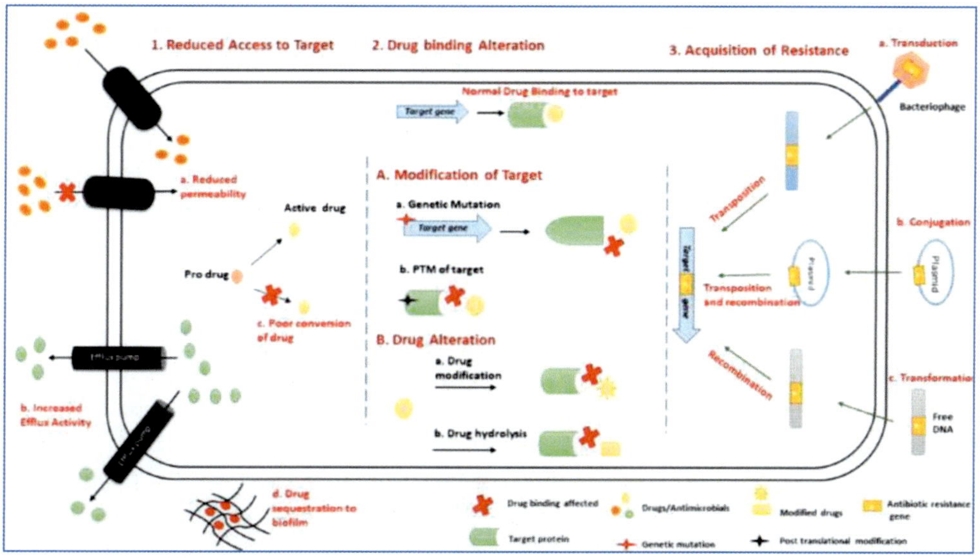

Fig. (2). Drug Resistance mechanism in microbes. Diverse mechanisms play a role in achieving MDR in microbes *viz.* **(1) Reduced access to the target.** This can occur due to **(a)** Reduced permeability of the drug through impermeable cell walls as in G-ve bacteria. **(b)** Increased Efflux Activity by the overexpression of drug resistance efflux pumps resulting in drug outflow from the cells. **(c)** Poor conversion of a drug by the mutations or absence of enzymes leading to the conversion of an inactive Pro-drug into its active form. **(d)** Drug Sequestration to biofilm, making it unavailable for the target within the cell; **(2) Drug Binding Alterations:** It involves loss of drug and target binding by either **(a)** Modification of target involving genetic mutations in genes expressing the target or by post-translational modification. **(b)** Drug Alteration by any form of modifications or its hydrolysis; **(3) Acquisition of Resistance:** It involves acquiring resistance genes by either recombination or transposition of mobile genetic elements by Transduction, Conjugation, or Transformation.

Genetic Basis for Drug Resistance

Genetic mechanisms for drug resistance mainly arise due to the transmission of resistant genes through mobile elements present on plasmids, transposons, and integrons. For example, in bacteria, such resistance elements are acquired by the drug-susceptible recipients *via* genetic events like transduction (by bacteriophages), transposition (mobile genetic elements), conjugation (by plasmids and conjugative transposons), transformation (by free DNA incorporation into the chromosome).

Biochemical and Mutational Basis for Drug Resistance

It is essential for a drug to be accessible and bind to the target site. Microbes have evolved several mechanisms, such as target modification, drug alterations, overexpression of efflux pumps, or cell wall impermeability, thus making the drug inaccessible or unable to reach the target at an appropriate concentration to inhibit its activity. We have discussed a few instances of these resistance mechanisms in the following sections, while these have been discussed along with respective microbes as well.

Target Modification

This is a common mode adopted by the microbes to develop resistance that helps evade antimicrobials. Resistance to antiviral drugs is predominantly achieved by mutations in the genes encoding the target protein and/or antiviral drug activator. For example, Amantadine resistant strains of influenza virus have developed resistance to the mutations in the gene encoding M9 protein, HIV develops a mutation in protease and reverses the transcriptase gene when in the presence of antiretrovirals [12, 13]. Few instances of antibiotic and antifungal resistance by target modification involve alteration of DNA gyrase resulting in quinolone resistance, modification of penicillin-binding protein leading to β-lactam resistance, the substitutional mutation in rpoB leading to rifampicin resistance, point mutations in genes encoding Erg11p makes the enzyme resistant to triazoles in *A. fumigatus*, *Candida* spp., mutations in *FKS1* reduces the susceptibility of yeast and *Candida* spp. against echinocandins [14 - 18].

Increased Activity of Energy-driven Efflux Pumps

Efflux transporters throw the drug out of the cell, making it inaccessible to the drug target. Efflux pumps are universally present in almost all kingdoms. They were first shown to be playing a role in tetracycline resistance in *E.coli* [19 - 21]. They are either selected for a single drug or may allow an array of structurally unrelated molecules to be transported, thus termed multidrug-resistant efflux transporters. The determinants for efflux pumps may be present extra-chromosomally on the mobile genetic element (plasmids) or chromosomes [22 - 24]. Therefore, the spread of resistance determinants may be contributed by either horizontal gene transfer and/or through a selection of resistance mutations, additionally resulting in a multi-resistance phenotype [11]. Efflux pumps are classified into five major categories based on features like sequence similarity, specificity for the substrate, number of components, membrane-spanning regions, and energy source. These include (a) ATP-binding cassette (ABC) superfamily; (b) Small Multidrug Resistance (SMR) family; (c) Multidrug and Toxic Extrusion (MATE) family; (d) Major Facilitator Superfamily (MFS); (e) Resistance-nodulation-division (RND) family. The ABC, SMR, MATE and MFS families are present in G+ve and G-ve bacteria, while the RND family is exclusively occurring in G-ve bacteria [11]. Fungus, on the other hand, usually possess either ABC transporters and/or MFS pumps. Drug resistance in fungi through efflux pumps is majorly contributed by either their constitutive overexpression or drug-responsive inducible response.

Other Biochemical and Mutational Events

Other significant resistance mechanisms of drug resistance in microbes include:

Decreased Cell Wall Permeability

The presence of an impermeable outer membrane usually makes G-ve bacteria resistant to drugs like vancomycin. Alterations of porins by modulating their expression and function also result in resistance in conjunction with increased activity of efflux pumps like OprD-mediated resistance in *P. aeruginosa* [25].

Bypass of the Target Site

Using this strategy, microbes can establish new targets mimicking the similar biochemical function of their original targets but fail to be inhibited by the anti-microbials. For example, the acquisition of PBP2a and peptidoglycan structure modification has resulted in the emergence of methicillin-resistant *S. aureus* (MRSA), and vancomycin-resistant *Enterococci* (VRE) strains, respectively [26].

Increased Expression Of Drug Modifying Enzymes Or Hydrolytic Enzymes

Mutations in the genes encoding drug activation enzymes have resulted in resistance in microbes and viruses. For instance, mutations in genes encoding thymidine kinase, an acyclovir activator, have led to resistance to the herpes simplex virus [27, 28].

EVOLUTION OF MULTIDRUG RESISTANCE

Multidrug Resistance in Bacteria

The emergence of MDR in bacteria (G+ve and G-ve) is a serious health challenge worldwide. Several global discussions and meetings like "68th meeting of World Health assembly" in May 2015, "United Nations General Assembly (UNGA)" in September 2016, "Rudolf-Schülke foundation" [29], *etc.*, involving many countries are timely held to discuss over the seriousness on resistant microbes. G+ve bacteria differ G-ve radically in the cell wall organization of the structures outside the plasma membrane but below the capsule [30]. G+ve cell wall comprises a thick peptidoglycan layer as the primary component with cell wall teichoic acid (WTA). Compared to G+ve, the G-ve bacteria additionally possesses a membrane (OM; lipopolysaccharide) outside the thin peptidoglycan layer, and this structure is known as the cell envelope (Fig. **3**). In the following sections, we intend to describe the evolution of multidrug resistance in G+ve and G-ve bacteria arising due to differences in their cell wall while briefly highlighting the concerns over the emergence of highly drug-resistant strains of *M.tuberculosis.*

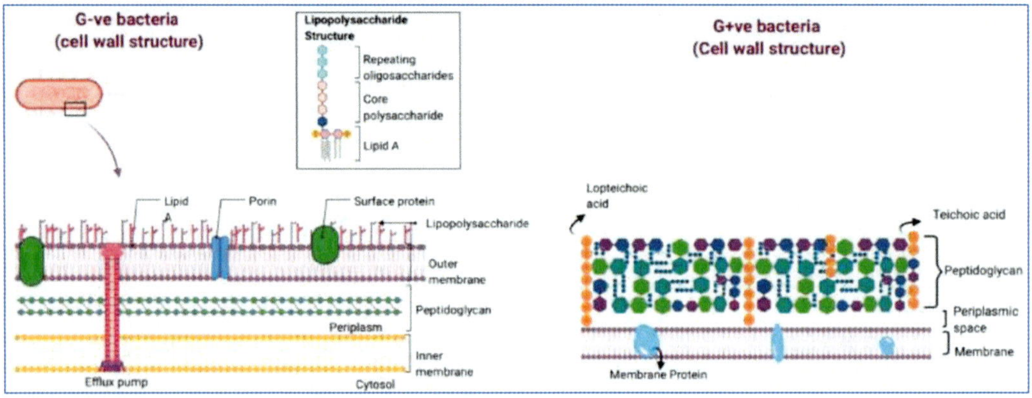

Fig. (3). Cell wall architecture of G+ve and G-ve bacteria. To better understand the drug resistance mechanism, described in this chapter, we highlight the major differences in the cell wall of G-ve (left panel) and G+ve bacteria (right panel). (Figure designed using BioRender.com).

MDR in Gram-negative Bacteria

The presence of an outer membrane obligates a permeability barrier to drugs in G-ve bacteria, attributing to intrinsic antibiotic resistance. Besides this, MDR evolved in G-ve bacteria by several other different mechanisms that are alarming to humans. Antimicrobial Resistance Threats Report by CDC (AR, 2019) has listed G-ve MDR bacteria under three different categories. The first class includes urgent threats, followed by the serious (2nd) and third—concerning threats [31]. The first class includes carbapenem-resistant *Acinetobacter*, carbapenem-resistant *Enterobacteriaceae* (CRE), and drug-resistant *Neisseria gonorrhoeae*; Serious threats include drug-resistant *Campylobacter*, extended-spectrum β-lactamase (ESBL)-producing *Enterobacteriaceae,* multidrug-resistant *P.aeruginosa,* drug-resistant nontyphoidal *Salmonella,* drug-resistant *Salmonella* serotype *Typhi,* and drug-resistant *Shigella.* Concerning threats and watch list include drug-resistant *Bordetella pertussis.* In Europe, there are different programs like ENABLE and MAGNET (European Gram-Negative Antibacterial Engine and Molecules Against Gram-Negative Infections, respectively; https://imi.europa.eu). In 2017,

WHO's priority list (to guide and promote research and development of new antibiotics) included G-ve bacteria in almost all the above categories [32]. Other MDR G-ve bacteria of medical importance include the following: *Acinetobacter* spp., *Citrobacter* spp., *Escherichia coli*, *Klebsiella* spp., *Serratia marcescens*, *Yersinia* spp., *Haemophilus influenzae*, *Helicobacter pylori*, *Legionella pneumophila*, and *Vibrio cholerae*. G-ve bacteria show MDR against all commonly used antibiotics used globally to control bacterial infections, including ureidopenicillins, fourth-generation cephalosporins (cefotaxime, ceftazidime), carbapenems, fluoroquinolones (ciprofloxacin), aminoglycosides, polymyxins, tetracyclines, chloramphenicol, sulphonamides (co-trimoxazole), and glycylcycline.

Besides these threats, the *ESBL* and *CRE* strains, along with strains with mcr-1 gene, pose serious global challenges in terms of their detection, control, and spread. The emergence of *ESBL* strains required the increased use of carbapenem antibiotics [33]. But due to the development of resistance against it, colistin (polypeptide) has been used to bind to the lipopolysaccharides and phospholipids in the OM. Unfortunately, bacteria have evolved to colistin too by the transferrable gene *mcr-1* and spread to different countries in the world (from China to the U.S.A.), affecting farm animals, and humans [34 - 37]. The most prominent strains amongst *CRE* are *K. pneumoniae carbapenemase* (KPC), New Delhi metallo-beta-lactamase (NDM), Oxacillin carbapenemases (OXA-48), Imipenemase metallo-beta-lactamase (IMP), and Verona integron-encoded metallo-beta-lactamase (VIM) [38, 39]. Different mechanisms have been opted by different bugs for evolving against the drugs to kill them to survive and spread infection. Some G-ve bacteria have innate resistance as their natural resistance to vancomycin due to the inability to penetrate the OM, innate ampicillin resistance in *Klebsiella* as well as chloramphenicol and trimethoprim resistance in *Pseudomonas* [29].

Clinically isolated strains act differently during antibiotic treatment. On drug treatment, the pathogenic bacteria usually get filamented without the division of chromosomes. In G-ve bacteria, (*e.g.*, *E. coli* strains upon ciprofloxacin treatment) development of resistance in the multi-chromosome filamented cell occurs mainly by mutations or recombination [40]. Other mechanisms include mutations in genes resulting in target or drug alterations and increased activity of efflux pumps [39]. The drug is also blocked by decreasing the number of porins in the cell wall. Different regulators affect efflux pump levels. Expression of *soxS* (a bacterial stress-response gene) increases ~ 3.5 times higher during antibiotic treatment. A similar change is also seen in the stress gene *rob* [41]. Other mechanisms, such as quorum sensing, biofilms, and bacterial two-component systems, also regulate the expression and activity of efflux pumps [42]. *V.cholerae* uses efflux mechanisms like MATE or integrons or SXT constant for developing resistance. MDR species is seen in Guinea Bissau with transmissible plasmid having a resistance gene for trimethoprim (dhfrXII) and aminoglycoside resistance [43, 44].

Mutations occur in the target protein-encoding gene giving resistance to the antibiotics. *Campylobacter* has mutations in *gyrA* (target genes for fluoroquinolones), 23s rRNA mutations, and rRNA methylase gene *erm(B),* conferring resistance to macrolides; *P. aeruginosa* bears mutations in genes encoding enzymes like DNA gyrases and type IV topoisomerases (GyrA, GyrB, and ParC, ParE) in TOPRIM domain leading to fluoroquinolone resistance and in elongation factor G (*fusA1*) for aminoglycoside resistance [45]. *N. gonorrhea,* the causative agent of sexually transmitted infections, infects 106 million cases annually worldwide [46]. Its infection affects the throat, rectum, and genitals. But MDR in *N. gonorrhea* has evolved to all known drugs used for its treatment. Moreover, its co-infection with *Chlamydia*, another major causative agent for sexually transmitted infections, is of concern and even transmits from the mother to the child during birth. Drug resistance in *Neiserria* arises due to mutations in genes. For instance, mutations in *gyrA* and *parC* genes for Quinolone, *penA* for Penicillin, *porB*, *mtrR*, and *ponA* genes for cefixime or ceftriaxone, *rpsJ* for tetracycline and *mtrR* and 23s RNA for azithromycin drugs [47]. Acquisition of resistance genes by transduction, transformation, and conjugation is also commonly seen in bacteria. The transfer of multiple resistance genes in *Salmonella* sp. consists of *blaTEM-1* (ampicillin resistance) and *dfrA7*, *sul1*, *sul2* (trimethoprim-sulfamethoxazole resistance). The presence of other genes, including the *catA1* for the chloramphenicol resistance and *strAB* for the streptomycin resistance, is associated with an IncHI1 plasmid. Ceftriaxone resistance is gained by acquiring an extended-spectrum beta-lactamase (ESBL; in Italy, Nepal, and France) gene [48]. The AcrAB-TolC efflux pump exports drugs (from the cell) used during the treatment. Overexpression of this pump occurs after exposure to fluoroquinolones, admitting the evolution of MDR [49]. *S. typhimurium* may have as many as 10 types of efflux pumps [50]. Horizontal gene transfer in *Campylobacter* acquired a plasmid harboring tetracycline and kanamycin-resistant genes, *tet(O)* and *aphA-3* [51, 52]. Other modes include transposons (mobile genetic elements, *e.g.*, Tn5053) or integrons like Verona integron encoded Metallo beta-lactamase, *e.g.*, VIMPA [53, 54]. Some multi-resistance regions (MRR) are made up of multiple types of mobile elements, as seen in *Klebsiella* and *Shigella* [55]. An evolving mechanism for developing MDR is also through exporting the resistance genes *via* Outer-inner membrane vesicles (O-IMVs), which involve protrusion from bacteria containing both the outer and the plasma membranes having the bacterial cytoplasmic components, too [56 - 59].

MDR in Gram-positive Bacteria

As discussed above, these bacteria have a thick cell wall and are exploited as a target for the most antibacterial to stop their growth. For example, penicillin has been widely used for treating *Staphylococci* and *Streptococci* infections. Unfortunately, resistance to penicillin was soon developed due to the emergence of the penicillinase enzyme, which resulted in the discovery of more β-lactams antibacterial drugs. But bacteria emerged with resistance to these too. Understanding the cell wall architecture is important to develop newer antibiotics required for bacterial killing. Penicillin-binding proteins (PBPs) are the bacterial catalytic proteins required for a new cell wall synthesis that has a glycosidic backbone consisting of polymerized lipid II by transglycosylase activity of PBPs (repeating units of N-acetylmuramic acid (MurNAc) and N-acetylglucosamine (GlcNAc). PBP's dd-transpeptidases activity carries out the indispensable step for cross-linking to get structural rigidity. The beta-lactam binds to the bacterial DD-transpeptidase enzyme and inhibits its activity, blocking new cell wall synthesis and hence bacterial cell death. G+ve bacteria have thick cell walls consisting thick peptidoglycan layer compared to G-ve bacteria, where OM has a role in blocking antibiotic entry into the cell. Hence, beta-lactams are most effective against G+ve bacteria with high transpeptidase activity. But these bacteria have developed several defense mechanisms in the form of beta-lactam sensors like BlaRs, which get acetylated and relay this sensing to the evolution of resistance in G+ve bacteria posing a significant public health threat. Hence, CDC has given a list of G+ve bacteria for R&D, and new antibiotics needed [31], while WHO has published a priority list for these resistant bacteria. CDC has classified these bacteria under different threat groups as Urgent Threats, including *Clostridium difficile*; Serious Threats, including vancomycin-resistant Enterococci (VRE), methicillin-resistant *Staphylococcus aureus* (MRSA), *Streptococcus pneumoniae*, and *Mycobacterium tuberculosis*; Concerning Threats for the erythromycin-resistant group A. *Streptococcus*, clindamycin resistant group A. *Streptococcus*; and a watch list for *Mycoplasma genitalium*.

We are discussing here, in detail, the top candidate, *S. aureus*. In 1959, methicillin (celbenin) was developed, and in 1961 methicillin resistance evolved in this bacterium against cephalosporins and carbapenems. The resistance is conferred by the modified PBP resulting in low binding of penicillin and other β-lactams. *S. aureus* is often present asymptomatically, with ~ 20% of individuals as a persistent nasal carriers and around 30% are intermittent carriers. MRSA emerged *via* horizontal gene transfer with penicillinase plasmids by bacteriophages. MRSA evolved with the *mecA* gene that generates PBP2 with a lower affinity to bind to beta-lactam antibiotics. The risk continues to be prolonged hospitalization, ICUs, long use of urinary catheters or central venous access, and other practices. MRSA outbreaks are known as both healthcare-associated (HA-MRSA) and community-associated (CA-MRSA). Many humans make antibodies against *S. aureus*. MRSA escapes the immune system by modifying the cell wall by glycosylation [60]. The vaccine development programs in the past have not been successful [61]. MRSA is a versatile pathogen and endures multiple resistance determinants and is found in food-producing animals to seafood (fish) due to extensive antibiotic use in these industries [62]. This issue gives a high chance to spread into the food chain too, both on land and sea. Many strains with resistance have zoonotic potential (moving between humans and animals), among livestock or pets, and wildlife [63]. MDR-MRSA transmission has been found among primates [64]. This bug is so smart that after extreme antibiotic treatment, a small subpopulation of bacteria as a result of phenotypic variation, differentiates into dormant, spore-like cells called persisters [65]. These species mutate highly, evolve against antibiotics used, and have small colony variants (SCVs), where many metabolic alterations result in MDR.

Access to the immune system is also limited by biofilm formation and allowing them to escape [66]. It is accompanied by the biosynthetic pathway downregulation, toxin/antitoxin expression to stop translation, *etc.* Many other genes like *Rel A, RelE, MazF, HipA kinase, ClP/Lon* (proteases), *DksA/ppGpp* (alarmone for TCA regulation), and persister genes *–glpD, plsB*, and *GlpF* (Glycerol metabolism), and by regulating counteraction for reactive oxygen species (ROS) [67 - 69] afford the cell with the ability to endure metabolic changes accepting persistence. MRSA undergoes mutations in the *bshA* and *fosB* genes encoding for bacillithiol (Glutathione counterpart, required for the ROS defense), reducing fosfomycin and H_2O_2 resistance, followed by a decrease in NADPH (required for thioredoxin antioxidant), altogether helping in the survival in the phagolysosome [70]. Some bacteria are resistant to glycopeptides (cell wall inhibitor), *e.g.*, in MDR *Enterococcus faecium* (owing to the presence of Van A-E, and VanG clusters), resulting in low binding affinity to antibiotics [71]. Also, the co-infection in patients (~30%) with VRE and MRSA resulting the transfer of drug-resistant genes. Due to the extensive use of vancomycin antibiotics for treating MRSA infections, the evolution of newer strains - vancomycin-intermediate *S. aureus* (VISA) and VRE resulted in the vancomycin-resistant *S. aureus* (VRSA) [72]. The D-Ala-D-Ala residues act as decoy targets to block the vancomycin antibiotic. Vancomycin physically blocks the D-alanyl-D-alanine residue (DDR) and interferes with the substrates of cell wall formation [73]. Transposon with genes for vancomycin resistance occurs in VRE, where it will replace the DDR with D-Ala-D-Lac [74]. *S. aureus* is resistant to quinolones (inhibits DNA gyrase and topoisomerase IV) by mutating genes of these two enzymes (grlA and gyrB) [75]. Other MDR species are *S. epidermidis* (resistant to exopolysaccharide matrix in biofilms against drug penetration), *S. saprophyticus* (UTI infections). In the case of *S. pneumonia,* the vaccination reduced the disease but also formed non-vaccine serotypes. MDR and XDR in serotype 15A is of concern [76, 77]. It is non-susceptible to many drugs (fluoroquinolones, penicillin, cephalosporins, macrolides, lincosamides, tetracyclines, and carbapenems).

Low-level MDR is reported in *Listeria monocytogenes* to streptomycin, chloramphenicol, trimethoprim, and macrolide. Sensitivity to drugs also decreased with the help of acquired active efflux pumps or transposons (against tetracyclines and fluoroquinolones) [78]. Its lipoteichoic acid (LTA) also plays a role in MDR. LTA is an important polymer of the bacterial cell wall (G+ve) (Fig. **3**). LTA forms an extracellular envelope with peptidoglycan and teichoic acids that protects from lysis. Also, the D-alanylation of LTA plays a major role in MDR [78]. *Mycoplasma genitalium* is sexually transmitted (20% of nongonococcal urethritis) and has increased its dissemination with HIV infections (human immunodeficiency virus). MDR in this organism is increasing as ~ 40% of the strains have resistant to azithromycin and fluoroquinolone. These MDR strains are prevalent in Australia, Japan, the US, and Europe [79].

In a healthcare setting, the MDR infection is also from *C. difficile* (CDI). These bugs produce the toxins - TcdA, TcdB, and CDT [80]. The life cycle of *C. difficile* involves spore formation, which endows it with persistence and resistance. MDR strains are found in the USA, China, and Asia. The strain found in Italy is resistant to rifampicin, imipenem, moxifloxacin, clindamycin, and chloramphenicol [81]. The *cfr* gene (encoding RNA methyltransferase) triggers multiple resistances (mainly to peptidyl transferase inhibitors) by methylation of the 23S rRNA [80, 82]. Similarly, spores formed in *Bacillus anthracis* and *Bacillus cereus* that cause infection by its toxin or capsule have gained resistance to penicillin, ampicillin, cephalosporins, and trimethoprim [73, 83]. Another toxin-mediated infection causative agent is *Corynebacterium diphtheria.* It causes diphtheria facilitated by A-B exotoxin (translational inhibitor). Resistance is acquired by the HGT of genes, but the non-toxigenic strains of *Corynebacterium* and *C. diphtheriae* (NTCD) are considered critical emerging bugs as they are not vaccine-preventable [84, 85].

MDR and XDR in M. tuberculosis

Apart from the above two categories of bacteria classified based on their cell wall, *M. tuberculosis* is considered as a separate class possessing an entirely different cell wall composition rich in mycolic acid. The presence of a highly impermeable outer membrane, along with other mechanisms like drug degrading and modifying enzymes as well as horizontally transferred chromosomal mutations, are responsible for conferring a high level of drug resistance. Moreover, the rapid emergence of multidrug-resistant *M. tuberculosis* (MDR-TB) is a growing public health concern that occurs due to resistance to front-line drugs, especially Rifampicin and Isoniazid. Management of MDR-TB thus required the use of second-line drugs that include fluoroquinolones and aminoglycosides. Geographically, MDR-TB is more challenging in Southeast Asian and African countries like India, China, the Russian Federation, Indonesia, Nigeria, and Pakistan, accounting for about 60% of the global burden [5]. Unfortunately, the advent of even more resistant strains of *M. tuberculosis* (known as extremely drug-resistant tuberculosis or XDR-TB), for the front-line drugs plus fluoroquinolones as well as one of the three injectables (kanamycin, capreomycin, or amikacin) has destabilized the global economy and WHO's END-TB program milestones. MDR- and XDR-TB is mainly arising due to inadequate treatment, the transmission of resistant bacteria, or infection with resistant strains. MDR-TB and XDR-TB are problematic as it involves difficult treatment with second-line drugs, low detection capabilities, a longer treatment regimen (18-24 months) with a lower treatment (55%), and a high death rate (15%). Whole-genome sequencing tools have facilitated the identification of drug-resistant strains for various drugs in use. Unfortunately, due to high costs and limited resources in developing countries, there is an increased spread of highly resistant strains in the community. Therefore, this demands an urgent development of collaborative aids, better and rapid diagnostic tools, identification of specific resistance markers as well as newer strategies and drugs that could significantly reduce the treatment time and complications associated with it.

Multidrug Resistance in Fungi

Infections caused due to fungi are a serious health concern around the world, posing a major threat to those with weakened immunity like HIV/AIDS, cancer, organ transplant, asthma, and corticosteroid therapies. The most commonly occurring fungal pathogens are *Candida, Aspergillus, Fusarium,* and *Cryptococcus* species. These fungal pathogens are responsible for an annual estimated case of over a billion and cause 1.4 million deaths worldwide. Like bacteria, fungi have also developed antifungal resistance. Antifungal resistance, especially MDR in fungi, arises when it becomes completely resistant to at least two classes of drugs simultaneously. Drug resistance in fungi is, therefore, a critical concern due to the limited treatment options available. Antifungal drugs can be classified into five major classes, namely polyenes, pyrimidine analogs, echinocandins, triazoles and allylamines, wherein each class can have several drugs. Azoles (ketoconazole, fluconazole, itraconazole, miconazole, voriconazole) are the most commonly used antimycotics inhibiting cytochrome P450 14α-lanosterol demethylase and block ergosterol synthesis [86]. Echinocandins (caspofungin, micafungin, and anidulafungin) inhibit the synthesis of the cell wall by targeting β-1,3-Glucan synthase [87]. Polyenes (including amphotericin B and Nystatin) bind to membrane sterols, mainly ergosterols and act by changing their permeability and cell death [88]. Pyrimidines (including 5-Flucytosine) inhibit nucleic acid synthesis by targeting thymidylate synthase [89]. Finally, Allylamines inhibit squalene epoxidase enzyme leading to squalene toxicity and thus affecting ergosterol synthesis.

Some fungi are naturally or intrinsically resistant to the drugs, while few develop resistance over time when exposed to antifungal agents. Several factors, like improper use of antifungal agents, extensive use of fungicides in agriculture, and the use of antibiotics for fungal treatments, lead to multidrug resistance in fungi. Drug resistance has been observed extensively in *Candida* and *Aspergillus* species to almost all drug options available. Drug resistance in fungi arises mainly due to (1) a decrease ineffective drug concentration which occurs by overexpression of efflux pumps and/or drug targets, poor pro-drug conversions, and drug sequestration in biofilms; (2) drug target alterations and (3) metabolic bypass. The following sections would briefly illustrate the evolution of multidrug resistance in pathogenic fungi, with special emphasis on *Candida* spp.

Drug Resistance in Candida albicans and NAC species

Among most of the invasive fungal infections occurring in immunocompromised patients, those caused by *Candida albicans* rank leading fungal pathogen of humans. But infections caused by other species of *Candida,* also known as non-albicans are also common, especially in neonates. Worryingly, the emergence of multidrug-resistant *Candida* species, which were naturally susceptible to the known drugs, is becoming another major worldwide concern. The evolution of drug resistance is a rapid phenomenon in *C. albicans*. Antifungal resistance in *Candida* occurs due to a variety of reasons, each of which is briefly detailed below:

(a) *Overexpression of efflux pumps:* Among various mechanisms, overexpression of drug efflux pumps contributes to the primary cause of resistance by reducing the intracellular accumulation of drugs. The genome of *C. albicans* possesses 28 ABC-type and 96 MFS transporters. Drug efflux caused due to either overexpression or transcriptional regulation of CDR1 and CDR2 (Candida Drug Resistance) transporters belonging to the PDR (pleiotropic drug resistance) class as well as MDR1 and FLU1 (Fluconazole resistance) belonging to MFS class are majorly involved in conferring drug resistance to *C. albicans* and *C. dubliniensis* [90 - 95]. Similarly, *C. glabrata* has 18 ABC and 33 MFS transporters.

(b) Alterations in the drug target enzyme: Another important mechanism responsible for drug resistance in *C. Albicans* involves modulation of the ergosterol biosynthetic pathway by the induction of 14α-demethylase expression or point mutations in the ERG11 gene. This altered expression results in reduced affinity to fluconazole and hence resistance [96, 97].

(c) Biofilm formation: Drug sequestration to the extracellular matrix is another very common mechanism of drug tolerance among most *Candida* species [98]. The presence of β-1,3- glucans (produced by enhanced expression of Fks1 encoded Glucan synthase) in the extracellular matrix of *C. albicans* biofilms plays a significant role in sequestering azole, echinocandins, pyrimidines, and polyenes, thus imparting resistance [17, 99 - 101].

(d) Loss of function mutation: Candida also circumvents the toxic effects of drugs by diverting the metabolic functions in the cell. For example, defective or loss of function mutation in *ERG3* gene encoding sterol $\Delta^{5,6}$ desaturase found in several species can bypass the otherwise toxic effects from 3,6-diol derivatives of 14α-methylated sterols. Such mutations have led to resistance to azole and AmB drugs [101].

(e) Cross Resistance and Co-resistance leading to Multidrug Resistance: Simultaneous mutations of *ERG11* (encoding sterol 14α-demethylase) and *ERG5* (encoding C22 desaturase) as seen in *C. Albicans* led to co-resistance for azole and amphotericin B [102]. Other incidences include selective pressures caused due to fluconazole resistance which led to cross-resistance to amphotericin B due to the defective sterol $\Delta^{5,6}$ desaturase [103]. Point mutations in *FKS2* encoding β-1,3-glucan synthase, as well as overexpression of ABC transporters, led to MDR to flucytosine, caspofungin, and azoles in *C. glabrata* [104].

C. auris is a new species identified in 2009 that exhibits features, not like other related fungi and has led to simultaneous exponential nosocomial outbreaks in healthcare facilities around the world. It has raised serious concern as it persists in colonization almost indefinitely and exhibits high transmissibility and shows resistance to multiple drug classes. It is the first fungal pathogen to be considered a public health threat by the Centers for Disease Control (CDC). *C. auris* has shown resistance to azoles, polyenes, and echinocandins. Currently, the mechanism for drug resistance remains less understood but is quite similar to that in *C. albicans*. For instance, mutations in the *ERG11* gene or increased efflux pump activity, or mutations in *FKS1* gene were found to be similar to *C. albicans* resulting in either fluconazole or echinocandin resistance. The resistance mechanism for amphotericin B is not clear yet. Though less complex, *C. auris* also forms a biofilm that pertains to its colonization over the surfaces and is responsible for multidrug resistance [6, 7, 105, 106].

MDR in Aspergillus Fumigatus

A. fumigatus is an invasive pathogen that is the primary cause of the disease aspergillosis. The rate of mortality often even increases to more than 88% in high-risk immunocompromised patients. Triazoles and AmB are the most effective treatment options for dealing with *A. fumigatus*. Unfortunately, their extensive use in agriculture as fungicides and accidental exposures on contact led to the emergence of resistant fungi. Drug-resistant strains of fungi were initially identified in Europe but have now been seen around the globe. Drug resistance to triazole is contributed by characteristic mutation TR_{34}/L98H in the gene encoding Cyp51A [7]. Other incidences include high expression of efflux pumps belonging to either ABC type (AfuMDR4) or MFS type (AfuMDR3) leading to resistance to itraconazole [107].

MDR in Fusarium

Fusarium is usually a plant pathogen but is also identified to cause human fusariosis. *Fusarium* is resistant to almost all classes of antifungals known. Few *Fusarium* spp. exhibits inherent primary resistance mechanisms without prior exposure to the antifungals. Secondary resistance develops for azole in previously susceptible strains. Unlike *A. fumigatus,* secondary resistance arises due to the deletion of *CYP51* encoding 14α-demethylase. *CYP51A* deletion makes *Fusarium graminearum* susceptible to azoles and other antifungals. Additionally, mutations in the *FKS1* gene (P647A and F639Y) might lead to intrinsic resistance against echinocandins in *Fusarium solani.* Efflux pumps are also predicted to be involved in incurring azole resistance in *Fusarium* spp. Cross-resistance for azoles, echinocandins, and polyenes are also commonly seen in *Fusarium.* The actual mechanism of resistance is yet to be understood [108].

Multidrug Resistance in Viruses

Similar to antibiotic resistance, the resistance against antiviral drugs is also well documented and increasing with the ever-growing mutational repertoire of the viral genome. The mitigation of drug resistance against viral infection is more challenging due to its obligate intracellular parasitic nature. The following sections explore some of the most prevalent forms of viruses and their current status of antiviral drug resistance.

MDR in Influenza Virus

Influenza virus is one the most prevalent form of virus found across the globe and is divided into classes of; type-A, type-B, type-C, and type-D. According to the Center for Disease Control and Prevention (CDC)-USA, type A and B cause seasonal epidemics every year in winter in the United States called flu—a form of the disease that causes acute febrile respiratory disease in young children, the elderly and immunocompromised individuals. Type-A has only the potential to emerge as a source of a global pandemic, however, type-C is considered a very mild form of influenza that does not qualify as an epidemic, and type-D is only limited to animals [109, 110].

For influenza A, two major classes of drugs are employed, 1) M2 Ion channel inhibitor (rimantadine and amantadine) and 2) neuraminidase inhibitor (NAIs; laninamivir, peramivir, oseltamivir, zanamivir). The M2 ion channels are tetrameric integral membrane protein that allows pH-sensitive gating on the host endosome surface [111], upon acidification, it leads to completion biogenesis, facilitated by the release of virion particles [112]. The drug adamantanes bind to the pore of the M2 Channel to block proton conductance and viral biogenesis [113]. The Influenza-virus (IAV) acquires resistance to the adamantanes by key mutations of pore-lining [114].

Adamantanes, and later included rimantadine [115], were highly successful drugs with an efficacy rate close to 90% [116], but just in the last two decades, the resistance increased significantly from 1%, before the year 2000 globally to; 73.8% in China, 22.7% in Taiwan, 69.6% in Hong Kong as a major contributor for overall 27% in the entire Asian continent, while Europe, North America, and South America had less than 5% that now encompasses overall 12.3% increase [117]. The single mutation of S31N is responsible for 98.2% of resistant cases against the drug adamantane [117].

These studies strongly support the notion of the development of drug resistance among IAVs with crucial mutations, and some of the key reasons for such adaptation are conferred on the nature of intermediate hosts—pigs, birds, and horses, like animals, act as reservoirs. Therefore, the development of antiviral medicine and universal vaccine, along with global surveillance of circulating intermediate hosts, becomes extremely important to restrict drug resistance.

MDR in Human Immunodeficiency Virus (HIV)

HIV infects the immune cells, causing acquired immune deficiency syndrome (AIDS) if it remains untreated. There are two forms of HIV viruses; HIV-1 is almost synonymous with HIV, while HIV-2 is very limited in west Africa and less than 0.01% in the population of the USA. The development of the combination of antiretroviral drugs leads to remarkable control and management of HIV [118]. Unfortunately, the development of resistance in HIV against these drugs is compromising their effectiveness [119].

HIV is a challenging viral infection for effective treatment due to; the very high genetic variability conferred by error-prone reverse transcriptase (RT) [120]. There are five commonly used classes of drugs for antiretroviral therapy (ART); 1) Protease inhibitors (PIs), 2) Nucleoside reverse transcriptase inhibitors (NRTIs), 3) Non-nucleoside reverse transcriptase inhibitors (NNRTIs), 4) Integrase strand transfer inhibitors (INSTIs) and 5) C-C motif chemokine receptor-5 (CCR5) antagonist [121]. Almost all drug resistance arises with the selection pressure of applied medicine; some drug-resistant mutation requires more than one site of alternation, while sometimes a single mutation confers the resistance. Usually, there is no cross-resistance; one class of resistant HIV remains completely susceptible to the other-unrelated class that patient has not been administered [122], however, significant resistance has been observed within the same class of ART drugs [123].

Protease inhibitors are key drugs applied for ART, stopping the proteases to act on polyproteins – responsible for the release of mature viral functional proteins. The key resistance associate mutation for this class may be found in over (Stanford University drug resistance database). Reverse transcriptase is targeted by two major classes of drugs: 1) Nucleoside reverse transcriptase inhibitors, and 2) Non-nucleoside reverse transcriptase. Moreover, the other class, Integrase, is an essential enzyme of HIV and an excellent drug target, which incorporates the viral genome into the host's cell chromatin. Integrase strand transfer inhibitors (INSTIs) are among the newest class of ART. The major mutations responsible for conferring resistance against these four classes are available in the Stanford University HIV drug resistance database (https://hivdb.stanford.edu/).

The "Maraviroc" is a special ART that prevents the binding of the tropical receptor of HIV-1 gp120 to seven transmembranes G protein-coupled R5 receptors [124]. The most common reason to acquire the resistance among the patients undergoing R5 receptor inhibitor is the elongation of the CXCR4 (X4) region on the surface of the virus [125], however, a rare binding of HIV-1 gp 120 to inhibitor-bound R5 receptor can also take place in some mutations [126].

There are significant classes of inhibitors that have been developed for ART, however, the error labile nature of viral genome amplification, always seeds the new mutation classes that compromise the efficacy to render the drug altogether ineffective. Therefore, a selective combination of unrelated classes has greater success in checking the viral load, along with this, a continuous understanding of fundamental principles always paves the way to discover new drug targets, better than previous generation drugs.

ACTION PLANS FOR COMBATING MULTIDRUG RESISTANCE AND THE GLOBAL SURVEILLANCE PROGRAMS

The main drivers of MDR microbes are the misuse and overuse of antibiotic drugs. Therefore, antimicrobial stewardship programs (ASP) are required to optimize the use of antimicrobials. ASP includes coordinated interventions aimed at achieving the appropriate use of antimicrobial drugs for enhanced health outcomes and reduced resistance by promoting the selection of the best procedures. Many steps, like education (improving awareness and understanding of resistance in microbes), deep surveillance, laboratory strategies (better detection and new equipment), pharmacy strategies (optimized effective dosing), new medicines and vaccines, *etc*., are included. Several organizations conduct ASP with education and advocacy programs for resistance control and access to effective antibiotics, *e.g.*, CDC [127], SHEA [128], EU (ESAC, ESCMID, ESGAP) [129], Do Bugs Need Drugs-CCAR [130], and WHO [131]. ASP is gaining thrust globally, Sweden (100% ASP), the United States (88% ASP), France (81% ASP), and the United Kingdom (77% ASP) [132]. Sharing of these best practices across the globe (including different institutions, regions, and countries) must continue for ASP to progress. It is well supported by the establishment of TATFAR [133] for the US and European activities and programs. To expedite the complete implementation, a stronger direction from WHO will play a big role; WHO has developed the global action plan (GAP) on microbial resistance [46, 134 - 136]. Many required steps like expanded coverage (including services like hygiene, preventative care, and vaccination to slow the spread of resistance), better oversight and quality of care (improve oversight in health care practices), smarter, fairer financing (in the case of treatable infections, to close existing access gaps), better information (surveillance, monitoring, and response), and improved stewardship and governance (improve awareness and understanding) are needed globally [137]. World Bank proposes to use the GAP where his recommendations as to the options for a country's action plan are structured in different sectors like health, agriculture, water, sanitation, and hygiene.

ADVANCES TO DEVELOP NEWER THERAPEUTICS FOR MULTIDRUG RESISTANCE

Recent studies demonstrated that the emergence of MDR bacterial infections, viral and fungal infections alarm our scientific communities, urging us to identify novel drugs with a different mechanism of action. We present a few of the advances that need to be exploited and implemented into real practice for treating MDR microbes. In the subsequent section, we also discuss alternatives to conventional therapeutics that have potential and found applications in the field of medicine.

Fecal Microbiota Transplantation (FMT)

FMT is the engraftment of fecal microbiota from a healthy donor to re-establish it in the gut of patients with MDR microbes. FMT has served to be a promising therapy for intestinal MDR bacterial decolonization while restoring the normal gut microbiota in patients with MDR. The success rate is 70–94% (*C. difficile*). FMT decolonized the vancomycin-resistant *Enterococci*, ESBL and carbapenemase-producing *Enterobacteriaceae*, and MRSA [138 - 143]. FMT has also worked well on immuno-compromised patients too. The safety of FMT is also tested in lentivirus-infected primates and showed increased numbers of Th17 and Th22 cells (which significantly decrease HIV or SIV infection) on performing FMT [144]. Successful FMT is also performed in dogs having diarrhea coupled with parvovirus infection [145].

CRISPR-Cas System

This is a DNA editing technology adaptive from bacteria and archaea where it provides a defense against viral attacks or plasmid challengers. The foreign nucleic acid fragments are integrated into the host chromosome at a repetitive element known as a CRISPR (clustered regularly interspaced short palindromic repeat). The other component is a protein associated with CRISPR called Cas9 that has nuclease activity and uses a small RNA guide to specify the site of cleavage [146]. Reprogrammed CRISPR-Cas-based genomic editing technique is very specific for the target and forms chances to manipulate MDR microbes [147]. It could be utilized to target sequence-specific MDR microbes while not affecting avirulent strains, for example, targeting plasmids harboring the resistance genes in *Corynebacterium, Staphylococcus,* and *M.tuberculosis* [148 - 151]. The antimicrobial function of CRISPR-Cas has already been documented in a mouse model to kill *S. aureus*. CRISPR has also been applied in *C.albicans* to delete the genes encoding efflux pumps and biofilm factors. This technology can also be extended to target other clinically relevant fungal pathogens like *C. auris*, *Cryptococcus*, *Aspergillus*, and *Mucorales* [152 - 155]. CRISPR-Cas system also finds its application to target different sequences of HIV. The small RNA guide is designed to match the sequences (stored in the patient's reservoir), conserved sites in the virus genome, and co-receptor CCR5 or CXCR4 [156, 157]. At the technical and ethical levels, the use of CRISPR-Cas in humans with MDR infections needs more advances to be implemented. But there lies a big hope in CRISPR-Cas.

Metallo-Drugs and Metallo-nanoparticles in MDR

MDR microbes (bacteria, viruses, and fungi) cause infections that are hard to control and are alarming to identify novel drugs with different mechanism of action [158]. The coordination of metal complexes, a combination of metal and organic compounds, has focused much attention on an alternative approach to antimicrobial therapy, which may overcome many issues such as solubility, toxicity, and resistance of individual metal ions and ligands. Several metal complexes such as bismuth-, gallium-, gold-, iron-, palladium-, silver-, ruthenium- and copper-based coordination complexes have attempted human clinical trials for curing cancer, malaria, and other neurodegenerative diseases [159 - 161]. The discovery of Salvarsan [162], Auranofin [163], and Cisplatin [164], *etc*., is the authentic evidence to dive deep into the search for metallodrugs against multidrug resistance. While several targets are exposed, nucleic acids (DNA or RNA) and membranes are the most vulnerable targets for antimicrobial drugs. The mode of action for the metal complexes is distinct, and the former is based on the intercalation between DNA base pairs or DNA cleavage, while the latter is based on either loss of membrane integrity or lipid peroxidation. Since the bioavailability of copper ions are prominent in the human body, the focus of this chapter is the interaction of copper (II) complexes with either nucleic acids or membranes. In the last few decades, copper intercalating complexes have been well-documented, creating a roadmap to discovering novel antimicrobial agents [165, 166]. The mechanism underlying those complexes is that they selectively target the DNA, cleaving into either single-stranded- (ssDNA) or double-stranded- (dsDNA) DNA, which also sheds light on the retrospective studies of old copper complexes that exhibited "Chemical-nuclease" activity [167]. The future direction toward the finding of antimicrobial drugs will include simple, modified, and mixed-ligand copper(II) complexes [168, 169]. The advantages of mixed ligand copper (II) complexes are to incorporate recognition elements such as amino acids, peptides, nucleotides, more specifically, antibiotics, along with intercalation ligands (simple, substituted-, and modified-1,10-phenanthrolines). To illustrate the mechanism of action in bacteria, we have devised several mutants to quantify the reactive oxygen species (ROS), iron-metabolism, generated by the copper (II) complexes as they are redox-active complexes, which will for sure create a huge impact on the discovery of drugs, alternative to multidrug resistance.

In recent years nanomaterials are playing a vital role in medicine and technology. One of the main applications is to deliver either metal ions or the metal-complexes to the specific target. For instance, cisplatin-based nanomaterials deliver cisplatin to the cancer cell specifically. Furthermore, the chemistry behind the metal-based nanocomplexes is also considered a prominent solution against multidrug resistance, and its mechanism of action is entirely different from traditional drugs, exhibiting resistance activity. The merits of metal-based nanoparticles over the metal complexes are size, shape, roughness, and surface energy [169], which are crucial for their *in vivo* activity. Several studies provide a detailed explanation of the copper nanoparticles (CuNP) that exhibited excellent antimicrobial activity against a wide range of microorganisms. However, the molecular level of understanding is still lacking, due to the limited knowledge of the interface between chemistry and microbiology. The future direction will be required to focus on the refurbished metallodrugs and their impact on the MDR microbes.

Use of Antibiotic Adjuvants

The adjuvants are compounds having the potential to enhance the bactericidal activities of antibiotics. It helps in killing resistant microbes in different ways, boosting host response, antibiotic accumulation, inhibiting signaling, *etc.*, and therefore also blocks the resistance [170]. For example, Zn^{2+} is used by the metallo beta-lactamases to cleave the antibiotic (carbapenems). But a natural fungal compound, Aspergillomarasmine A (AMA), sequesters the Zn^{2+} metal and enhances the meropenem's antibiotic activity on NDM-1 or VNM. Interestingly, other metals like Cu^{+2} ions also act as carbapenem adjuvants by metal exchange [171]. Similarly, adjuvants (after screening compound libraries) could be used against efflux pumps or biofilm to stop diverse MDR microbes.

Herbal Medicine

Herbal extracts have a wide spectrum of activity with minimal side effects (*e.g.*, Apocynaceae family, used much in the Ayurvedic system of medicine). Many plant secondary metabolites have been used for broad antimicrobial activities, *e.g.*, flavonoids, alkaloids, essential oils, *etc*. [172]. Anti-microbial activities are shown against many pathogenic microbes like *Acinetobacter, Staphylococcus, Streptococcus, Bacillus, E. coli, M. tb, S. typhi, Candida, Fusarium, etc*. Therefore, a metabolic engineering approach toward the secondary metabolites of medicinal plants or marine microbes could be applied for MDR control [173, 174].

Quorum Sensing (QS) Inhibitors

Quorum sensing is a bacterial communication system based on cell number using small molecules to regulate virulence. For QS, G-ve bacteria use N-acyl homoserine lactones and G+ve bacteria use ribosomally produced peptides. Several molecules have been identified that inhibit QS in many bacteria. A few of these include M64 which inhibits QS of *P. aeruginosa* and 5'-methylthi--DADMe-ImmucillinAs, inhibiting QS of *E. coli* and *Vibrio cholerae*. More chemical libraries with similar actions need to be screened and tested for use as effective antimicrobials.

Phage Therapy

Phages are gaining an advantage over antibiotics because of their stability in the environment. They have found a positive impact on their use in the development of new therapeutics to control MDR microbes [175, 176]. *Pseudomonas* phage inhibits *Candida* and *Aspergillus* by blocking the fungal metabolism and also possesses potent antiviral activity against human adenovirus [177, 178]. Phage phi11 endolysin (*via* endopeptidase) shows anti-biofilm activity in *Staphylococcus* [179]. Another phage with lytic endolysin inhibits biofilms in *Streptococcus* [180].

Antimicrobial Peptides (AMP)

These are the peptides (~ 5-100 amino acid residues) used by microbes as a defense against invasion by other species. AMP directly kills a vast class of pathogens, bacteria, fungi, and viruses [181, 182]. They act on membranes as they have a positive charge (Arg/Lys residues) and hydrophobicity too. A few of the examples include P5 and P9 peptides that displayed activity against MRSA [183]; cationic peptides having antibacterial activity against VRE strains [184].

Therapies for Oxidant: Antioxidant Balance

Oxidants like reactive oxygen or nitrogen species (superoxide, hydrogen peroxide, peroxynitrite ions, and radicals) are produced inside the aerobic cells continuously (accidental oxygen collision at the flavin center or by Fenton chemistry) and comprise threats to the proper cell growth [185 - 187]. To counter this effect, several redox molecules, enzymes, or mechanisms exist in the cell [188]. Even macrophages produce reactive oxygen (ROS) or reactive nitrogen species (NO) to kill microbes [189]. In this regard, quantum dot therapeutics (QD therapy) could use oxygen for oxidant species generation and targeting MDR cells. Using the redox chemistry, many types of oxidant species ($O2^-$, $OH^\bullet$, H_2O_2, and NO or ONOO−) can be formed inside the MDR microbes resulting in multiple cell damages [190]. In the case of *M. tuberculosis*, several front-line drugs are prodrugs and need activation by the oxidants from the host cell. Unfortunately, the bacilli have evolved to neutralize these oxidants making the drugs ineffective. Therefore, it is an urgent need to develop methods to increase/proper use of oxidants to make the drug active and stop bug growth.

Vaccines

The use of vaccines to prevent the spread of antimicrobial resistance is at a boom and requires global implementation. They have an advantage over traditional antibiotics as they are sustainable without causing resistance. There is a minimal likelihood for microbes to develop resistance to vaccines because of various reasons, like (1) vaccines are used as prophylaxis when the microbial load is less to develop resistance; (2) multiple modes of action of vaccines which would require several mutations to generate resistance; (3) even if resistance is observed, disease reduction is quite effective [191]. Moreover, in contrast to the discovery of antibiotic options coming up, numerous vaccines are being developed having activity against antimicrobial-resistant pathogens [192]. Next, the use of vaccines has been shown to reduce the occurrence of disease and has also reduced the use of antibiotics. For instance, the introduction of the pneumococcal conjugate vaccine for pneumococcal infection has indicated a 64% drop in antibiotic use and reduced incidences of infection [193]. Finally, vaccines preserve the microbiome, unlike the disruptions observed with the use of antibiotics [194].

Monoclonal Antibodies

Other promising alternatives to antibiotics are pathogen-specific monoclonal antibodies (MAbs), which are safe and potent molecules having the ability to reduce the spread of resistance by reducing the antibiotic load. They have been increasingly used for rheumatologic and inflammatory diseases and oncological practices. Currently, only three mABs have been licensed for bacterial infections, namely raxibacumb and obiltoxaximab for inhalational anthrax and bezlotoxumab against the *C.difficile* toxin B. In contrast, the only licensed mAB for viral infection is Palivizumab, which is required for the treatment of respiratory syncytial virus (RSV). Numerous other mAB candidates are in the clinical trials pipeline to be used against HIV, *C.difficile*, rabies prophylaxis, and *S.aureus* [195, 196].

CONCLUSION

Drug resistance is an ever-evolving problem creating serious health concerns throughout the world. Wide exposure to antimicrobials in all sectors of life, like in agriculture, hospitals, the pharma industry, or hospital wastewater treatment, has led to selection pressure and the evolution of resistance in almost all microbes to adapt to these drugs. The rate of emergence of resistant microbes far exceeds the frequency at which newer therapeutic options are developed. Moreover, co-infection with more than one MDR microbe in patients is tough to control. For instance, one-third of deaths by MDR microbes are due to MDR-TB co-infected with HIV. It has been seen in the majority that drug resistance is mainly arising and persistent in developing nations and henceforth travels to other parts of the world, thus making it a global health problem. Furthermore, factors like lack of resources, inability to understand the dynamicity of disease resistance, high cost as well as improper treatments and diagnosis have led to the rapid spread of resistance within the population. To circumvent the problem of drug resistance, there is an urgent need to think and act multi-dimensionally. Development of rapid diagnostic measures to detect MDR strains of microbes, proper treatment, controlled use of antimicrobials, and involvement of newer alternative approaches. Finally, educating and spreading awareness within the rural and urban societies as well as collaborative socio-political commitments for the proper use of antimicrobials would aid to curb the disease.

CONSENT OF PUBLICATION

Declared None

CONFLICT OF INTEREST

The author declares no conflict of interest, financial or otherwise.

ACKNOWLEDGEMENTS

We acknowledge all the authors for contributing equally to writing this book chapter. SKR and NM conceptualized the idea for designing the outline of the chapter. SKR, NM, PK, and RS wrote the manuscript. Funding was provided in part by the Division of Intramural Research, National Institute of Allergy and Infectious Diseases and National Institutes of Health.

REFERENCES

[1] Pérez-Rodríguez F, Mercanoglu Taban B. A State-of-Art Review on Multi-Drug Resistant Pathogens in Foods of Animal Origin: Risk Factors and Mitigation Strategies. Front Microbiol 2019; 10: 2091. [PMID: 31555256]

[2] Pendleton JN, Gorman SP, Gilmore BF. Clinical relevance of the ESKAPE pathogens. Expert Rev Anti Infect Ther 2013; 11(3): 297-308. [PMID: 23458769]

[3] Ghafur A, Shankar C, GnanaSoundari P, *et al.* Detection of chromosomal and plasmid-mediated mechanisms of colistin resistance in *Escherichia coli* and *Klebsiella pneumoniae* from Indian food samples. J Glob Antimicrob Resist 2019; 16: 48-52. [PMID: 30244040]

[4] Khan AU, Maryam L, Zarrilli R. Structure, Genetics and Worldwide Spread of New Delhi Metallo--lactamase (NDM): a threat to public health. BMC Microbiol 2017; 17(1): 101. [PMID: 28449650]

[5] WHO. WHO TB Report 2019; 7.

[6] Lockhart SR. *Candida auris* and multidrug resistance: Defining the new normal. Fungal Genet Biol 2019; 131: 103243. [http://dx.doi.org/10.1016/j.fgb.2019.103243] [PMID: 31228646]

[7] Forsberg K, Woodworth K, Walters M, *et al. Candida auris*: The recent emergence of a multidrug-resistant fungal pathogen. Med Mycol 2019; 57(1): 1-12. [PMID: 30085270]

[8] Marimani M. Combination therapy against multidrug resistance. Combination Therapy Against Multidrug Resistance 2020; 39-64.

[9] Nikaido H. Prevention of drug access to bacterial targets: Permeability barriers and active efflux. Science (80-) 1994; 264: 382-8.

[10] Cantón R, Ruiz-Garbajosa P. Co-resistance: an opportunity for the bacteria and resistance genes. Curr Opin Pharmacol 2011; 11(5): 477-85. [PMID: 21840259]

[11] Blanco P, Hernando-Amado S, Reales-Calderon JA, *et al.* Bacterial multidrug efflux pumps: much more than antibiotic resistance determinants. Microorganisms 2016; 4(1): 14. [PMID: 27681908]

[12] Belshe RB, Smith MH, Hall CB, Betts R, Hay AJ. Genetic basis of resistance to rimantadine emerging during treatment of influenza virus infection. J Virol 1988; 62(5): 1508-12. [PMID: 3282079]

[13] Hirsch MS, Conway B, D'Aquila RT, *et al.* Antiretroviral Drug Resistance Testing in Adults With HIV InfectionImplications for Clinical Management. JAMA 1998; 279: 1984-91. [PMID: 9643863]

[14] Mellado E, Garcia-Effron G, Alcázar-Fuoli L, *et al.* A new Aspergillus fumigatus resistance mechanism conferring *in vitro* cross-resistance to azole antifungals involves a combination of cyp51A alterations. Antimicrob Agents Chemother 2007; 51(6): 1897-904. [PMID: 17371828]

[15] Marichal P, Koymans L, Willemsens S, *et al.* Contribution of mutations in the cytochrome P450 14α-demethylase (Erg11p, Cyp51p) to azole resistance in *Candida albicans*. 1999.

[16] Ferrero L, Cameron B, Crouzet J. Analysis of gyrA and grlA mutations in stepwise-selected ciprofloxacin-resistant mutants of *Staphylococcus aureus*. Antimicrob Agents Chemother 1995; 39(7): 1554-8.
[PMID: 7492103]

[17] Nett JE, Crawford K, Marchillo K, Andes DR. Role of Fks1p and matrix glucan in *Candida albicans* biofilm resistance to an echinocandin, pyrimidine, and polyene. Antimicrob Agents Chemother 2010; 54(8): 3505-8.
[PMID: 20516280]

[18] Floss HG, Yu TW. Rifamycin-mode of action, resistance, and biosynthesis. Chem Rev 2005; 105(2): 621-32.
[PMID: 15700959]

[19] McMurry L, Petrucci RE Jr, Levy SB. Active efflux of tetracycline encoded by four genetically different tetracycline resistance determinants in *Escherichia coli* . Proc Natl Acad Sci USA 1980; 77(7): 3974-7.
[PMID: 7001450]

[20] Martinez JL, Sánchez MB, Martínez-Solano L, *et al.* Functional role of bacterial multidrug efflux pumps in microbial natural ecosystems. FEMS Microbiol Rev 2009; 33(2): 430-49.
[PMID: 19207745]

[21] Cannon RD, Lamping E, Holmes AR, *et al.* Efflux-mediated antifungal drug resistance. Clin Microbiol Rev 2009; 22(2): 291-321.
[PMID: 19366916]

[22] George AM, Levy SB. Amplifiable resistance to tetracycline, chloramphenicol, and other antibiotics in *Escherichia coli* : involvement of a non-plasmid-determined efflux of tetracycline. J Bacteriol 1983; 155(2): 531-40.
[PMID: 6348022]

[23] Piddock LJV. Clinically relevant chromosomally encoded multidrug resistance efflux pumps in bacteria. Clin Microbiol Rev 2006; 19(2): 382-402.
[PMID: 16614254]

[24] Davies J. Inactivation of Antibiotics and the Dissemination of Resistance genes. Science 1994; 264(5157): 375-82.

[25] Quinn JP, Dudek EJ, DiVincenzo CA, Lucks DA, Lerner SA. Emergence of resistance to imipenem during therapy for *Pseudomonas aeruginosa* infections. J Infect Dis 1986; 154(2): 289-94.
[PMID: 3088133]

[26] Munita JM, Arias CA. Mechanisms of Antibiotic Resistance. Microbiol Spectr 2016; 4(2).
[http://dx.doi.org/10.1128/microbiolspec.VMBF-0016–2015] [PMID: 27227291]

[27] Coen DM, Schaffer PA. Two distinct loci confer resistance to acycloguanosine in herpes simplex virus type 1. Proc Natl Acad Sci USA 1980; 77(4): 2265-9.
[PMID: 6246531]

[28] Morfin F, Thouvenot D. Herpes simplex virus resistance to antiviral drugs. J Clin Virol 2003; 26(1): 29-37.

[29] Exner M, Bhattacharya S, Christiansen B, *et al.* Antibiotic resistance: What is so special about multidrug-resistant Gram-negative bacteria? GMS Hyg Infect Control 2017.

[30] Salton MRJ, Kim K-S. Medical Microbiology 4th ed., 1996.

[31] Centers for Disease Control and Prevention. Biggest Threats and Data | Antibiotic/Antimicrobial Resistance | CDC. 2019.

[32] WHO priority pathogens list for R&D of new antibioticse. http://www.who.int/ en/news-room/detail/27-02-2017-who-publishes-list-of-bacteria-for-which-new-antibiotics-are-urgently-needed

[33] Gharbi M, Moore LSP, Gilchrist M, *et al.* Forecasting carbapenem resistance from antimicrobial consumption surveillance: Lessons learnt from an OXA-48-producing *Klebsiella pneumoniae* outbreak in a West London renal unit. Int J Antimicrob Agents 2015; 46(2): 150-6. [http://dx.doi.org/10.1016/j.ijantimicag.2015.03.005] [PMID: 25979640]

[34] Giufrè M, Monaco M, Accogli M, Pantosti A, Cerquetti M. Emergence of the colistin resistance mcr-1 determinant in commensal *Escherichia coli* from residents of long-term-care facilities in Italy. J Antimicrob Chemother 2016; 71(8): 2329-31. [http://dx.doi.org/10.1093/jac/dkw195] [PMID: 27261262]

[35] Irrgang A, Roschanski N, Tenhagen BA, *et al.* Prevalence of mcr-1 in E. coli from livestock and food in Germany, 2010-2015. PLoS One 2016; 11(7): e0159863. [http://dx.doi.org/10.1371/journal.pone.0159863] [PMID: 27454527]

[36] Liu YY, Wang Y, Walsh TR, *et al.* Emergence of plasmid-mediated colistin resistance mechanism MCR-1 in animals and human beings in China: a microbiological and molecular biological study. Lancet Infect Dis 2016; 16(2): 161-8. [http://dx.doi.org/10.1016/S1473-3099(15)00424-7] [PMID: 26603172]

[37] Skov RL, Monnet DL. Plasmid-mediated colistin resistance (mcr-1 gene): three months later, the story unfolds. Euro Surveill 2016; 21(9): 30155. [http://dx.doi.org/10.2807/1560-7917.ES.2016.21.9.30155] [PMID: 26967914]

[38] Nordmann P, Naas T, Poirel L. Global spread of carbapenemase producing Enterobacteriaceae. Emerg Infect Dis 2011; 17(10): 1791-8. [http://dx.doi.org/10.3201/eid1710.110655] [PMID: 22000347]

[39] Ruppé É, Woerther PL, Barbier F. Mechanisms of antimicrobial resistance in Gram-negative bacilli. Ann Intensive Care 2015; 5(1): 61. [http://dx.doi.org/10.1186/s13613-015-0061-0] [PMID: 26261001]

[40] Bos J, Zhang Q, Vyawahare S, Rogers E, Rosenberg SM, Austin RH. Emergence of antibiotic resistance from multinucleated bacterial filaments. Proc Natl Acad Sci USA 2015; 112(1): 178-83. [http://dx.doi.org/10.1073/pnas.1420702111] [PMID: 25492931]

[41] Viveiros M, Dupont M, Rodrigues L, *et al.* Antibiotic stress, genetic response and altered permeability of E. coli. PLoS One 2007; 2(4): e365. [http://dx.doi.org/10.1371/journal.pone.0000365] [PMID: 17426813]

[42] Amaral L, Martins A, Spengler G, Molnar J. Efflux pumps of Gram-negative bacteria: what they do, how they do it, with what and how to deal with them. Front Pharmacol 2014; 4: 168. [http://dx.doi.org/10.3389/fphar.2013.00168] [PMID: 24427138]

[43] Dalsgaard A, Forslund A, Petersen A, *et al.* Class 1 integron-borne, multiple-antibiotic resistance encoded by a 150-kilobase conjugative plasmid in epidemic vibrio cholerae O1 strains isolated in Guinea-Bissau. J Clin Microbiol 2000; 38(10): 3774-9. [http://dx.doi.org/10.1128/jcm.38.10.3774-3779.2000] [PMID: 11015401]

[44] Narendrakumar L, Gupta SS, Johnson JB, Ramamurthy T, Thomas S. Molecular Adaptations and Antibiotic Resistance in *Vibrio cholerae*: A Communal Challenge. Microb Drug Resist 2019; 25(7): 1012-22. [http://dx.doi.org/10.1089/mdr.2018.0354] [PMID: 31021308]

[45] Mehla K, Ramana J. Molecular Dynamics Simulations of Quinolone Resistance-Associated T86I and P104S Mutations in Campylobacter jejuni gyrA: Unraveling Structural Repercussions. Microb Drug Resist 2018; 24(3): 232-43. [http://dx.doi.org/10.1089/mdr.2017.0020] [PMID: 28759328]

[46] WHO. Sexually transmitted infections (STIs): key facts. 2016. [http://dx.doi.org/10.1016/j.jiec.2015.10.029]

[47] Yang Y, Yang Y, Martin I, *et al.* NG-STAR genotypes are associated with MDR in Neisseria

gonorrhoeae isolates collected in 2017 in Shanghai. J Antimicrob Chemother 2020; 75(3): 566-70. [http://dx.doi.org/10.1093/jac/dkz471] [PMID: 31713620]

[48] Narasanna R, Chavadi M, Chandrakanth K. Prevalence of multidrug-resistant Salmonella typhi in typhoid patients and detection of blaCTX-M2 and blaCTX-M9 genes in cefetoxime-mediated extended spectrum β-lactamase-producing Salmonella typhi isolates. Biomed Res (Aligarh) 2018; 29(14).
[http://dx.doi.org/10.4066/biomedicalresearch.29-18-794]

[49] Grimsey EM, Weston N, Ricci V, Stone JW, Piddock LJV. Overexpression of RamA, which regulates production of the multidrug resistance efflux pump AcrAB-TolC, increases mutation rate and influences drug resistance phenotype. Antimicrob Agents Chemother 2020; 64(4): e02460-19. [http://dx.doi.org/10.1128/AAC.02460-19] [PMID: 31988103]

[50] Yamasaki S, Nagasawa S, Fukushima A, Hayashi-Nishino M, Nishino K. Cooperation of the multidrug efflux pump and lipopolysaccharides in the intrinsic antibiotic resistance of Salmonella enterica serovar Typhimurium. J Antimicrob Chemother 2013; 68(5): 1066-70. [http://dx.doi.org/10.1093/jac/dks528] [PMID: 23378414]

[51] Tang Y, Dai L, Sahin O, Wu Z, Liu M, Zhang Q. Emergence of a plasmid-borne multidrug resistance gene cfr(C) in foodborne pathogen Campylobacter. J Antimicrob Chemother 2017; 72(6): 1581-8. [http://dx.doi.org/10.1093/jac/dkx023] [PMID: 28186558]

[52] Crespo MD, Altermann E, Olson J, Miller WG, Chandrashekhar K, Kathariou S. Novel plasmid conferring kanamycin and tetracycline resistance in the turkey-derived Campylobacter jejuni strain 11601MD. Plasmid 2016; 86: 32-7.
[http://dx.doi.org/10.1016/j.plasmid.2016.06.001] [PMID: 27268853]

[53] Partridge SR, Kwong SM, Firth N, Jensen SO. Mobile genetic elements associated with antimicrobial resistance. Clin Microbiol Rev 2018; 31(4): e00088-17.
[http://dx.doi.org/10.1128/CMR.00088-17] [PMID: 30068738]

[54] Tijet N, Macmullin G, Lastovetska O, *et al.* Verona integron-encoded metallo-β-lactamase 1 in Enterobacteria, Ontario, Canada. Emerg Infect Dis 2013; 19(7): 1156-8.
[http://dx.doi.org/10.3201/eid1907.121294] [PMID: 23763841]

[55] Partridge SR, Hall RM. Complex multiple antibiotic and mercury resistance region derived from the r-det of NR1 (R100). Antimicrob Agents Chemother 2004; 48(11): 4250-5.
[http://dx.doi.org/10.1128/AAC.48.11.4250-4255.2004] [PMID: 15504849]

[56] Dorward DW, Garon CF, Judd RC. Export and intercellular transfer of DNA *via* membrane blebs of Neisseria gonorrhoeae. J Bacteriol 1989; 171(5): 2499-505.
[http://dx.doi.org/10.1128/jb.171.5.2499-2505.1989] [PMID: 2496108]

[57] Pérez-Cruz C, Delgado L, López-Iglesias C, Mercade E. Outer-inner membrane vesicles naturally secreted by gram-negative pathogenic bacteria. PLoS One 2015; 10(1): e0116896.
[http://dx.doi.org/10.1371/journal.pone.0116896] [PMID: 25581302]

[58] Rumbo C, Fernández-Moreira E, Merino M, *et al.* Horizontal transfer of the OXA-24 carbapenemase gene *via* outer membrane vesicles: a new mechanism of dissemination of carbapenem resistance genes in Acinetobacter baumannii. Antimicrob Agents Chemother 2011; 55(7): 3084-90.
[http://dx.doi.org/10.1128/AAC.00929-10] [PMID: 21518847]

[59] Yaron S, Kolling GL, Simon L, Matthews KR. Vesicle-mediated transfer of virulence genes from *Escherichia coli* O157:H7 to other enteric bacteria. Appl Environ Microbiol 2000; 66(10): 4414-20. [http://dx.doi.org/10.1128/AEM.66.10.4414-4420.2000] [PMID: 11010892]

[60] Gerlach D, Guo Y, De Castro C, *et al.* Methicillin-resistant *Staphylococcus aureus* alters cell wall glycosylation to evade immunity. Nature 2018; 563(7733): 705-9.
[http://dx.doi.org/10.1038/s41586-018-0730-x] [PMID: 30464342]

[61] Missiakas D, Schneewind O. *Staphylococcus aureus* vaccines: Deviating from the carol. J Exp Med

2016; 213(9): 1645-53.
[http://dx.doi.org/10.1084/jem.20160569] [PMID: 27526714]

[62] Vaiyapuri M, Joseph TC, Rao BM, Lalitha KV, Prasad MM. Methicillin-Resistant *Staphylococcus aureus* in Seafood: Prevalence, Laboratory Detection, Clonal Nature, and Control in Seafood Chain. J Food Sci 2019; 84(12): 3341-51.
[http://dx.doi.org/10.1111/1750-3841.14915] [PMID: 31769517]

[63] Heaton CJ, Gerbig GR, Sensius LD, Patel V, Smith TC. *Staphylococcus aureus* epidemiology in wildlife: A systematic review. Antibiotics (Basel) 2020; 9(2): E89.
[http://dx.doi.org/10.3390/antibiotics9020089] [PMID: 32085586]

[64] Soge OO, No D, Michael KE, *et al.* Transmission of MDR MRSA between primates, their environment and personnel at a United States primate centre. J Antimicrob Chemother 2016; 71(10): 2798-803.
[http://dx.doi.org/10.1093/jac/dkw236] [PMID: 27439524]

[65] Keren I, Kaldalu N, Spoering A, Wang Y, Lewis K. Persister cells and tolerance to antimicrobials. FEMS Microbiol Lett 2004; 230(1): 13-8.
[http://dx.doi.org/10.1016/S0378-1097(03)00856-5] [PMID: 14734160]

[66] Lewis K. Multidrug tolerance of biofilms and persister cells. Curr Top Microbiol Immunol 2008; 322: 107-31.
[http://dx.doi.org/10.1007/978-3-540-75418-3_6] [PMID: 18453274]

[67] Germain E, Roghanian M, Gerdes K, *et al.* Stochastic induction of persister cells by HipA through (p)ppGpp-mediated activation of mRNA endonucleases. Proceedings of the National Academy of Sciences of the United States of America 2015.

[68] Han J, He L, Shi W, *et al.* Glycerol uptake is important for L-form formation and persistence in *Staphylococcus aureus*. PLoS One 2014; 9(9): e108325.
[http://dx.doi.org/10.1371/journal.pone.0108325] [PMID: 25251561]

[69] Seke Etet PF, Nwabo Kamdje AH, Vecchio L, *et al.* Signaling Pathways Sustaining Antibiotic Resistance: Lessons From Methicillin-Resistant *Staphylococcus aureus*. Lessons From Methicillin-Resistant *Staphylococcus aureus*. In: Antibiotic Resistance: Mechanisms and New Antimicrobial Approaches 2016.
[http://dx.doi.org/10.1016/B978-0-12-803642-6.00003-4]

[70] Posada AC, Kolar SL, Dusi RG, *et al.* Importance of bacillithiol in the oxidative stress response of *Staphylococcus aureus*. Infect Immun 2014; 82(1): 316-32.
[http://dx.doi.org/10.1128/IAI.01074-13] [PMID: 24166956]

[71] Sujatha S, Praharaj I. Glycopeptide resistance in gram-positive cocci: a review. Interdiscip Perspect Infect Dis 2012; 2012: 781679.
[http://dx.doi.org/10.1155/2012/781679] [PMID: 22778729]

[72] Chambers HF, Deleo FR. Waves of resistance: *Staphylococcus aureus* in the antibiotic era. Nat Rev Microbiol 2009; 7(9): 629-41.
[http://dx.doi.org/10.1038/nrmicro2200] [PMID: 19680247]

[73] Breijyeh Z, Jubeh B, Karaman R. Resistance of gram-negative bacteria to current antibacterial agents and approaches to resolve it. Molecules 2020; 25(6): E1340.
[http://dx.doi.org/10.3390/molecules25061340] [PMID: 32187986]

[74] Cui L, Iwamoto A, Lian JQ, *et al.* Novel mechanism of antibiotic resistance originating in vancomycin-intermediate *Staphylococcus aureus*. Antimicrob Agents Chemother 2006; 50(2): 428-38.
[http://dx.doi.org/10.1128/AAC.50.2.428-438.2006] [PMID: 16436693]

[75] Tanaka M, Wang T, Onodera Y, Uchida Y, Sato K. Mechanism of quinolone resistance in *Staphylococcus aureus*. J Infect Chemother 2000; 6(3): 131-9.
[http://dx.doi.org/10.1007/s101560070010] [PMID: 11810552]

[76] Arushothy R, Ramasamy H, Hashim R, *et al.* Multidrug-resistant Streptococcus pneumoniae causing invasive pneumococcal disease isolated from a paediatric patient. Int J Infect Dis 2020; 90: 219-22. [http://dx.doi.org/10.1016/j.ijid.2019.10.037] [PMID: 31682962]

[77] Yang Baek J, Kim SH, Kang CI, *et al.* Emergence of an extensively drug-resistant (XDR) Streptococcus pneumoniae serotype 15A by capsular switching. Int J Med Microbiol 2018; 308(8): 986-9.
[http://dx.doi.org/10.1016/j.ijmm.2018.08.004] [PMID: 30143394]

[78] Tadmor K, Pozniak Y, Burg Golani T, *et al.* Listeria monocytogenes MDR transporters are involved in LTA synthesis and triggering of innate immunity during infection. Front Cell Infect Microbiol 2014; 4: 16.
[http://dx.doi.org/10.3389/fcimb.2014.00016] [PMID: 24611134]

[79] Braam JF, van Dommelen L, Henquet CJM, van de Bovenkamp JHB, Kusters JG. Multidrug-resistant Mycoplasma genitalium infections in Europe. Eur J Clin Microbiol Infect Dis 2017; 36(9): 1565-7. [http://dx.doi.org/10.1007/s10096-017-2969-9] [PMID: 28361246]

[80] Isidro J, Mendes AL, Serrano M, *et al.* Overview of Clostridium difficile Infection: Life Cycle, Epidemiology, Antimicrobial Resistance and Treatment. In: Clostridium Difficile - A Comprehensive Overview 2017 Epub ahead of print 2017.
[http://dx.doi.org/10.5772/intechopen.69053]

[81] Freeman J, Vernon J, Morris K, *et al.* Pan-European longitudinal surveillance of antibiotic resistance among prevalent Clostridium difficile ribotypes. Clin Microbiol Infect 2015; 21(3): 248.e9-248.e16. [http://dx.doi.org/10.1016/j.cmi.2014.09.017] [PMID: 25701178]

[82] Hansen LH, Vester B. A cfr-like gene from Clostridium difficile confers multiple antibiotic resistance by the same mechanism as the cfr gene. Antimicrob Agents Chemother 2015; 59(9): 5841-3. [http://dx.doi.org/10.1128/AAC.01274-15] [PMID: 26149991]

[83] Loesche WJ. Microbiology of Dental Decay and Periodontal Disease 1996.

[84] Indumathi VA, Shikha R, Suryaprakash DR. Diphtheria-like illness in a fully immunised child caused by Corynebacterium pseudodiphtheriticum. Indian J Med Microbiol 2014; 32(4): 443-5. [http://dx.doi.org/10.4103/0255-0857.142250] [PMID: 25297035]

[85] Sharma NC, Efstratiou A, Mokrousov I, Sharma NC, Efstratiou A, Mokrousov I, *et al.* Diphtheria. Nat Rev Dis Prim 2019; 5(1): 81.
[http://dx.doi.org/10.1038/s41572-019-0131-y]

[86] Sheehan DJ, Hitchcock CA, Sibley CM. Current and emerging azole antifungal agents. Clin Microbiol Rev 1999; 12(1): 40-79.
[PMID: 9880474]

[87] Sucher AJ, Chahine EB, Balcer HE. Echinocandins: the newest class of antifungals. Ann Pharmacother 2009; 43(10): 1647-57.
[PMID: 19724014]

[88] Kathiravan MK, Salake AB, Chothe AS, *et al.* The biology and chemistry of antifungal agents: a review. Bioorg Med Chem 2012; 20(19): 5678-98.
[PMID: 22902032]

[89] Mayers DL, Lerner SA, Ouellette M, *et al.* Antimicrobial drug resistance: mechanisms of drug resistance. Springer 2nd Ed.., 2017.

[90] Gaur M, Puri N, Manoharlal R, *et al.* MFS transportome of the human pathogenic yeast *Candida albicans*. BMC Genomics 2008; 9: 579.
[PMID: 19055746]

[91] Prasad R, Goffeau A. Yeast ATP-binding cassette transporters conferring multidrug resistance. Annu Rev Microbiol 2012; 66: 39-63.

[PMID: 22703054]

[92] Sanglard D, Coste A, Ferrari S. Antifungal drug resistance mechanisms in fungal pathogens from the perspective of transcriptional gene regulation. FEMS Yeast Res 2009; 9(7): 1029-50. [PMID: 19799636]

[93] Calabrese D, Bille J, Sanglard D. A novel multidrug efflux transporter gene of the major facilitator superfamily from Candida albicans (FLU1) conferring resistance to fluconazole 2000.

[94] Coste AT, Karababa M, Ischer F, Bille J, Sanglard D. TAC1, transcriptional activator of CDR genes, is a new transcription factor involved in the regulation of *Candida albicans* ABC transporters CDR1 and CDR2. Eukaryot Cell 2004; 3(6): 1639-52. [PMID: 15590837]

[95] Morschhäuser J, Barker KS, Liu TT, BlaB-Warmuth J, Homayouni R, Rogers PD. The transcription factor Mrr1p controls expression of the MDR1 efflux pump and mediates multidrug resistance in *Candida albicans*. PLoS Pathog 2007; 3(11): e164-4. [PMID: 17983269]

[96] Favre B, Didmon M, Ryder NS. Multiple amino acid substitutions in lanosterol 14α-demethylase contribute to azole resistance in Candida albicans 1999.

[97] Kelly SL, Lamb DC, Loeffler J, *et al.* The G464S Amino Acid Substitution in Candida albicans Sterol 14-Demethylase Causes Fluconazole Resistance in the Clinic through Reduced Affinity 1999.

[98] Mitchell KF, Zarnowski R, Sanchez H, *et al.* Community participation in biofilm matrix assembly and function. Proc Natl Acad Sci USA 2015; 112(13): 4092-7. [PMID: 25770218]

[99] Nett J, Lincoln L, Marchillo K, *et al.* Putative role of beta-1,3 glucans in *Candida albicans* biofilm resistance. Antimicrob Agents Chemother 2007; 51(2): 510-20. [PMID: 17130296]

[100] Vediyappan G, Rossignol T, d'Enfert C. Interaction of *Candida albicans* biofilms with antifungals: transcriptional response and binding of antifungals to beta-glucans. Antimicrob Agents Chemother 2010; 54(5): 2096-111. [PMID: 20194705]

[101] Sanglard D. Emerging Threats in Antifungal-Resistant Fungal Pathogens. Front Med (Lausanne) 2016; 3: 11. [PMID: 27014694]

[102] Martel CM, Parker JE, Bader O, *et al.* A clinical isolate of *Candida albicans* with mutations in ERG11 (encoding sterol 14alpha-demethylase) and ERG5 (encoding C22 desaturase) is cross resistant to azoles and amphotericin B. Antimicrob Agents Chemother 2010; 54(9): 3578-83. [PMID: 20547793]

[103] Kelly SL, Lamb DC, Kelly DE, *et al.* Resistance to fluconazole and cross-resistance to amphotericin B in *Candida albicans* from AIDS patients caused by defective sterol delta5,6-desaturation. FEBS Lett 1997; 400(1): 80-2. [PMID: 9000517]

[104] Chapeland-Leclerc F, Hennequin C, Papon N, *et al.* Acquisition of flucytosine, azole, and caspofungin resistance in *Candida glabrata* bloodstream isolates serially obtained from a hematopoietic stem cell transplant recipient. Antimicrob Agents Chemother 2010; 54(3): 1360-2. [PMID: 20038613]

[105] Lockhart SR, Etienne KA, Vallabhaneni S, *et al.* Simultaneous emergence of multidrug-resistant *Candida auris* on 3 continents confirmed by whole-genome sequencing and epidemiological analyses. Clin Infect Dis 2017; 64(2): 134-40. [PMID: 27988485]

[106] Dominguez EG, Zarnowski R, Choy HL, *et al.* Conserved Role for Biofilm Matrix Polysaccharides in

Candida auris Drug Resistance. MSphere 2019; 4(1): e00680-18.
[PMID: 30602527]

[107] Nascimento AM, Goldman GH, Park S, *et al.* Multiple resistance mechanisms among Aspergillus fumigatus mutants with high-level resistance to itraconazole. Antimicrob Agents Chemother 2003; 47(5): 1719-26.
[PMID: 12709346]

[108] Al-Hatmi AMS, Meis JF, de Hoog GS. Fusarium: Molecular Diversity and Intrinsic Drug Resistance. PLoS Pathog 2016; 12(4): e1005464.
[http://dx.doi.org/10.1371/journal.ppat.1005464] [PMID: 27054821]

[109] Peasah SK, Azziz-Baumgartner E, Breese J, Meltzer MI, Widdowson MA. Influenza cost and cost-effectiveness studies globally--a review. Vaccine 2013; 31(46): 5339-48.
[http://dx.doi.org/10.1016/j.vaccine.2013.09.013] [PMID: 24055351]

[110] Gasparini R, Amicizia D, Lai PL, Panatto D. Clinical and socioeconomic impact of seasonal and pandemic influenza in adults and the elderly. Hum Vaccin Immunother 2012; 8(1): 21-8.
[http://dx.doi.org/10.4161/hv.8.1.17622] [PMID: 22252007]

[111] Nayak DP, Balogun RA, Yamada H, Zhou ZH, Barman S. Influenza virus morphogenesis and budding. Virus Res 2009; 143(2): 147-61.
[http://dx.doi.org/10.1016/j.virusres.2009.05.010] [PMID: 19481124]

[112] Pinto LH, Lamb RA. Influenza virus proton channels. Photochem Photobiol Sci 2006; 5(6): 629-32.
[http://dx.doi.org/10.1039/b517734k] [PMID: 16761092]

[113] Stouffer AL, Acharya R, Salom D, *et al.* Structural basis for the function and inhibition of an influenza virus proton channel. Nature 2008; 451(7178): 596-9.
[PMID: 18235504]

[114] Cady SD, Schmidt-Rohr K, Wang J, Soto CS, Degrado WF, Hong M. Structure of the amantadine binding site of influenza M2 proton channels in lipid bilayers. Nature 2010; 463(7281): 689-92.
[PMID: 20130653]

[115] Hay AJ, Wolstenholme AJ, Skehel JJ, Smith MH. The molecular basis of the specific anti-influenza action of amantadine. EMBO J 1985; 4(11): 3021-4.
[PMID: 4065098]

[116] Dolin R, Reichman RC, Madore HP, Maynard R, Linton PN, Webber-Jones J. A controlled trial of amantadine and rimantadine in the prophylaxis of influenza A infection. N Engl J Med 1982; 307(10): 580-4.
[PMID: 7050702]

[117] Bright RA, Medina MJ, Xu X, *et al.* Incidence of adamantane resistance among influenza A (H3N2) viruses isolated worldwide from 1994 to 2005: a cause for concern. Lancet 2005; 366(9492): 1175-81.
[PMID: 16198766]

[118] Palella FJ Jr, Delaney KM, Moorman AC, *et al.* Declining morbidity and mortality among patients with advanced human immunodeficiency virus infection. N Engl J Med 1998; 338(13): 853-60.
[PMID: 9516219]

[119] DeGruttola V, Dix L, D'Aquila R, *et al.* The relation between baseline HIV drug resistance and response to antiretroviral therapy: re-analysis of retrospective and prospective studies using a standardized data analysis plan. Antivir Ther 2000; 5(1): 41-8.
[PMID: 10846592]

[120] Abram ME, Ferris AL, Shao W, Alvord WG, Hughes SH. Nature, position, and frequency of mutations made in a single cycle of HIV-1 replication. J Virol 2010; 84(19): 9864-78.
[PMID: 20660205]

[121] Clutter DS, Jordan MR, Bertagnolio S, Shafer RW. HIV-1 drug resistance and resistance testing. Infect Genet Evol 2016; 46: 292-307.

[PMID: 27587334]

[122] Larder BA. Interactions between drug resistance mutations in human immunodeficiency virus type 1 reverse transcriptase. J Gen Virol 1994; 75(Pt 5): 951-7.
[PMID: 7513745]

[123] Tang MW, Shafer RW. HIV-1 antiretroviral resistance: scientific principles and clinical applications. Drugs 2012; 72(9): e1-e25.
[PMID: 22686620]

[124] Tan Q, Zhu Y, Li J, *et al.* Structure of the CCR5 chemokine receptor-HIV entry inhibitor maraviroc complex. Science 2013; 341(6152): 1387-90.
[PMID: 24030490]

[125] Gulick RM, Lalezari J, Goodrich J, *et al.* Maraviroc for previously treated patients with R5 HIV-1 infection. N Engl J Med 2008; 359(14): 1429-41.
[PMID: 18832244]

[126] Jiang X, Feyertag F, Meehan CJ, *et al.* Characterizing the Diverse Mutational Pathways Associated with R5-Tropic Maraviroc Resistance: HIV-1 That Uses the Drug-Bound CCR5 Coreceptor. J Virol 2015; 89(22): 11457-72.
[PMID: 26339063]

[127] Centers for Disease Control and Prevention. Core elements of hospital antibiotic stewardship programs. US Dep Heal Hum Serv CDC 2014.

[128] Chiotos K, Rock C, Schweizer ML, *et al.* Current infection prevention and antibiotic stewardship program practices: A survey of the Society for Healthcare Epidemiology of America (SHEA) Research Network (SRN). Infect Control Hosp Epidemiol 2019; 40(9): 1046-9.
[http://dx.doi.org/10.1017/ice.2019.172] [PMID: 31311610]

[129] Allerberger F, Gareis R, Jindrák V, Struelens MJ. Antibiotic stewardship implementation in the EU: the way forward. Expert Rev Anti Infect Ther 2009; 7(10): 1175-83.
[http://dx.doi.org/10.1586/ERI.09.96] [PMID: 19968511]

[130] Conly JM, McEwen S, Hutchinson J, Boyd N, Callery S, Bryce E. Canadian Committee on Antibiotic Resistance report. Can J Infect Dis Med Microbiol 2004; 15(5): 257-60.
[http://dx.doi.org/10.1155/2004/582574] [PMID: 18159502]

[131] WHO. Global Antimicrobial Resistance Surveillance System (GLASS) Report. 2017.
[http://dx.doi.org/ISBN 978-92-4-151344-9]

[132] Trivedi KK, Dumartin C, Gilchrist M, Wade P, Howard P. Identifying best practices across three countries: hospital antimicrobial stewardship in the United Kingdom, France, and the United States. Clin Infect Dis 2014; 59 (Suppl. 3): S170-8.
[http://dx.doi.org/10.1093/cid/ciu538] [PMID: 25261544]

[133] European Centre for Disease Prevention and Control. Transatlantic Taskforce on Antimicrobial Resistance – TATFAR. ECDC Website

[134] WHO. Global action plan on antimicrobial resistance 2017.

[135] WHO. WHO | Country progress in the implementation of the global action plan on antimicrobial resistance: WHO, FAO and OIE global tripartite database 2018.

[136] FAO. The FAO Action Plan on Antimicrobial Resistance 2016. BMJ 2020.

[137] World Bank Group. Drug-resistant infections: A Threat to Our Economic Future. World Bank Rep. Epub ahead of print 2017.
[http://dx.doi.org/10.1007/s11947-009-0181-3]

[138] A.R. M. T.S. S, A.J. W. Fecal microbiota transplantation for the intestinal decolonization of extensively antimicrobial-resistant opportunistic pathogens: A review. Infect Dis (Auckl) 2016; 48(8): 587-92.

[http://dx.doi.org/10.1080/23744235.2016.1177199]

[139] Freedman A, Eppes S. Use of Stool Transplant to Clear Fecal Colonization with Carbapenem-Resistant Enterobacteraciae (CRE): Proof of Concept. Open Forum Infect Dis 2014; 1(Suppl 1): S65. [http://dx.doi.org/10.1093/ofid/ofu051.177]

[140] Stripling J, Kumar R, Baddley JW, *et al.* Loss of vancomycin-resistant enterococcus fecal dominance in an organ transplant patient with Clostridium difficile colitis after fecal microbiota transplant. Open Forum Infect Dis 2015; 2(2): ofv078. [http://dx.doi.org/10.1093/ofid/ofv078] [PMID: 26180828]

[141] Pardi DS, Kelly C, Khanna S, Pomerantz RJ, Hohmann EL. Ser-109, An Oral, Microbiome-based Therapeutic, Is Efficacious For The Treatment Of Recurrent C. Difficile And Eliminates Enterobacteriacea And Vancomycin-resistant Enterococci Colonizing The Gut. Interscience Conference on Antimicrobial Agents and Chemotherapy.

[142] van Nood E, Vrieze A, Nieuwdorp M, *et al.* Duodenal infusion of donor feces for recurrent Clostridium difficile. N Engl J Med 2013; 368(5): 407-15. [http://dx.doi.org/10.1056/NEJMoa1205037] [PMID: 23323867]

[143] Wei Y, Gong J, Zhu W, *et al.* Fecal microbiota transplantation restores dysbiosis in patients with methicillin resistant *Staphylococcus aureusenterocolitis*. BMC Infect Dis 2015; 15: 265. [http://dx.doi.org/10.1186/s12879-015-0973-1] [PMID: 26159166]

[144] Hensley-McBain T, Zevin AS, Manuzak J, *et al.* Effects of Fecal Microbial Transplantation on Microbiome and Immunity in Simian Immunodeficiency Virus-Infected Macaques. J Virol 2016; 90(10): 4981-9. [http://dx.doi.org/10.1128/jvi.00099-16] [PMID: 26937040]

[145] Pereira GQ, Gomes LA, Santos IS, Alfieri AF, Weese JS, Costa MC. Fecal microbiota transplantation in puppies with canine parvovirus infection. J Vet Intern Med 2018; 32(2): 707-11. [http://dx.doi.org/10.1111/jvim.15072] [PMID: 29460302]

[146] Sorek R, Lawrence CM, Wiedenheft B. CRISPR-mediated adaptive immune systems in bacteria and archaea. Annu Rev Biochem 2013; 82: 237-66. [http://dx.doi.org/10.1146/annurev-biochem-072911-172315] [PMID: 23495939]

[147] Lima R, Del Fiol FS, Balcão VM. Prospects for the use of new technologies to combat multidrug-resistant bacteria. Front Pharmacol 2019; 10: 692. [http://dx.doi.org/10.3389/fphar.2019.00692] [PMID: 31293420]

[148] Choudhary E, Thakur P, Pareek M, Agarwal N. Gene silencing by CRISPR interference in mycobacteria. Nat Commun 2015; 6: 6267. [http://dx.doi.org/10.1038/ncomms7267] [PMID: 25711368]

[149] Cobb LH, Park J, Swanson EA, *et al.* CRISPR-Cas9 modified bacteriophage for treatment of *Staphylococcus aureus* induced osteomyelitis and soft tissue infection. PLoS One 2019; 14(11): e0220421. [http://dx.doi.org/10.1371/journal.pone.0220421] [PMID: 31756187]

[150] David B, Euler Chad W, Wenyan J, *et al.* Exploiting CRISPR-Cas nucleases to produce sequence-specific antimicrobials. Nat Biotechnol 2014; 32: 1146-50.

[151] Liu J, Wang Y, Lu Y, Zheng P, Sun J, Ma Y. Development of a CRISPR/Cas9 genome editing toolbox for Corynebacterium glutamicum. Microb Cell Fact 2017; 16(1): 205. [http://dx.doi.org/10.1186/s12934-017-0815-5] [PMID: 29145843]

[152] Grahl N, Demers EG, Crocker AW, Hogan DA. Use of RNA-Protein Complexes for Genome Editing in Non-*albicans Candida* Species. MSphere 2017; 2(3): e00218-17. [http://dx.doi.org/10.1128/msphere.00218-17] [PMID: 28657070]

[153] Kujoth GC, Sullivan TD, Merkhofer R, *et al.* CRISPR/Cas9-mediated gene disruption reveals the importance of zinc metabolism for fitness of the dimorphic fungal pathogen Blastomyces dermatitidis.

MBio 2018; 9(2): e00412-18.
[http://dx.doi.org/10.1128/mBio.00412-18] [PMID: 29615501]

[154] Satish L, Shamili S, Muthubharathi BC. Satish L, Shamili S, Muthubharathi BC, *et al.* CRISPR-Cas9 system for fungi genome engineering toward industrial applications. In: Genome Engineering *via* CRISPR-Cas9 System. 2020.
[http://dx.doi.org/10.1016/b978-0-12-818140-9.00006-4]

[155] Wensing L, Sharma J, Uthayakumar D, Proteau Y, Chavez A, Shapiro RS. A CRISPR Interference Platform for Efficient Genetic Repression in *Candida albicans*. MSphere 2019; 4(1): e00002-19.
[http://dx.doi.org/10.1128/msphere.00002-19] [PMID: 30760609]

[156] Liao HK, Gu Y, Diaz A, *et al.* Use of the CRISPR/Cas9 system as an intracellular defense against HIV-1 infection in human cells. Nat Commun 2015; 6: 6413.
[http://dx.doi.org/10.1038/ncomms7413] [PMID: 25752527]

[157] Xiao Q, Guo D, Chen S. Application of CRISPR/Cas9-based gene editing in HIV-1/AIDS therapy. Front Cell Infect Microbiol 2019; 9: 69.
[http://dx.doi.org/10.3389/fcimb.2019.00069] [PMID: 30968001]

[158] Boucher HW, Talbot GH, Bradley JS, *et al.* Bad bugs, no drugs: no ESKAPE! An update from the Infectious Diseases Society of America. Clin Infect Dis 2009; 48(1): 1-12.
[http://dx.doi.org/10.1086/595011] [PMID: 19035777]

[159] Monro S, Colón KL, Yin H, *et al.* Transition Metal Complexes and Photodynamic Therapy from a Tumor-Centered Approach: Challenges, Opportunities, and Highlights from the Development of TLD1433. Chem Rev 2019; 119(2): 797-828.
[PMID: 30295467]

[160] Zeng L, Gupta P, Chen Y, *et al.* The development of anticancer ruthenium(ii) complexes: from single molecule compounds to nanomaterials. Chem Soc Rev 2017; 46(19): 5771-804.
[PMID: 28654103]

[161] Kenny RG, Marmion CJ. Toward Multi-Targeted Platinum and Ruthenium Drugs-A New Paradigm in Cancer Drug Treatment Regimens? Chem Rev 2019; 119(2): 1058-137.
[PMID: 30640441]

[162] Lloyd NC, Morgan HW, Nicholson BK, Ronimus RS. The composition of Ehrlich's salvarsan: resolution of a century-old debate. Angew Chem Int Ed 2005; 44(6): 941-4.
[PMID: 15624113]

[163] Gasser G. Metal complexes and medicine: A successful combination. Chimia (Aarau) 2015; 69: 442-6.

[164] Johnstone TC, Suntharalingam K, Lippard SJ. The Next Generation of Platinum Drugs: Targeted Pt(II) Agents, Nanoparticle Delivery, and Pt(IV) Prodrugs. Chem Rev 2016; 116(5): 3436-86.
[PMID: 26865551]

[165] Rajalakshmi S, Fathima A, Rao JR, *et al.* Antibacterial activity of copper(ii) complexes against *Staphylococcus aureus*. RSC Advances 2014; 4: 32004-12.

[166] Ng NS, Wu MJ, Aldrich-Wright JR. The cytotoxicity of some phenanthroline-based antimicrobial copper(II) and ruthenium(II) complexes. J Inorg Biochem 2018; 180: 61-8.
[PMID: 29247868]

[167] Sigman DS, Mazumder A, Perrin DM. Chemical Nucleases. Chem Rev 1993; 93: 2295-316.

[168] Ramakrishnan S, Rajendiran V, Palaniandavar M, *et al.* Induction of cell death by ternary copper(II) complexes of L-tyrosine and diimines: role of coligands on DNA binding and cleavage and anticancer activity. Inorg Chem 2009; 48(4): 1309-22.
[PMID: 19140687]

[169] Ramakrishnan S, Shakthipriya D, Suresh E, Periasamy VS, Akbarsha MA, Palaniandavar M. Ternary dinuclear copper(II) complexes of a hydroxybenzamide ligand with diimine coligands: the 5,6-dmp

ligand enhances DNA binding and cleavage and induces apoptosis. Inorg Chem 2011; 50(14): 6458-71.
[PMID: 21671566]

[170] Liu Y, Li R, Xiao X, Wang Z. Antibiotic adjuvants: an alternative approach to overcome multi-drug resistant Gram-negative bacteria. Crit Rev Microbiol 2019; 45(3): 301-14.
[http://dx.doi.org/10.1080/1040841X.2019.1599813] [PMID: 30985240]

[171] Djoko KY, Achard MES, Phan MD, *et al.* Copper ions and coordination complexes as novel carbapenem adjuvants. Antimicrob Agents Chemother 2018; 62(2): e02280-17.
[http://dx.doi.org/10.1128/AAC.02280-17] [PMID: 29133551]

[172] Anand U, Nandy S, Mundhra A, Das N, Pandey DK, Dey A. A review on antimicrobial botanicals, phytochemicals and natural resistance modifying agents from Apocynaceae family: Possible therapeutic approaches against multidrug resistance in pathogenic microorganisms. Drug Resist Updat 2020; 51: 100695.
[http://dx.doi.org/10.1016/j.drup.2020.100695] [PMID: 32442892]

[173] Shin J, Prabhakaran VS, Kim KS. The multi-faceted potential of plant-derived metabolites as antimicrobial agents against multidrug-resistant pathogens. Microb Pathog 2018; 116: 209-14.
[http://dx.doi.org/10.1016/j.micpath.2018.01.043] [PMID: 29407230]

[174] Liu X, Chen C, He W, *et al.* Exploring anti-TB leads from natural products library originated from marine microbes and medicinal plants. Antonie van Leeuwenhoek 2012; 102(3): 447-61.
[http://dx.doi.org/10.1007/s10482-012-9777-0] [PMID: 22814612]

[175] Bagińska N, Pichlak A, Górski A, Jończyk-Matysiak E. Specific and Selective Bacteriophages in the Fight against Multidrug-resistant Acinetobacter baumannii. Virol Sin 2019; 34(4): 347-57.
[http://dx.doi.org/10.1007/s12250-019-00125-0] [PMID: 31093881]

[176] Taati Moghadam M, Khoshbayan A, Chegini Z, Farahani I, Shariati A. Bacteriophages, a new therapeutic solution for inhibiting multidrug-resistant bacteria causing wound infection: Lesson from animal models and clinical trials. Drug Des Devel Ther 2020; 14: 1867-83.
[http://dx.doi.org/10.2147/DDDT.S251171] [PMID: 32523333]

[177] Nazik H, Joubert LM, Secor PR, *et al.* Pseudomonas phage inhibition of Candida albicans. Microbiol (United Kingdom). 2017. 63(11): 1568-77.
[http://dx.doi.org/10.1099/mic.0.000539]

[178] Górski A, Bollyky PL, Przybylski M, *et al.* Perspectives of phage therapy in non-bacterial infections. Front Microbiol 2019; 9: 3306.
[http://dx.doi.org/10.3389/fmicb.2018.03306] [PMID: 30687285]

[179] Donovan DM, Lardeo M, Foster-Frey J. Lysis of staphylococcal mastitis pathogens by bacteriophage phi11 endolysin. FEMS Microbiol Lett 2006; 265(1): 133-9.
[http://dx.doi.org/10.1111/j.1574-6968.2006.00483.x] [PMID: 17054440]

[180] Shen Y, Köller T, Kreikemeyer B, Nelson DC. Rapid degradation of it. *Streptococcus pyogenes* biofilms by PlyC, a bacteriophage-encoded endolysin. J Antimicrob Chemother 2013; 68(8): 1818-24.
[http://dx.doi.org/10.1093/jac/dkt104] [PMID: 23557924]

[181] Mwangi J, Hao X, Lai R, Zhang ZY. Antimicrobial peptides: new hope in the war against multidrug resistance. Zool Res 2019; 40(6): 488-505.
[http://dx.doi.org/10.24272/j.issn.2095-8137.2019.062] [PMID: 31592585]

[182] Pacios O, Blasco L, Bleriot I, *et al.* Strategies to combat multidrug-resistant and persistent infectious diseases. Antibiotics (Basel) 2020; 9(2): E65.
[http://dx.doi.org/10.3390/antibiotics9020065] [PMID: 32041137]

[183] Li C, Zhu C, Ren B, *et al.* Two optimized antimicrobial peptides with therapeutic potential for clinical antibiotic-resistant *Staphylococcus aureus*. Eur J Med Chem 2019; 183: 111686.
[http://dx.doi.org/10.1016/j.ejmech.2019.111686] [PMID: 31520928]

[184] Liu Y, Jia Y, Yang K, *et al.* Antagonizing vancomycin resistance in enterococcus by surface localized antimicrobial display-derived peptides. ACS Infect Dis 2019. 6(5): 761-767. [http://dx.doi.org/10.1021/acsinfecdis.9b00164] [PMID: 31505930]

[185] Wardman P, Candeias LP. Fenton chemistry: an introduction. Radiat Res 1996; 145(5): 523-31. [http://dx.doi.org/10.2307/3579270] [PMID: 8619017]

[186] Imlay JA. Where in the world do bacteria experience oxidative stress? Environ Microbiol 2019; 21(2): 521-30. [http://dx.doi.org/10.1111/1462-2920.14445] [PMID: 30307099]

[187] Kumar S, Kanudia P, Karthikeyan S, Chakraborti PK. Identification of crucial amino acids of bacterial Peptide deformylases affecting enzymatic activity in response to oxidative stress. J Bacteriol 2014; 196(1): 90-9. [http://dx.doi.org/10.1128/JB.00916-13] [PMID: 24142250]

[188] Imlay JA, Sethu R, Rohaun SK. Evolutionary adaptations that enable enzymes to tolerate oxidative stress. Free Radic Biol Med 2019; 140: 4-13. [http://dx.doi.org/10.1016/j.freeradbiomed.2019.01.048] [PMID: 30735836]

[189] Kawahara T, Quinn MT, Lambeth JD. Molecular evolution of the reactive oxygen-generating NADPH oxidase (Nox/Duox) family of enzymes. BMC Evol Biol 2007; 7: 109. [http://dx.doi.org/10.1186/1471-2148-7-109] [PMID: 17612411]

[190] Levy M, Chowdhury PP, Nagpal P. Quantum dot therapeutics: a new class of radical therapies. J Biol Eng 2019; 13: 48. [http://dx.doi.org/10.1186/s13036-019-0173-4] [PMID: 31160923]

[191] Kennedy DA, Read AF. Why the evolution of vaccine resistance is less of a concern than the evolution of drug resistance. Proc Natl Acad Sci USA 2018; 115(51): 12878-86. [PMID: 30559199]

[192] Baker SJ, Payne DJ, Rappuoli R, De Gregorio E. Technologies to address antimicrobial resistance. Proc Natl Acad Sci USA 2018; 115(51): 12887-95. [PMID: 30559181]

[193] Klugman KP, Black S. Impact of existing vaccines in reducing antibiotic resistance: Primary and secondary effects. Proc Natl Acad Sci USA 2018; 115(51): 12896-901. [PMID: 30559195]

[194] Relman DA, Lipsitch M. Relman DA, Lipsitch M. Microbiome as a tool and a target in the effort to address antimicrobial resistance. Proc Natl Acad Sci 2018.115: 12902 LP – 12910.

[195] Babb R, Pirofski L-A. Help is on the way: Monoclonal antibody therapy for multi-drug resistant bacteria. Virulence 2017; 8(7): 1055-8. [PMID: 28306387]

[196] McConnell MJ. Where are we with monoclonal antibodies for multidrug-resistant infections? Drug Discov Today 2019; 24(5): 1132-8. [PMID: 30853568]

CHAPTER 3

Basics of the Drug Development Process

Salim Faruk Bashir[1,2], **Shivani Meena**[1] and **Gaurav Kumar**[1,*]

[1] *Department of Microbiology, School of Bioengineering and Biosciences, Lovely Professional University, Phagwara, 144411, Punjab, India*

[2] *Department of Microbiology and Biotechnology, Faculty of Science, Federal University Dutse, Jigawa State, Nigeria*

Abstract: Drug discovery and development is a complex and lengthy process aimed at producing therapeutic substances that can be both effective in terms of pharmacological activity, specificity, good affinity to its target molecule, and safe for humans. It is a necessary step due to many emerging diseases of microbial, parasitic and genetic diseases affecting the entire world so that effective prophylaxis and treatment could be provided. The successful process of discovering a new drug relies on proper discovery and characterization of the lead compound followed by the preclinical studies that ascertain the safety and efficacy of the newly discovered compound. A number of information gathered from preclinical studies that, include information about the formulation, dosage, delivery, pharmacokinetic, pharmacodynamic, mode of action of the drug as well as its relation with other drugs when they interacted, could determine the fate of the new drug's approval by the regulatory agency for a clinical trial on humans. Human clinical trials with the new drug under investigation are carried out on volunteers in different phases with a common goal to ascertain the new drug's safety, efficacy, and possible side effect in the actual environment. Since the human body is more dynamic, optimal dosage and effect of other substances on the drug itself are determined so as to ensure better treatment; satisfactory results from the human trial could pave the way for application and approval for a human trial in phase IV where the drug may subsequently go for commercialization but with strict monitoring for any unforeseen side effect most especially in a vulnerable group. Although this is an expensive, tedious and risky process for the pharmaceutical industry and volunteers, which takes many years, it is necessary. This chapter discusses the necessary steps for developing a new drug from the initial discovery from bench-top up to human trial and commercialization as an over-the-counter drug.

Keywords: Clinical trial, Discovery, Drug, Efficacy, Lead compound, Pharmacodynamic, Pharmacokinetic, Toxicity.

* **Corresponding authors Gaurav Kumar:** Department of Microbiology, School of Bioengineering and Biosciences, Lovely Professional University, Phagwara, 144411, Punjab, India; Phone: +91- 8847669615; E-mail: gaurav.19454@lpu.co.in, gau_rav@yahoo.com

M. Amin-ul Mannan & Gaurav Kumar (Eds.)

INTRODUCTION

Drug discovery and development is the process that takes years to serve fruitful results as a potential drug. The advent of modern medicine and challenges in drug development has dramatically changed the process of drug discovery. Emerging pandemic viral infections remain a constant threat, with many entering the human population due to contact with animals. The most recent such infection includes the recent SARS-Covid-19 (severe acute respiratory syndrome) that affected different countries worldwide [1]. The mortality rate and speedy transmission of coronavirus have influenced the need to discover and develop potential candidate drug molecules. Several research studies are ongoing to develop antiviral therapy. Tan *et al.* tested *in vitro* inhibition of SARS-CoV with 10 commercially available antiviral drugs and have found complete inhibition of cytopathic effects of SARS-CoV in culture for subtypes of interferon, β-1b, α-n1, α-n3, and human leukocyte interferon α [2]. The discovery of new therapeutic drugs has increased the survival of the population with varieties of infectious and non-infectious diseases [3]. Progress has been observed in reducing the mortality rate in HIV-infected children with the advent of HAART therapy and the case of the world's leading cause of death from the infectious agent *Mycobacterium tuberculosis* [4]. Significant improvements in tuberculosis treatment occurred over 2-3 decades after the new drug regime was discovered [5, 6]. Effective prophylaxis, treatment regimens, and chemotherapy have evolved over decades of research, dramatically reducing tuberculosis morbidity and mortality [7].

The discovery of a drug with the desired pharmacological activity, specificity, and affinity with the target molecule involves a series of processes after target identification and validation [8]. This series of processes involves lead discovery, lead optimization, preclinical studies, clinical studies, and post-marketing surveillance, along with the involvement of regulatory bodies in each step of drug development [9]. Each step in drug discovery and development plays a considerable role and thus determines the quality of a clinical candidate molecule [10].

The development and discovery of drugs remain lengthy, complicated processes affected by different factors. The advent of technology provides a set of tools and algorithms to carry out the process of drug discovery and development smoothly without complexity [11]. In recent times, very good progress has been made in new techniques and approaches to discover drugs for everyday use. Techniques like quantum chemistry, virtual target profiling, and *in silico* drug designing through computational methods have made the process of drug development easier than the traditional method [12 - 14].

LEAD DISCOVERY

Once a target is identified and validated, the drug design and development process moves toward the lead discovery against an identified target. Lead discovery is regarded as screening active molecules known as hits that show desired antimicrobial activity against an infectious microorganism. There are different varieties of sources for lead molecules; these sources include natural products (Microbial metabolites, plant extracts, and animal isolates), chemical libraries, and computational medicinal chemistry. Screening strategies are employed to get lead molecules for several hits, thus, these strategies play a vital role to discover a lead molecule against the target protein or enzyme [15]. These Screening techniques range from conventional strategies of screening to modern strategies like computer-aided drug designing and fragment-based lead discovery, thus, the selection of the strategies itself presents a challenge for the generation of quality leads for novel drug development [16]. Several fragment-based drug discoveries of novel drug molecules have been reported in recent years with antibacterial and antiviral activity [17, 18]. Table **1** demonstrates a few lead molecules with antimicrobial activity isolated from different sources with their mode of action.

Besides the availability of potent sources of leads, natural products serve as a viable source for lead discovery. Natural products research keeps investigating an assortment of lead molecules with a desired biological spectrum that can be utilized as a layout for new medication by the drug industries [19].

Microbial products are one of the most promising sources that have long been reported as potent pieces of evidence for lead discovery. Singh S.B. *et al.* reported the discovery of Lucensimycins A and B from *Streptomyces licenses MA7349* using an antisense strategy that showed activity against a well-established target known as protein synthesis [20]. Plant extracts have been regarded as the second most useful natural source for lead generation due to their desired pharmacological activities, most of which are not investigated yet [21]. Gadamsetty and his group have reported cytotoxic, anti-inflammatory, and antioxidant activity of extracts from *Drypetes Sepiaria.* The screening of extracts from *Drypetes Sepiaria* was performed by standard methods and further tested by *in vitro* methods for cytotoxicity, anti-inflammatory response, and free radical scavenging activity, which have shown more activity than standard compounds [22]. Whereas microbial and plant sources serve as continuing sources for lead generation, animal sources have also been reported for drug development. There are a variety of animal sources that serve as a path for the development of drugs used against diseases. For example, compounds from leech have shown potent characteristics as anticoagulants and serve as an antibiotic [23]. Recent studies on extracts of saliva have reported antimicrobial activity of theromacin and theromyzin [24].

Table 1. Lead compounds with their therapeutic values and sources

Source Name	Lead Compound	Pathogen Specificity	Mode of Action	References
Camellia sinensis	Catechins	Broadly active against G+ and G- MRSA	Suppression of virulence factors and reversion of resistance	[25]
Clusiaceae, Umbelliferae* and *Rutaceae	Coumarin	Hepatitis, influenza, HIV	Inhibits protein involved in transcription and translation	[26]
Nectandra leucantha	Dehydrodieugenol B Neolignans	*Trypanosoma cruzi*	Inibition of *Trypanosoma. cruzi* amastigotes	[27]
Actinomycete Verrucosispora strain AB 18-032	Abyssomicins	Methicillin-resistant *Staphylococcus aureus* strains (MRSA),	Inhibit the biosynthesis of pABA	[28]
Tethya crypta	Nucleoside Ara-A (vidarabine)	Herpes, vaccinica and varicella zoster viruses.	Inhibits viral DNA polymerase and DNA synthesis	[29]
Marinactinospora thermotolerans SCSIO 00652	Marthiapeptide A	Antibacterial activity against a panel of Gram-positive bacteria	-	[30]
Actinoplanes philippinensis MA7347	Philipimycin	*S. aureus*	Inhibits protein synthesis	[31]
Daphne gnidium L.	Daphnane diterpenes	HIV type-1 and coxsackievirus	Interference with the expression of HIV co-receptors, CCR5, and CXCR4	[32]
Aspergillus versicolor	Anthraquinone	Vibrio MRSA ATCC and MRSA CGMCC	topoisomerase IV and AmpC β-lactamase enzymes	[33]
Aigialus parvus	Aigialomycins D	*in vitro* antimalarial	_	[34]

LEAD OPTIMIZATION AND CHARACTERIZATION

The discovery and identification of new lead compounds with the potential to be developed into a potent drug is just an early or first step in the complex process of drug discovery; proper optimization and characterization of such molecules need to be carried out, which can play a significant role in the success of developing the compound into an active drug for preliminary studies of safety, efficacy in the preclinical trial [35], such characterization and optimization can help check the presence of desirable interactions against the target and thus may lead to producing a potent pre-clinical drug candidate with required attributes and minimize the rate of failure of drug discoveries [36]. This step is also taken to further understand the lead molecule for a structure-activity relationship, biophysical characteristics, as well as other compounds properties that define its activity [37].

The activities of many compounds have a lot of correlation between their structure and their binding sites, with a focus on their affinity to the target protein. High throughput screenings by technological advances, such as nuclear magnetic resonance (NMR), liquid chromatography-Mass spectrometry (LCMS), liquid chromatography-Nuclear magnetic resonance (LC-NMR), X-ray crystallography, high-performance liquid chromatography (HPLC), and gas chromatography-mass spectrometry (GCMS) in collaboration with *in silico* or computer-based mathematical models are used widely in understanding the new compound structure in relation to the target protein or ligand so that it helps in predicting the possible structure-function relationships through compound-target affinity strength [38, 39].

NMR technique proved to be an invaluable method for optimization and characterizing a newly discovered lead compound with respect to its activity [40]. This is a technique that is highly sensitive even with the most complex mixture of compounds, such as crude extracts of natural entities like plants crude extract when compared to other techniques. It is the most used method in the structure-based approach of biomolecule activity identification; depending on the type of target for the new lead compound, different types of NMR are employed, such as ligand-based NMR, protein-based NMR, *etc.* [41, 42]. Some of these techniques give rise to other screening methods, such as fragment-based ones that identify the sequence of the compound residues and thus give clear structure [43].

The mass spectrometric techniques, such as GCMS and LCMS, are the advanced methods that usually measure the mass of the newly discovered compound and compare it with that of known compounds databases so that structure homology can be assessed. Since similarity in structure can mostly depict similar functionality, these methods have a great tendency to detect metabolites or compounds at very low concentrations; hence widely used for phytochemical analysis, the choice of the method depends on the type and volatile nature of the compound under investigation [42].

HPLC is one of the most advanced methods of chromatographic techniques with a good promise of separating and characterizing complex biomolecules present within a biological system. This technique is adopted widely in pharmaceutical industries and research centers for characterizing potential drug molecules due to its excellent specificity, accuracy, and precision [44]. The combination of HPLC and NMR recently has proved to be an invaluable approach in investigating complex compounds, such as through stop-flow HPLC-NMR; this is successfully used in characterizing and analyzing lead compounds during drug discovery (Fig. **1**) [45].

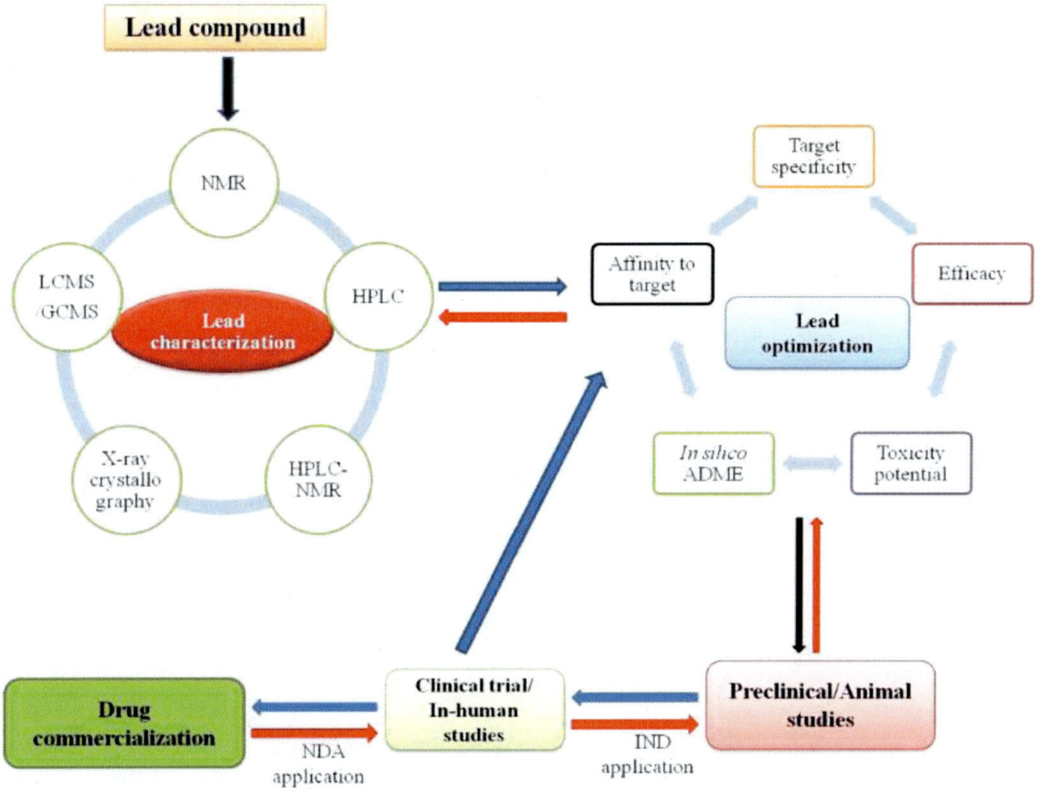

Fig. (1). Schematic representation of drug discovery and development process.

Suitable lead compound characterizations that have the potential to be developed into a final approved drug are expected to have certain attributes obtained from the various number of the above techniques. Such attributes include less than g/Mol molecular weight, the number of hydrogen donors, and acceptors to be maximum 5 and more than 10 respectively, lower than 5 partition coefficient, these properties are, however, not tentative as many excellent compounds were proved to be worthy of drug candidates without fulfilling all [46]. Some lead compounds may have a good attribute but may show lower affinity toward their target protein or ligand, which may hamper their potential as better drug candidates. These types of compounds are thus optimized for better affinity and activity through what is called the hit-to-lead approach, which may lead to forming a potent and efficient compound [47].

The process of optimization may involve mutating the molecule into another in a smoother passion so that the new one can have better attributes of molecule-ligand affinity as compared to the original molecule; this is usually achieved with the aid of computer-based simulations, such as molecular mechanics structure design [48, 49]. Another method of optimizing a compound is by simplifying its overall structure for better binding, which is achieved by reducing the complexity of its scaffold and/or chiral center removal and maintaining proper conformation; these are then subjected to various docking approaches to assess the binding affinity to the ligand or target macromolecules [50, 51].

PRECLINICAL STUDIES

The preclinical studies of any new drug to be developed are designed and undertaken to assess the initial aspects of the drug, such as its efficacy, safety/toxicity, pharmacodynamics, and pharmacokinetics on non-human subjects before it is given or tested on human subjects. This step is crucial to the pharmaceutical industry because it is a deciding factor as to whether or not the new compound or drug possesses any benefit of clinical relevance to be considered worthy of developing a new drug for human use or development into an investigational new drug. In preclinical studies, it begins with an unrestricted or unspecified dose, and it is left to the investigator to decide what a starting dose should be [52, 53].

Preclinical studies or testing involve testing the new drug candidate or compounds using two approaches that, are *in vitro* and *in vivo* methods, however, a third approach, called *in silico* method, is recently gaining momentum. The *in vitro* method involves testing the new drug candidate in a test tube, plate/petri dishes, or in cell culture, while the *in vivo* method involves testing the drug candidate in a living system such as animal models and other human tissues [52]. Preclinical studies do not usually take a very long period of time, most especially the *in vitro* aspect, and the detailed information gathered determines whether or not the investigational new drug application could be filed for approval for testing in human subjects in phase I clinical studies from the concerned regulatory agencies or authorities, it is important to note that a large number of new compounds that undergoes this preclinical studiesfailed to qualify for approval to the next phase of investigation with an average of 1 new molecule accepted out of 5000 under investigation [54].

Preclinical studies though an initial aspect of a new drug investigation, is very important to any new drug and thus cover the following aspects with respect to the new compound under investigation:

- Formulation, dosage, delivery.
- Absorption, distribution, metabolization, and excretion information.
- Mode of action of the drug.
- Toxicity/side effects.
- Compression with other similar drugs.
- Interaction with other drug molecules.
- Packaging development.

These and other factors help the researcher or the regulatory agency decide whether such a new compound or drug can be considered for approval to proceed to the next phase of development [55].

Formulation, Dosage, and Delivery

The study of the formulation of a potential drug, its intended dosage, and its delivery system is important in the preclinical studies of any new drug. These aspects significantly affect the way a drug act in the body of its consumer (pharmacodynamic) as well as how the body act on the drug itself (pharmacokinetic), the data obtained from this study can thus be used efficiently in designing further drug trials. The formulation, dosage, and delivery are most importantly determined when *in vivo* or animal studies are undertaken since this is where the effect of the living system on the drug can be seen; nevertheless, these parameters are also that important even during *in vitro* studies [56].

Formulation

Many factors are placed under consideration during the formulation process because not only do they affect the new drug's potential action in the body of the consumers, but they also could significantly influence its acceptability to the targeted population, a change in any one of these factors may influence the drug action in the body of the consumer.

In formulating a new drug, some of the main factors considered are the stability of both the compound and its base (to temperature, humidity, light, solvents, and in the presence of other compounds), organoleptic properties (odor, taste, and color), physical property (solid, liquid or gas), solubility (in both water, acid or basic medium), characterization studies on its bulk material (crystallization, polymorphism, its fine particles, hygroscopic property, material density) and interactions with other body components, such as proteins and plasma [57].

Proper knowledge of the stability of the drug to temperature can show the suitability of the drug to be administered to the target consumers. A suitable drug must not be destroyed by a certain range of temperatures, both high and low, for the purpose of consumption or storage. Many compounds within a new drug can be changed to another form at a certain temperature which may automatically alter the initial drug properties and thus deviate from its intended purpose. Proper testing under different temperature ranges is essential under both *in vitro* and *in vivo* tests [58]. Factors such as humidity and light are equally important parameters tested as they may influence drug stability, especially during storage. The stability is also affected by the type of solvent used or which the drug may come in contact with during administration; some compounds dissolve better in a given solvent such as water than others hence inappropriate solvent may likely give less activity even on a standard dose; other organic compounds when present may alter the new drug stability if such compounds were found to react with some of the new drugs. All these can be tested on the new drug before and/or during *in vitro* and *in vivo* study [59].

The taste, odor, and color (organoleptic properties) of the drug is another important aspect that is considered while formulating a drug; not all new compounds isolated and intended to be used in developing a new drug have an acceptable taste, odor, or color to the consumers. The new drug may be unpalatable or unappealing to the consumers in its original form. The taste and odor could be influenced by using a suitable vehicle, excipients or coating for easy deliverance into the body (most especially for oral formulations); depending on the target population (children or adults for instance), the formulation should be made such that the new drug can have wider acceptability [60].

The physical property or intended form of the drug, such as the solid-liquid or gas, should be put into account; this largely depends on the intended route of administration, such as the oral or intravenous route, since this may affect the absorption of the drug as well as its bioavailability in the body [61]. The crystalline and polymorphic nature of the intended drug can also define its activity in the body as such, it has to be appropriately determined during the preclinical studies. Some of the earlier known drugs were shown to exhibit different activity and stability when presented in different polymorphic forms, for instance, the well-known chloramphenicol palmitate exists in 3 crystalline forms A, B, C and amorphous D form, among which B form was more active and preferred as well based on its higher stability in the body of the consumers such as stomach acid [62].

Dosage

During preclinical studies, it is imperative to design properly the recommended dosage of the new drug to be administered in each target group. This is because the new compounds nowadays are hardly taken in their natural forms. They are usually mixed with other substances that serve as stabilizers, vehicles or aides for administration and thus referred to by many names such as excipients, adjuncts, *etc.*; these substances has no medical value whatsoever, but it is proper to be taken into account as they may influence the final required dose when the new drug is taken by the consumer.

Most of the drug dosages are usually in a very minute amount however, when designing a drug during the preclinical stage, a large dosage must be prepared, for instance, when a tablet or capsule is to be made for oral administration such that the resultant dose is achieved when the patient/consumer take the drug or when it is absorbed into the body. The dosage of the drug must be carefully chosen such that the drug required amount is obtained before being destroyed by various body conditions such as an acidic environment or eliminated, and the vehicle for the drug chosen to protect the drug from its unwanted organoleptic property such as bitterness and odor does not alter the final required dose. Proper design of the appropriate vehicle is made for administering liquid drugs and gives optimal drug action when it reaches its intended target, *etc.* [63].

The dosage design of the new drug may differ based on the intended mode of administration, such as the oral, intravenous, nasal, or topical routes. These and many more factors can play a significant role in deciding which dose should be used for investigational new drug application for starting trials on human subjects after establishing its safety level using *in vivo* study. Many natural products are poisonous if used without control as such, and the new drug dosing could be that required amount to give optimum therapeutic value with little or no side effects possible. Some factors considered for dosing include the ability of the drug to be absorbed once administered, body parameters of the intended consumer such as weight, age, *etc.*, drug binding ability to receptors, the rate at which it is metabolized, *etc.* [64].

Delivery

The new drug delivery design is closely related to the formulation design whereby the approaches selected for the drug delivery must be ones that may result in maximal therapeutic value at a relatively low dosage. Many delivery routes, such as the oral, intravenous, topical, ophthalmic, nasal, *etc.*, play a significant role in deciding the basic approach to be adopted for the delivery, poor choice of delivery approach may not directly affect the drug's pharmacodynamics, but rather it will change its pharmacokinetic nature which in turn place an impact on the drug pharmacodynamic thus may lead to various pharmacology-related problems [65].

The overall idea of designing delivery approaches is to ensure maximum drug bioavailability in the body with respect to various solvents used and, most likely environment in which the drug may be localized when consumed. For drugs that are meant for oral administration, for instance, some of the parameters considered are those related to the drug solubility, such as the effect of co-solvents, reducing the size of the particles, crystallization, pH adjustments, *etc.* Nowadays, some novel ways are used for efficient drug delivery design, even during preclinical studies, such as targeted drug delivery, local drug delivery, delayed drug delivery, *etc.* [66].

Absorption, Distribution, Metabolization, and Excretion Information (ADME)

Preclinical studies help an investigator compile information about the potential new drug pharmacokinetic data; pharmacokinetics describes how the new drug is processed from the point of entry or route of administration up to the point where it is eliminated from the body. This pharmacokinetic information can be seen generally in two ways that are absorption and elimination, with the latter encompassing the distribution, metabolism, and excretion from the body [67]. The study of new drug pharmacokinetics during preclinical studies is carried out under both *in vitro* and *in vivo* methods, however when taken by *in vitro* method, the term ADME is commonly employed; when the study is carried out by *in vivo* technique, the term that is mostly used is pharmacokinetics even though both terms refers to the same thing. Apart from the aforementioned two methods, nowadays, an additional third method called *in silico* is gaining popularity which employs the use of mathematical models and computer simulation methods to estimate or predict the way a new drug under investigation could be absorbed and eliminated from the body [68].

The pharmacokinetic information provides a great benefit in every drug discovery and development process since a lot of other processes that follow may directly or indirectly be defendants on this information before the investigational new drug application (IND) could be filed [69]. Proper understanding of the new drug attributes about its absorption, distribution, metabolism, and excretion can be part of the most considerable factors while designing the final dose regimens for all particular groups of potential consumers such as the adult, children, pregnant, and nursing mothers [70]. In this part of the study, most works are associated with the oral route of administration as it is one of the most preferred ways of administering drugs due to the convenience associated with it, information gathered from the mathematical model or *in silico* analysis together with that of the *in vitro* analysis are used to give an idea on the *in vivo* analysis to be undertaken since the *in vitro* and *in silico* analyses showed data or result under the fixed condition as compared to *in vivo* which is more dynamic in nature [71].

All pharmacokinetics study is carried out as per the principles of the declaration of Helsinki and its amendments, and the study is designed based on the type and nature of the drug under investigation, for instance, the study may be designed for the single formulation, double formulation with interval or a third approach if the new drug has a very long half-life [72]. The absorption, distribution, metabolization, and excretion are assessed by estimating an area under a concentration-time graph or by a kinetic model by determining the drug's plasma concentrations [14, 68].

Mode of Action of Drug

Every new drug under discovery and development must undergo what is referred to as a mechanism of action test during a preclinical stage to help direct the researcher on how the drug may work in its consumers. This will also clear a path on the toxicological study aspect of the preclinical testing or study. Drugs act differently in the way they exhibit their actions, however, most of the action occurs at the molecular level, with proper understanding being a key to setting the appropriate dose. This, together with pharmacokinetic information, can give a solid base for dose-effect-related study [73]. The mechanism of action study on a new drug provides a deep insight understanding of how the drug binds with different receptors or target sites as well as how its activity is similar to other previously known standard drugs. These processes are usually carried out nowadays through the bioinformatics-based mathematical model at the molecular level [74].

The new drug under discovery and development is assessed for a defined mechanism of action, including its specificity and potential high affinity. In this approach, two molecular tests are usually carried out with regard to the potential targets. These are high-throughput test which is based on the new molecule or drug interaction with the target protein or cell lines in shortest possible time (*e.g.*, 24 hours) and compared with standard compound databases. At the same time, the second is a low throughput which is carried over a long period of time on tissue, organs, and animals (*in vivo*) [52].

The drugs are known to act by an array of methods such as inhibiting a particular pathway of a particular metabolite, or hormones, or preventing the synthesis of some key components. However, the main aspect is shown by earlier researchers such as Ehrlich, who indicates that a substance can never work unless it binds to a particular receptor. Modern scientists consider this a gold standard in defining a new drug potential activity even though it might not be clear-cut evidence due to other factors associated with it [75].

One of the best *in vitro* approaches for determining the mode of action nowadays is target-based screens which are regarded as efficient based on the fact that it is based on assays that are developed to identify compounds with specificity to a particular molecule in the body. This is closely followed by a phenotypic screen used on cellular, tissue or organ level, even though cells are the most used [76]. The target-based approach is regarded however as the most cost-effective and efficient but has a certain setback as it requires an in-depth understanding of the disease itself [77]. Whichever means is chosen, the objective always remains the same, and the result or data generated are crucial toward the success of the newly discovered drug development during both filing for approval for human trial as well as optimizing the drug performance with regard to the targeted disease.

Toxicity/Side Effects

The toxicity tests on the new drug under development are rigorously taken to ascertain whether or not the new molecule or drug poses some health risk or is safe for consumption. Just like other parameters discussed above, a toxicity study is of paramount importance because it is one of the largest determinants on which the dose formulation of the new drug is made. This is because despite the fact that the new drug may possess a reasonable activity, efficacy or potency on a particular type of disease, it may, however, be disregarded if it is found to be too toxic at smaller doses thus, safety determination matters [78]. It is a well-known fact that most usable substances may indeed become poison at a given concentration or dose. In fact, a large number of therapeutic substances have toxicity potential, but with careful selection of suitable dosing, most of their side effects are avoided hence the need for toxicity studies or testing during preclinical studies before investigational new drug application (IND) [79].

Toxicity studies of new drugs are carried out under both *in vitro* and *in vivo* methods however, more recently, an *in silico* approach is also adopted. *In vitro* methods mostly employs testing different doses of the drug on cells to determine which concentration of the drug is lethal enough to kill half of the cells used for the study, the higher that concentration of the drug is observed, the better or safer is the new drug [80, 81]. *In vivo* studies are usually carried out with animals that mimic the human body systems so that less variation could be experienced when it comes to human trial in terms of both treatment and dosage, such model animals are chosen based on their metabolic pathways and genetic similarities with humans, of these, the most commonly used models are rodents such as the mice, rats and guinea pigs [82]. Some of the most common tests carried out to determine the toxicity or safety of a new drug during preclinical studies include repeated doses, acute toxicity, reproductive toxicity, carcinogenic toxicity, genetic toxicity, and toxicokinetic testing.

The repeated dose toxicity studies carried out on the test animals involve testing a given dose of the new drug over a certain period of time. This type of toxicity testing is also known by other names such as chronic toxicity test, subchronic or subacute toxicity testing. This test gives an idea of the possible toxicity or safety of the new drug when used over a long period of time, thus guiding for proper instruction about the duration of use. On the other hand, acute toxicity testing is used to show an effect of a particular dose, usually, a high dose of the drug, over a short period, such as one day, so that safe dosing can be made [83].

The reproductive toxicity testing is aimed at determining how the new drug under investigation could affect the test animal fertility; this is further tested for its effect on pregnancy, such as on fetuses and newborn infants, data from this study is used in ascertaining whether or not the new drug may be safe for pregnant and nursing mothers when the human trial is taken [84]. Based on the reproductive toxicity study, the new drug may be recommended or not to a such group, or a certain dose may be recommended at a particular stage of their pregnancy for a short period of time only [85]. The carcinogenic toxicity study is a study that is taken for a long period of time and thus may prolong the preclinical studies' time. This study is usually carried out if the drug under discovery and development is designed for such chronic diseases or those that are recurring; most of the time, such studies are designed and started as early as possible [86].

Genetic toxicity testing is aimed at assessing the drug's potential to alter the genetic constitution of the test subject over a given period of time, and this is thus focused on the molecular composition of the host animal for possible mutation, or the cancer-causing ability of the drug, such test is carried under both *in vitro* and *in vivo* conditions [87]. The toxicokinetic studies are similar to that of studies on the absorption, distribution, metabolism, and excretion (ADME), but here, a very high dose is used so that information can be acquired about the high dosage relation with the ADME parameters. This will help the regulatory authorities, and the investigation set a safety margin for the new drug under clinical conditions [83, 88].

The preclinical toxicology studies, despite being efficient in assessing the safety of the new drug under investigation, have their shortcomings as well, despite *in vivo* testing; some of the immunomodulation in humans are poorly predicted, in addition to limitations that are associated with the experiments [89].

Compression with Other Similar Drugs

The new drug compression properties in relation to other similar drugs are studied under preclinical study, such properties such as the new molecule elastic behavior, plasticity, the punch filming propensity, and its fragment ability are intensively studied as they play a key role in making a decision about choosing materials for the drug formulation, indicate how these properties could affect the rate of absorption, distribution, metabolism, and excretion and also how the new drug may behave in the presence of other drugs when used as standard prescription drugs or during clinical trial [90].

The compression properties of the new drug are compared with that of the other known similar drug, and the information gathered can help significantly in the formulation of the new drug since such properties such as elasticity and plasticity can affect the drug's mechanical stability during manufacturing as well as oral stability, flowability, *etc.* less plasticity in a new drug can lead to high dosage formulation which may ultimately increase possible plasma concentration and toxicity [91]. Compression of the most powdered drug into tablets used for oral administration is of paramount importance under suitable properties so that dosing can be made with high precision during the manufacturing process, knowledge of the compression with a similar drug during preclinical trial can help file an application for human trial indicating a precise dose to be used that will be safe for human consumption [92].

The new drug that may come as tablets are usually formulated from a wide variety of starting materials as their vehicles or excipients, which range from granules, powders or pellets, such materials must have good physical properties related to their compressibility that may help in predicting the new drug performance in the body when made into a tablet with their properties such as surface area, porosity, compaction and flow property put in consideration [93, 94]. Understanding these properties during preclinical studies can help prepare optimal drug formulation in suitable size without affecting the original chemical property of the new drug under investigation or that of other drugs that may be co-administered with it [95].

Interaction with Other Drug Molecules

A study of the interaction of the new drug under investigation is crucial during preclinical studies. Due to a large number of drugs available to consumers with increasingly new ones developed every year, many concerns are raised as a result of the potential of the newly developed drug to interact with other drug molecules, some of these interactions may results in the loss of activity to one of the drug or development of adverse reactions in the consumer which may lead to a number of consequences that are life-threatening sickness, physical disability, genetic alteration or even death [96].

The safety assessment of a new drug *in vitro* or *in vivo* describes the behavior of the new drug alone in the body of the test models only, which is not enough for the drug's overall safety until its interaction profile with other drugs is assessed in relation to altering the new drug bioavailability [97]. Some drug-drug interactions can lead to toxicity in the body when interacting with other drugs and increase their plasma concentration due to anticoagulation failure, such as that between phenytoin and warfarin coadministration [98]. Different drug-drug interaction-related clinical manifestations have been documented, including but not limited to renal and hepatic failures, seizures, heart problems, visual impairments and coma [99].

The new drug interaction study with other drugs during the preclinical stage is of paramount importance in consideration to those vulnerable groups such as the elderly that are using a large number of medications daily, thus a particular or different formulation form of the same drug may be used or recommended for them if it is discovered to be safe. Drug interaction with other molecules is nowadays shown to be on the rise due to the large use of herbal medicines as an alternative therapy which necessitates the new drug regulatory agencies to set it as mandatory practices during preclinical testing of any potential new drug to be tested for possible drug-drug related-interaction which may lead to the adverse reaction before approval for human trial is given [100].

Most of the drug interactions occur during absorption and elimination (distribution, metabolism, and excretion), whereby drugs may bind to a particular receptor and prevent the work of another. A typical example is seen with such anti-acid drugs that prevent other drug's activity like ketoconazole, which needs an acidic environment, some drugs like tetracycline may form complexes with other drugs that have multivalent cations such as magnesium or aluminum (or even food like milk that is rich in such ions) and thus prevented from been absorbed into the body, some drugs may simply prevent others from binding to a particular receptor due to competition if they happen to have a same binding site and thus increase the other drug's concentration in the body which may lead to increased toxicity [100 - 102].

Packaging Development

During the preclinical stage, packaging development is another important aspect considered for a new drug under development before it is approved for human clinical trial. Such development may come with different challenges, leading to the rejection of the investigational new drug application. Since this is used for application and approval purposes, package design and development of the new drug under investigation must be done in such a way that it meets the basic requirements of the concerned regulatory agencies and thus may not necessarily be exactly as it would be for commercialization purpose where certain aspects such as graphics can be added. In packaging development, the following parameters are placed into consideration: Labeling, biocompatible materials, and quantity or volume of the drug. Proper package design may ensure consistency in information between batches of each drug for commercial purposes [88, 103].

Labeling

This is one of the most important aspects of any packaging development; it gives information about the drugs' initial finding, such as those related to safety, labeling must contain a well-recognizable text and symbols that are standard as per the regulatory agency guidelines and region. One of the most important things is a warning label that clearly shows that it is an investigational new drug so that it will be known that the respective regulatory authority does not clear it. In addition, the detail of the manufacturer and or distributor, batch number, other warnings, the quantity of the drug, intended route of administration *etc.* must all be present or proper justification for omission should be given [103].

Biocompatible Materials

Consideration of the packaging design/material of a new drug under investigation is very important due to the fact that most of the drug activity on the test subjects relies on the purity of its compounds. When choosing a packaging material, such materials must be biocompatible with the content (the new drug) and thus should not be reactive, capable of absorbing moisture, and thus less chance of contaminating the new drug, which may alter its intended purpose and safety. The choice of proper packaging material includes the primary container, covers, seals, and outer package; selecting packaging material at the proper time can enhance or help in getting approval from the regulatory agencies during investigational new drug applications (IND) [104].

Quantity/Volume

The quantity or volumes of the investigational new drug play a significant role mostly in terms of the proposed cost of the new drug, a package design that is made to hold a low volume of the drug tends to lead to a high-cost drug due to mostly budget challenge. Many new drugs, most especially those intended for intravenous administration, usually are packaged in lower quantities for single use only to prevent contamination; hence may likely be relatively expensive, low volume packaging may also be used for the purpose of the drug trial only after which may be changed for commercial purpose in larger volume (for drugs that are supported) provided that it did not change the approval of the regulatory agencies [105].

CLINICAL TRIALS (PHASE 1-3)

Clinical trials are steps taken by scientists to ensure long-term and short-term safety as well as the acceptability of novel drugs that are going to be introduced into the population. This is imperative as new drugs may have both benefits and risks associated with them. Clinical trials are carried out in phases (phase I-IV), which are monitored by the funding agency such as pharmaceutical companies and the regulatory authority such as the United States Food and Drugs Administration (FDA) or India Central Drugs Standard Control Organization (CDSCO), the process is strictly monitored by the concerned authorities and reports analyzed to measure the potential benefit/risk ratio on the tested population in both the long and short term tests before finally approved for clinical usage [106]. A large amount of money in billions of U. S dollars is spent every year on trials related to new pharmaceutical drugs and interventions, and for a drug to be finally ready and fully accepted, it may take an average of 10 to 12 years before such drug reaches the completion of its trial phases [107, 108].

Phase I Clinical Trial

This phase of the clinical trial is aimed at providing safety, optimum dosage, and side effect information related to the new drug, where it is tested on healthy volunteers or patients. The trial is usually carried out in a small number of healthy volunteers, such as 20-100 or 15-30 patients whose conventional treatment failed or are in terminal illness due to severe life-threatening diseases such as cancer [109]. This trial phase is taken to assess the safety of the new drug rather than its efficacy and thus is referred to as first-in-human studies. The data generated may support the possibility of further testing in the next phase [110]. Phase I clinical trial is not carried out in a randomized pattern as such, it may be prone to selection bias with regard to test subjects [111, 112].

In Phase I, the new drug to be tested is administered to a few volunteers at a very low dose usually carried by contract research organizations (CRO) in a clinical trial clinic; the dosage used is usually a fraction of the toxic dose that is obtained from the data of pre-clinical studies in animal models, and the volunteers will be kept under 24-hour surveillance by trained professionals until half-life of the drugs is over [113]. The dosage is increased gradually in other patients to the point where the drug side effects are considered too high to bear, or the required result is obtained on the test subjects. A piece of good safety information qualifies the test drug to be tested in a phase II clinical trial. Since this phase is conducted in a controlled setting, most healthy volunteers are often inconvenienced by spending time in such an environment, and as such, they may be given some incentives to cater to the inconvenience [114]. With healthy volunteers, control groups are usually kept while in life-threatening disease patients as volunteers (*e.g.*, cancer patients); the control groups are usually ignored [113].

Phase I clinical trial can be taken or divided into different parts, which include the single dose (Phase Ia), multiple doses (Phase Ib), and food effects on the drug; these divisions can help generate a piece of pharmacokinetic information about the new drug in question such as its rates of absorption and elimination under both functional and impaired hepatic and renal conditions.

Phase Ia- Single Ascending Dose

In this phase, volunteers are divided into fewer groups of about 3 people, which are taken sequentially and given the lower dose of the new drug followed by their behavioral observation for its safety, when there is a lack of any side effects and the result is within expected value then a higher dose is given to the next group and so on till unacceptable toxicity is observed in a given group [115]. Another group will then be treated with the same dose until side effect is observed that is too severe to be tolerated; this is then considered as the highest tolerable dose or maximum tolerable dose for the new drug, which is usually considered when one-third of the volunteers showed the unacceptable toxicity, the dose is then considered the maximal tolerable dose when other group showed similar result and thus the trial with increasing dose is terminated [116].

Phase Ib- Multiple Ascending Doses

In this test, the test subjects are subjected to multiple doses of the new drug to assess its tolerable and safe dose through constant investigation of the pharmacokinetics and pharmacodynamics properties from samples taken such as blood and other body of the participants or volunteers. The doses are increased gradually in other groups till a given predetermined level of observable toxicity, after which the test is stopped, and available data is analyzed [115].

Food Effect on Drug

This is usually carried out as a short crossover study that is aimed to analyze the effect of food on the absorption rate of the new drug; the test is carried out on volunteers in any of the above testing scenarios whereby some of the volunteers are tested with a given dose of the drug after eating. In contrast, others were given the same dose after they had fasted, and differences in the bioavailability of the new drug in various body fluids of the volunteers were assessed. Its rate of absorption is determined under both conditions [115].

Phase I clinical trials usually have a success rate of about 70% on the tested volunteers, however, where the trial result in this phase showed undesirable or unexpected results due to adverse toxicity, consideration for the next treatment might be reassessed or dropped [109].

Phase II Clinical Trial

In this phase, the drug under investigation is assessed for its efficacy, effectiveness, and its further side effect on the trial participants or volunteers; phase II is carried out once the dose range and preliminary safety are assessed and found to be within the acceptable range in phase I trial, in addition to the aforementioned aim, phase II drug trial also is carried out with reference to specificity on a particular type of disease or outcome as such different combinations of the drug are tested on the volunteers for better efficacy and effectiveness [117].

Phase II drug trial is usually carried out on 100-300 volunteers that harbor a particular disease, however, the drug is assumed to have no curative value as such, it is not compared to a standard cure [109, 118]. To some extent, this phase is also regarded as a further extension of phase I in terms of safety assessment, but unlike the previous phase, this phase is carried out as a randomized controlled trial in treatment centers such as hospitals by professional clinical researchers, thus less prone to selection bias as compared to phase I trial, this phase account for most of the drug discovery and development failures when it failed its efficacy test or showed to have high toxic effects under both standard and genetic testing [116, 119].

In the study of the efficacy, the assessment is focused on the drug's ability to give a result of interest on a particular disease in the tested group, for example, the fever, tumor size, reduction of a particular hormone or metabolite, *etc.* within a specified period of time [120]. On the other hand, when the drug is to be tested for effectiveness, the focus on the drug is based on how it influences the disease itself rather than its outcome; as such, the trial volunteers are treated similar to the way they would be with the standard drug under standard medical practice; In both these tests, the participants under investigation must have a specific disease to which the drug is tested for and have no any other ongoing ailment [121].

The trial participants are treated with either the new drug, standard drug or a placebo, and the outcome is assessed as per the volunteer's conditions; those that receive the standard drug or placebo serve as a control for the trial. The clinical researcher usually observes the participants for months or years to ascertain the overall safety, efficacy, and effectiveness of the new drug for its acceptability as compared to the placebo or the standard drug [122]. The phase also paves the way for the investigator or clinical researcher to design methods for the next phase of the investigation, although studies have shown that the nature of reporting pattern about clinical trial may differ and leads to many data losses along the way [123, 124]. The success rate of phase II drug trials is around 33% and has the highest rate of drug trial failures as stated above, the success with regard to the drug effects such as lesser hospital visits, longer life, and general body well-being is more preferred than its efficacy such as less cell count of the causative agents [125, 126].

Phase III Clinical Trial

This is one of the most valuable phases regarding new drug clinical trials. It is designed based on the outcome of the phase II trial. It is also a randomized controlled trial conducted at various clinical centers with many participants involved. The number of participants taken for the phase III trial depends largely on the type of disease the researchers are dealing with, but generally, the number ranges between 300-3000 participants or even higher [109]. Although the trial is randomized, the participants are usually grouped based on their similarities to ensure more acceptable results, such that the data generated from them should be a result of the drug's effectiveness rather than their differences. The groupings are not influenced by the clinicians or participants until the trial period is over and the result is assessed [114].

Phase III clinical account for the most time taken during drug trials and is also the most expensive due to the larger number of participants. It is also the most difficult aspect of the new drug trial as many variables have to be considered by each participant [127]. This phase is aimed at further assessment of the drug's effectiveness and can be terminated in any given group if a problem arises, such as severe side effects observed in the participants. Phase III clinical trial is usually conducted pre-approval of the new drug for commercial purposes by the concerned or regulatory authorities such as the Food and Drug Administration of the United States (FDA) hence regarded as the pre-marketing phase, and one of the most important and critical stage since its outcome determines the new drug fate in terms of filing for approval [128].

In some instances, which mostly depend on the requirement and urgency during the trial phase, an adaptive design may be employed whereby the previous trial phases may be altered to end an unsuccessful design which may greatly reduce the time taken for a drug trial, as seen during the most recent Covid-19 pandemic [118, 129].

A couple of successful or satisfactory data generated from phase III clinical trials are gathered and used for application for approval; detailed information is shown from both animal studies, safety or toxicity, optimum dosage, formulation, and shelf life. This information is used during the marketing phase for calling off the newly approved drug should certain side effects emerge post-approval.

This phase has been shown to have an average success rate of 25-30%, with a disparity among industry-sponsored (40%) and non-industry-sponsored (7%) candidates [130]. A ten-year survey of phase II clinical trial drug success rate showed that it was found to be between 5-14% with a result of more defendants on the type of disease treated; hence some maybe even at a lower range [125]. Most of the time, however, almost half of the drug candidates failed to produce a satisfactory result in this phase or meet the requirements for approval into the marketing phase [131].

NEW DRUG APPLICATION FILING AND APPROVALS (REGULATORY APPROVALS FOR CLINICAL TRIALS)

New drug application is a step taken by new drug regulatory agencies before giving approval to any newly developed drug to be released into the market for commercialization. This application has a primary role of ensuring general safety to the public upon which they will serve as consumers of the test drug. The regulatory authority or agency can thus make an informed decision about whether or not the approval can be given based on the information submitted about the initial test such as the *in vitro* test, animal test, Phase I-III clinical trials, *etc.* [132]. All this information can give an overall idea of the following:

• Risk/benefit analysis of the new drug safety and efficacy

• Whether its labeling reflects drug contents, precautions,and warnings

• Good manufacturing ethics and standards are adhered to

• Quality control measures are taken to ensure maintaining the original drug's nature and shelf life.

All the above-mentioned are required not only for the new drug but also for an investigational drug application [133].

Different regulatory authorities of countries or regions have their specific guidelines, some of which may be similar, information submitted about the new drug is preliminary reviewed, and a decision about its acceptance or otherwise is communicated to the applicant within about two months. A standard review of the application is then carried out, which can last about 10 months or 6 months for a priority review. It is only when the drug is approved that it can be legally sold as an over-the-counter drug. It is indicated, however, that some of the review processes of agencies such as the FDA lack transparency, although efforts are underway for improvement [134].

It is important to note that the new drug application (NDA) differs from an investigational new drug application (IND), whereby the former is filed at the end of phase III clinical trial before releasing the drug for commercialization while the latter is filed just after animal studies is completed before the human trial began and thus its requirement is based on non-clinical data [135]. Similar product applications are made under different names and regulations, for instance, the application for a generic drug is termed an Abbreviated New Drug Application (ANDA), vaccines are termed a Biologic License Application (BLA), while drugs made for animal use (veterinary purpose) are applied as New Animal Drug

Application (NADA), but all the purpose remain the same; for generic drugs (drugs that are chemically similar to a previously patented new drug), the requirements are not as strict as that of new drug application whereby clinical trial data is exempted from its application [115, 135].

Different countries have a different sets of regulations that govern their new drug applications; in India, the Central Drugs Standard Control Organization (CDSCO) is responsible for all regulations related to drugs, medical devices, and cosmetics for the well-being of the general public, which is ensured by different sets of rules such as ethical guidelines for biomedical research that involve human volunteers, act on drugs and cosmetics as well as good clinical practices; many regulations relevant to clinical trial in India keep changing especially from 2005-2016 as such can be regarded as under development [136].

In a general sense, however, the regulatory guidelines for all concerned authorities are aimed primarily at enforcing the laws of the country about patient's/consumers health and prevention from exploitation by the pharma industries; different people also have shown to have different attitudes toward clinical trials in terms of both safety and privacy which varies widely between countries as well as from urban and rural areas based on behavioral and cultural diversity [137]. Some of the requirements of the new drug application include the following, among others:

- Registration with the ethics committee for study approval.

- Registration for initiation of clinical trial.

- Obtaining the volunteers informed consent.

- Reporting any adverse situation that happens during the course of the trial.

- Detailed evidence about the new drug's effectiveness for clinical purposes.

- Toxicological information of the new drug based on non-clinical data.

- All microbiological aspects about the application of the new drug.

- Detailed information about the labeling and packaging material of the new drug.

- Full guidelines on the drug usage, including dosage and timing.

- The drug bioavailability in the body, shelf-life, and metabolic aspects.

All the above, along with others, can pave the way to the successful approval of the new drug for commercial purposes or the next line of trial by drug developers [133, 138].

CLINICAL TRIAL (PHASE 4)

This phase is taken only after approval is sought from the concerned authorities to be sold onto the public domain, which is based on the data and information obtained from previous phases (Phase I-III) and pre-clinical/animal studies; it is usually taken after the newly approved drug is released into the market thus referred to as post-marketing study or surveillance. This phase is a final phase that is concerned or aimed at assessing the final safety of the drug in the general public, commonly called pharmacovigilance [139]. This phase is crucial as it is focused on the long-term effect of the drug on the general population, usually due to compatibility issues with other drugs, effect on some populations not covered in previous phases, such as pregnant and nursing mothers, as well as certain effects it may incur on babies during breastfeeding [109].

Side effects or abnormalities reported due to the approved drug may lead to a certain change in policy about its approval given, some of which may include termination of the drug use when severe side effects are reported, changing the drug use to a particular population instead of the general population or restricted to be used only under certain conditions such as last resort choice or under severe clinical disease to which there are no any available alternatives [140].

Much information is generated in Phase IV clinical trials such as the drug safety and side effects related to a larger population of different ethnicities, social statuses, and other underlying medical conditions. In other words, this phase tests the final performance of the drug in real-world; by this, it signifies the taking of what is referred to as non-interventional studies where the new drug is administered as per the normal clinical setting of other approved drugs in accordance to standard clinical practice, adverse or unseen side effects that are seriously observed during this phase is constantly reported to the concerned regulatory authorities for further assessment as to whether or not the approval of the drug remains or certain changes need to be made, on the other hand, non-serious side effects are also documented and reported at certain time intervals [141].

It is important to note that despite the fact that phase IV clinical trial follows a randomized sampling pattern; a study has shown that the degree of randomization taken during phase IV clinical trial of drugs differs significantly between the type of disease under investigation; drugs related to mental health have the highest randomization pattern while that of cancer-related ones showed to have fewer randomization, this is largely associated with the number of funding from authorities such as National Institute of Health (NIH) in the United States which is showed to be higher on mental health-related drugs than on those designed for oncology purposes [142].

POST-MARKET MONITORING

Postmarketing monitoring or surveillance is the study that is designed to ensure the well-being or safety of the patients as consumers of the newly approved drug, and it is also carried out to check on any disparities such as changes in the originally-approved drug that may be either voluntary or involuntary, many of the information can be gathered from a variety of source such as databases and publications [143]. Despite a ton of information generated during the course of a new drug trial from the initial phase to the last one, still, additional information can be obtained about the true nature of the drug when released to the general population, most of the information gathered is dependent on the time taken by the drug post-approval into the market or in the general public, the report made is reviewed, and certain modifications are made if necessary with respect to initial approval given to the drug [109].

Postmarketing monitoring or surveillance thus serves many or diverse purposes during both short term and long term of the new drug use, however, certain general objectives can be mentioned, some of which include assessing the newly approved drug's true efficacy and safety in standard conditions within populations, nature of the drug in the presence of other prescription drugs, assessment between actual and potential impact of the drug intended use, generate data that can be used in its cost-benefit analysis [144].

It is mostly a common perception by many that the regulatory authority-approved drugs such as FDA are generally safe, but that is not the case in reality. Despite the randomized trials strategies that were adopted in the course of testing the drugs, many adverse reactions are reported with respect to drugs approved for commercial purposes, which can range from minor to major ones, including death in some instances, adverse reactions that were reported during post-market surveillance/monitoring of approved drugs have been showing to be rising from the last two decades with about 50% of the approved ones showed to have severe side effects [145].

Post-market monitoring/surveillance is not practiced the same way across the globe, different countries have a different set of rules and guidelines on how this surveillance is taken hence a variability in findings and approach to the study design about the approved drugs, but in the general sense, the objective remains the same which is the plan for risk management, most of this surveillance is carried out in basically two ways which include:

1. Monitoring all populations that are using the approved drug.

2. Monitoring vulnerable populations such as older people, children, pregnant women, *etc*. While countries like The United States, Japan, China, Korea, and those under European Union have their established guidelines, others like India have their guidelines still under development and thus less strict as compared to the rest [146].

The post-marketing surveillance is carried out not only by the respective country's regulatory authority but also by the manufacturing pharmaceutical companies themselves, some withdrawal of drugs from the market has been reported by some of the major pharmaceutical companies across the globe after they discover some of their drugs severe side effects with respect to a certain group of patients, these unforeseen circumstances are what makes post-marketing monitoring important aspect of any drug development process [108]. Errors that may lead to withdrawal or termination of a particular new drug approval may sometimes be far from common human errors but rather based on diverse collective factors related to the environmental, genetic, and socioeconomic profile of the consumers [147].

As there are many ways of generating data about post-market drug surveillance, many problems are reported due to many disparities in information, and thus post-marketing monitoring needs to be further improved to ensure the proper well-being of the consumers of such newly approved drugs [148]. Post-marketing monitoring is constantly evolving since drug regulatory authorities, and pharma industries record all information and learn from previously evolved problems related to drug development [149].

CONCLUSION

New drug development is the tedious, expensive and complicated process taken before any new drug reaches potential consumers. However, despite all these challenges, the process is necessary due to the risk involved when drug safety is assessed wrongly, which may lead to an array of mild and severe side effects or even death in critical situations. Clinical trials ensure that all the benefits, such as the efficacy and effectiveness of the new drug, is obtained and compared with its risks and side effect. Ensuring an unbiased and standard trial of the drug by its investigators and regulatory authorities can result in producing a qualitative drug that may save a life. There is no doubt that a clinical trial can be both beneficial and counter-productive to both the volunteers, investigators, and pharmaceutical companies, but the proper design of such a clinical trial may lead to a high benefit/risk ratio and thus approval of the new drug if the regulatory process is carried under professional, transparent and unbiased manner or with reevaluation of different regulatory rules and regulation should the need arises.

CONSENT FOR PUBLICATION

Not applicable.

CONFLICT OF INTEREST

The author declares no conflict of interest, financial or otherwise.

ACKNOWLEDGEMENT

Declared none.

REFERENCES

[1] Wu Y-S, Lin W-H, Hsu JT, Hsieh HP. Antiviral drug discovery against SARS-CoV. Curr Med Chem 2006; 13(17): 2003-20. [http://dx.doi.org/10.2174/092986706777584988] [PMID: 16842194]

[2] Tan ELC, Ooi EE, Lin CY, *et al.* Inhibition of SARS coronavirus infection *in vitro* with clinically approved antiviral drugs. Emerg Infect Dis 2004; 10(4): 581-6. [http://dx.doi.org/10.3201/eid1004.030458] [PMID: 15200845]

[3] Holmes KK, Bertozzi S, Bloom BR, *et al.* Major infectious diseases: Key messages from disease control priorities. In: Major Infectious Diseases: Disease Control Priorities (Volume 6), 3rd ed. The World Bank 2017; 1-27.

[4] Brady MT, Oleske JM, Williams PL, *et al.* Declines in mortality rates and changes in causes of death in HIV-1-infected children during the HAART era. J Acquir Immune Defic Syndr 2010; 53(1): 86-94. [http://dx.doi.org/10.1097/QAI.0b013e3181b9869f] [PMID: 20035164]

[5] Sloan D, Davies G, Khoo S. Recent advances in tuberculosis: New drugs and treatment regimens. Curr Respir Med Rev 2013; 9(3): 200-10. [http://dx.doi.org/10.2174/1573398X113099990017] [PMID: 24683386]

[6] Rabahi MF, Silva Júnior JLR, Ferreira ACG, Tannus-Silva DGS, Conde MB. Tuberculosis treatment. J Bras Pneumol 2017; 43(6): 472-86. [http://dx.doi.org/10.1590/s1806-37562016000000388] [PMID: 29340497]

[7] Dye C, Watt CJ, Bleed DM, Hosseini SM, Raviglione MC. Evolution of tuberculosis control and prospects for reducing tuberculosis incidence, prevalence, and deaths globally. JAMA 2005; 293(22): 2767-75. [http://dx.doi.org/10.1001/jama.293.22.2767] [PMID: 15941807]

[8] Rang HP. The development of the pharmaceutical industry. In: Hill R, Rang H, Eds. Drug Discovery and Development 2012; 3-18.

[9] Mohs RC, Greig NH. Drug discovery and development: Role of basic biological research. Alzheimers Dement (N Y) 2017; 3(4): 651-7. [http://dx.doi.org/10.1016/j.trci.2017.10.005] [PMID: 29255791]

[10] Chen J, Luo X, Qiu H, Mackey V, Sun L, Ouyang X. Drug discovery and drug marketing with the critical roles of modern administration. Am J Transl Res 2018; 10(12): 4302-12. [PMID: 30662672]

[11] Petrova E. Innovation in the pharmaceutical industry: The process of drug discovery and development. In: Din M, Eliashberg J, Stepfan Strre (Eds.) Innovation and marketing in the pharmaceutical industry. Springer, New York, 2014; 19-81.

[12] Zheng M, Liu X, Xu Y, Li H, Luo C, Jiang H. Computational methods for drug design and discovery: focus on China. Trends Pharmacol Sci 2013; 34(10): 549-59. [http://dx.doi.org/10.1016/j.tips.2013.08.004] [PMID: 24035675]

[13] Chan HCS, Shan H, Dahoun T, Vogel H, Yuan S. Advancing drug discovery *via* artificial intelligence. Trends Pharmacol Sci 2019; 40(8): 592-604. [http://dx.doi.org/10.1016/j.tips.2019.06.004] [PMID: 31320117]

[14] van de Waterbeemd H, Gifford E. ADMET *in silico* modelling: towards prediction paradise? Nat Rev Drug Discov 2003; 2(3): 192-204. [http://dx.doi.org/10.1038/nrd1032] [PMID: 12612645]

[15] Lahlou M. The success of natural products in drug discovery. Pharmacol Pharm 2013; 4(3): 17-31. [http://dx.doi.org/10.4236/pp.2013.43A003]

[16] Wu G, Zhao T, Kang D, *et al.* Overview of recent strategic advances in medicinal chemistry. J Med Chem 2019; 62(21): 9375-414. [http://dx.doi.org/10.1021/acs.jmedchem.9b00359] [PMID: 31050421]

[17] Yamada Y, Takashima H, Walmsley DL, *et al.* Fragment-based discovery of novel non-hydroxamate LpxC inhibitors with antibacterial activity. J Med Chem 2020; 63(23): 14805-20. [http://dx.doi.org/10.1021/acs.jmedchem.0c01215] [PMID: 33210531]

[18] Benmansour F, Trist I, Coutard B, *et al.* Discovery of novel dengue virus NS5 methyltransferase non-nucleoside inhibitors by fragment-based drug design. Eur J Med Chem 2017; 125: 865-80. [http://dx.doi.org/10.1016/j.ejmech.2016.10.007] [PMID: 27750202]

[19] de Moraes J. Natural products with antischistosomal activity. Future Med Chem 2015; 7(6): 801-20. [http://dx.doi.org/10.4155/fmc.15.23] [PMID: 25996071]

[20] Singh SB, Zink DL, Huber J, *et al.* Discovery of lucensimycins A and B from *Streptomyces lucensis* MA7349 using an antisense strategy. Org Lett 2006; 8(24): 5449-52. [http://dx.doi.org/10.1021/ol062041r] [PMID: 17107044]

[21] Balandrin MF, Kinghorn AD, Farnsworth NR. Plant-derived natural products in drug discovery and development. Human Medicinal Agents from Plants 1993; 2-12. [http://dx.doi.org/10.1021/bk-1993-0534.ch001]

[22] Gadamsetty G, Maru S, Tyagi A, Chakravarthula SN. Anti-inflammatory, cytotoxic and antioxidant effects of methanolic extracts of Drypetes Sepiaria (Euphorbiaceae). Afr J Tradit Complement Altern Med 2013; 10(5): 274-82. [http://dx.doi.org/10.4314/ajtcam.v10i5.9] [PMID: 24311836]

[23] Abdualkader AM, Ghawi AM, Alaama M, Awang M, Merzouk A. Leech therapeutic applications. Indian J Pharm Sci 2013; 75(2): 127-37. [PMID: 24019559]

[24] Sig AK, Guney M, Uskudar Guclu A, Ozmen E. Medicinal leech therapy—an overall perspective. Integr Med Res 2017; 6(4): 337-43. [http://dx.doi.org/10.1016/j.imr.2017.08.001] [PMID: 29296560]

[25] Taylor PW, Hamilton-Miller JMT, Stapleton PD. Antimicrobial properties of green tea catechins. Food Sci Technol Bull 2005; 2(7): 71-81. [http://dx.doi.org/10.1616/1476-2137.14184] [PMID: 19844590]

[26] Mishra S, Pandey A, Manvati S. Coumarin: An emerging antiviral agent. Heliyon 2020; 6(1): e03217. [http://dx.doi.org/10.1016/j.heliyon.2020.e03217] [PMID: 32042967]

[27] Sear CE, Pieper P, Amaral M, *et al.* Synthesis and structure-activity relationship of dehydrodieugenol B neolignans against *Trypanosoma cruzi.* ACS Infect Dis 2020; 6(11): 2872-8. [http://dx.doi.org/10.1021/acsinfecdis.0c00523] [PMID: 33047947]

[28] Bister B, Bischoff D, Ströbele M, *et al.* Abyssomicin C-A polycyclic antibiotic from a marine Verrucosispora strain as an inhibitor of the p-aminobenzoic acid/tetrahydrofolate biosynthesis pathway. Angew Chem Int Ed 2004; 43(19): 2574-6. [http://dx.doi.org/10.1002/anie.200353160] [PMID: 15127456]

[29] Sagar S, Kaur M, Minneman KP. Antiviral lead compounds from marine sponges. Mar Drugs 2010; 8(10): 2619-38. [http://dx.doi.org/10.3390/md8102619] [PMID: 21116410]

[30] Zhou X, Huang H, Chen Y, *et al.* Marthiapeptide A, an anti-infective and cytotoxic polythiazole cyclopeptide from a 60 L scale fermentation of the deep sea-derived *Marinactinospora thermotolerans* SCSIO 00652. J Nat Prod 2012; 75(12): 2251-5. [http://dx.doi.org/10.1021/np300554f] [PMID: 23215246]

[31] Zhang C, Occi J, Masurekar P, *et al.* Isolation, structure, and antibacterial activity of philipimycin, a thiazolyl peptide discovered from *Actinoplanes philippinensis* MA7347. J Am Chem Soc 2008; 130(36): 12102-10. [http://dx.doi.org/10.1021/ja803183u] [PMID: 18698773]

[32] Vidal V, Potterat O, Louvel S, *et al.* Library-based discovery and characterization of daphnane diterpenes as potent and selective HIV inhibitors in *Daphne gnidium.* J Nat Prod 2012; 75(3): 414-9. [http://dx.doi.org/10.1021/np200855d] [PMID: 22148316]

[33] Wang W, Chen R, Luo Z, Wang W, Chen J. Antimicrobial activity and molecular docking studies of a novel anthraquinone from a marine-derived fungus *Aspergillus versicolor.* Nat Prod Res 2018; 32(5): 558-63. [http://dx.doi.org/10.1080/14786419.2017.1329732] [PMID: 28511613]

[34] Isaka M, Suyarnsestakorn C, Tanticharoen M, Kongsaeree P, Thebtaranonth Y. Aigialomycins A-E, new resorcylic macrolides from the marine mangrove fungus *Aigialus parvus.* J Org Chem 2002; 67(5): 1561-6. [http://dx.doi.org/10.1021/jo010930g] [PMID: 11871887]

[35] Keserű GM, Makara GM. Hit discovery and hit-to-lead approaches. Drug Discov Today 2006; 11(15-16): 741-8. [http://dx.doi.org/10.1016/j.drudis.2006.06.016] [PMID: 16846802]

[36] Todd A, Anderson R, Groundwater PW. Rational drug design - Identifying and characterising a target. Pharm J 2009; 283: 19-20.

[37] Moore JM. NMR techniques for characterization of ligand binding: Utility for lead generation and optimization in drug discovery. 1999; 221-43.

[38] Wolfender JL, Queiroz EF, Hostettmann K. The importance of hyphenated techniques in the discovery of new lead compounds from nature. Expert Opin Drug Discov 2006; 1(3): 237-60. [http://dx.doi.org/10.1517/17460441.1.3.237] [PMID: 23495845]

[39] Janero DR. The future of drug discovery: enabling technologies for enhancing lead characterization and profiling therapeutic potential. Expert Opin Drug Discov 2014; 9(8): 847-58. [http://dx.doi.org/10.1517/17460441.2014.925876] [PMID: 24965547]

[40] Pellecchia M, Bertini I, Cowburn D, *et al.* Perspectives on NMR in drug discovery: a technique comes of age. Nat Rev Drug Discov 2008; 7(9): 738-45. [http://dx.doi.org/10.1038/nrd2606] [PMID: 19172689]

[41] Shortridge MD, Powers R. NMR screening methods for drug discovery. Adv Biomed Spectrosc 2011; 3: 381-412.
[http://dx.doi.org/10.3233/978-1-60750-695-9-381]

[42] Baghel US, Singh A, Singh D, Sinha M. Application of mass spectroscopy in pharmaceutical and biomedical analysis. Spectroscopic Analyses - Developments and Applications 2017; 105-21.
[http://dx.doi.org/10.5772/intechopen.70655]

[43] Dias DM, Ciulli A. NMR approaches in structure-based lead discovery: Recent developments and new frontiers for targeting multi-protein complexes. Prog Biophys Mol Biol 2014; 116(2-3): 101-12.
[http://dx.doi.org/10.1016/j.pbiomolbio.2014.08.012] [PMID: 25175337]

[44] Siddiqui MR, AlOthman ZA, Rahman N. Analytical techniques in pharmaceutical analysis: A review. Arab J Chem 2017; 10: S1409-21.
[http://dx.doi.org/10.1016/j.arabjc.2013.04.016]

[45] Sidelmann UG, Bjørnsdottir I, Shockcor JP, Hansen SH, Lindon JC, Nicholson JK. Directly coupled HPLC-NMR and HPLC-MS approaches for the rapid characterisation of drug metabolites in urine: application to the human metabolism of naproxen. J Pharm Biomed Anal 2001; 24(4): 569-79.
[http://dx.doi.org/10.1016/S0731-7085(00)00482-9] [PMID: 11272313]

[46] Hefti FF. Requirements for a lead compound to become a clinical candidate. BMC Neurosci 2008; 9(S3) (Suppl. 3): S7.
[http://dx.doi.org/10.1186/1471-2202-9-S3-S7] [PMID: 19091004]

[47] Hoffer L, Voitovich YV, Raux B, *et al.* Integrated strategy for lead optimization based on fragment growing: The diversity-oriented-target-focused-synthesis approach. J Med Chem 2018; 61(13): 5719-32.
[http://dx.doi.org/10.1021/acs.jmedchem.8b00653] [PMID: 29883107]

[48] Jorgensen WL. Efficient drug lead discovery and optimization. Acc Chem Res 2009; 42(6): 724-33.
[http://dx.doi.org/10.1021/ar800236t] [PMID: 19317443]

[49] Maruca A, Ambrosio FA, Lupia A, *et al.* Computer-based techniques for lead identification and optimization I: Basics. Physical Sciences Reviews 2019; 4(6): 1-14.
[http://dx.doi.org/10.1515/psr-2018-0113]

[50] Wang S, Dong G, Sheng C. Structural simplification: an efficient strategy in lead optimization. Acta Pharm Sin B 2019; 9(5): 880-901.
[http://dx.doi.org/10.1016/j.apsb.2019.05.004] [PMID: 31649841]

[51] Hardianto A, Yusuf M, Liu F, Ranganathan S. Structure-based drug design workflow. Encycl Bioinforma Comput Biol ABC Bioinforma 2019; 1–3: 273-82.
[http://dx.doi.org/10.1016/B978-0-12-809633-8.20104-0]

[52] Brodniewicz T, Grynkiewicz G. Preclinical drug development. Acta Pol Pharm 2010; 67(6): 578-85.
[http://dx.doi.org/10.1016/S0959-8049(01)80184-2] [PMID: 21229871]

[53] New drug application : preclinical research. https://www.fda.gov/patients/drug-developmen--process/step-2-preclinical-research

[54] Atanasov AG, Waltenberger B, Pferschy-Wenzig EM, *et al.* Discovery and resupply of pharmacologically active plant-derived natural products: A review. Biotechnol Adv 2015; 33(8): 1582-614.
[http://dx.doi.org/10.1016/j.biotechadv.2015.08.001] [PMID: 26281720]

[55] Kulkarni G, Gowthamarajan K, Suresh B. Stability testing of pharmaceutical products: An overview. Indian J Pharm Educ 2004; 38: 194-8.

[56] Mitragotri S, Burke PA, Langer R. Overcoming the challenges in administering biopharmaceuticals: formulation and delivery strategies. Nat Rev Drug Discov 2014; 13(9): 655-72.
[http://dx.doi.org/10.1038/nrd4363] [PMID: 25103255]

[57] Chaurasia G. A review on pharmaceutical preformulation studies in formulation and development of new drug molecules. Int J Pharma Sci Res 2016. 13(12): 2313-0

[58] Hoag SW, Hussain AS. The impact of formulation on bioavailability: summary of workshop discussion. J Nutr 2001; 131(4) (Suppl.): 1389S-91S. [http://dx.doi.org/10.1093/jn/131.4.1389S] [PMID: 11285360]

[59] Vilegave K, Vidyasagar G, Chandankar P. Preformulation studies of pharmaceutical new drug molecule & products: An overview. Am J Pharm Heal Res 2013; 1: 1-20.

[60] Thabet Y, Klingmann V, Breitkreutz J. Drug formulations: Standards and novel strategies for drug administration in pediatrics. J Clin Pharmacol 2018; 58 (Suppl. 10): S26-35. [http://dx.doi.org/10.1002/jcph.1138] [PMID: 30248193]

[61] Tibalinda P, Pius D, Shedafa R, Masota N, Temu M, Kaale E. Pre-formulation development of lamivudine 300 mg and tenofovir disoproxil fumarate (TDF) 300 mg fixed dose combination tablets. Pharmacol &amp. Pharmacol Pharm 2016; 7(7): 247-54. [http://dx.doi.org/10.4236/pp.2016.77031]

[62] Shah SM, Jain AS, Kaushik R, Nagarsenker MS, Nerurkar MJ. Preclinical formulations: insight, strategies, and practical considerations. AAPS PharmSciTech 2014; 15(5): 1307-23. [http://dx.doi.org/10.1208/s12249-014-0156-1] [PMID: 24920522]

[63] Alexander K. Dosage forms and their routes of administration 1st ed., 2009. [http://dx.doi.org/10.1016/B978-0-12-369521-5.00002-6]

[64] Honek J. Preclinical research in drug development. Med Writ 2017; 26: 5-8.

[65] Schwartz RS, Edelman E, Virmani R, *et al.* Drug-eluting stents in preclinical studies: updated consensus recommendations for preclinical evaluation. Circ Cardiovasc Interv 2008; 1(2): 143-53. [http://dx.doi.org/10.1161/CIRCINTERVENTIONS.108.789974] [PMID: 20031669]

[66] Wen H, Jung H, Li X. Drug delivery approaches in addressing clinical pharmacology-related issues: Opportunities and challenges. AAPS J 2015; 17(6): 1327-40. [http://dx.doi.org/10.1208/s12248-015-9814-9] [PMID: 26276218]

[67] Ruiz-Garcia A, Bermejo M, Moss A, Casabo VG. Pharmacokinetics in drug discovery. J Pharm Sci 2008; 97(2): 654-90. [http://dx.doi.org/10.1002/jps.21009] [PMID: 17630642]

[68] Toomula N, Kumar D S, Kumar A, M P. Role of pharmacokinetic studies in drug discovery. J Bioequivalence Bioavailab 2011; 3(11): 263-7. [http://dx.doi.org/10.4172/jbb.1000097]

[69] Kola I, Landis J. Can the pharmaceutical industry reduce attrition rates? Nat Rev Drug Discov 2004; 3(8): 711-6. [http://dx.doi.org/10.1038/nrd1470] [PMID: 15286737]

[70] Fostel J. Predictive ADME-Tox. Expert Opin Drug Metab Toxicol 2005; 1(3): 565-70. [http://dx.doi.org/10.1517/17425255.1.3.565] [PMID: 16863462]

[71] Lin L, Wong H. Predicting oral drug absorption: Mini review on physiologically-based pharmacokinetic models. Pharmaceutics 2017; 9(4): 41. [http://dx.doi.org/10.3390/pharmaceutics9040041] [PMID: 28954416]

[72] O'Brien SE, de Groot MJ. Greater than the sum of its parts: combining models for useful ADMET prediction. J Med Chem 2005; 48(4): 1287-91. [http://dx.doi.org/10.1021/jm049254b] [PMID: 15715500]

[73] Swinney DC. Drug discoveries and molecular mechanism of action. In: Successful Drug Discovery. Wiley-VCH Verlag GmbH & Co. KGaA, Weinheim, Germany, 2015, 9–34. [http://dx.doi.org/10.1002/9783527678433.ch2]

[74] Iorio F, Bosotti R, Scacheri E, *et al.* Discovery of drug mode of action and drug repositioning from transcriptional responses. Proc Natl Acad Sci USA 2010; 107(33): 14621-6. [http://dx.doi.org/10.1073/pnas.1000138107] [PMID: 20679242]

[75] Tillotson GS, Zinner SH. Paul Ehrlich, a prescient pioneer in the field of antimicrobial chemotherapy: what did he foresee a century ago? Expert Rev Anti Infect Ther 2013; 11(2): 113-5. [http://dx.doi.org/10.1586/eri.12.166] [PMID: 23409816]

[76] Kiriiri GK, Njogu PM, Mwangi AN. Exploring different approaches to improve the success of drug discovery and development projects: a review. Future Journal of Pharmaceutical Sciences 2020; 6(1): 27. [http://dx.doi.org/10.1186/s43094-020-00047-9]

[77] Davis RL. Mechanism of action and target identification: A matter of timing in drug discovery. iScience 2020; 23(9): 101487. [http://dx.doi.org/10.1016/j.isci.2020.101487] [PMID: 32891054]

[78] Herter-Sprie GS, Kung AL, Wong KK. New cast for a new era: preclinical cancer drug development revisited. J Clin Invest 2013; 123(9): 3639-45. [http://dx.doi.org/10.1172/JCI68340] [PMID: 23999436]

[79] Faqi AS. Introduction. A Comprehensive Guide to Toxicology in Preclinical Drug Development 2013; 1-2. [http://dx.doi.org/10.1016/B978-0-12-387815-1.00001-0]

[80] Amberg A. *In silico* methods. In: Vogel HG, Maas J, Hock FJ, Mayer D (eds) Drug Discovery and Evaluation: Safety and Pharmacokinetic Assays, Second Edition. Springer Berlin Heidelberg, Berlin, Heidelberg, 2013; 1-1402.

[81] Everitt JI. The future of preclinical animal models in pharmaceutical discovery and development: a need to bring in cerebro to the *in vivo* discussions. Toxicol Pathol 2015; 43(1): 70-7. [http://dx.doi.org/10.1177/0192623314555162] [PMID: 25351920]

[82] Vandamme T. Use of rodents as models of human diseases. J Pharm Bioallied Sci 2014; 6(1): 2-9. [http://dx.doi.org/10.4103/0975-7406.124301] [PMID: 24459397]

[83] Singh SS. Toxicokinetics: An integral component of preclinical toxicity Studies. John Wiley & Sons, Inc, Hoboken, NJ, USA, 2010.

[84] Parkinson C, Thomas KE, Lumley CE. Reproductive toxicity testing of pharmaceutical compounds to support the inclusion of women in clinical trials. Hum Exp Toxicol 1997; 16(5): 239-46. [http://dx.doi.org/10.1177/096032719701600501] [PMID: 9192201]

[85] Parkinson C, Thomas KE, Lumley CE. The timing of preclinical toxicological studies: pharmaceutical company approaches to toxicity testing in support of initial clinical investigations. Regul Toxicol Pharmacol 1996; 23(2): 162-72. [http://dx.doi.org/10.1006/rtph.1996.0038] [PMID: 8661335]

[86] Day CP, Merlino G, Van Dyke T. Preclinical mouse cancer models: a maze of opportunities and challenges. Cell 2015; 163(1): 39-53. [http://dx.doi.org/10.1016/j.cell.2015.08.068] [PMID: 26406370]

[87] Custer L, Sweder K. The role of genetic toxicology in drug discovery and optimization. Curr Drug Metab 2008; 9(9): 978-85. [http://dx.doi.org/10.2174/138920008786485191] [PMID: 18991595]

[88] Steinmetz KL, Spack EG. The basics of preclinical drug development for neurodegenerative disease indications. BMC Neurol 2009; 9 (Suppl. 1): S2. [http://dx.doi.org/10.1186/1471-2377-9-S1-S2] [PMID: 19534731]

[89] Polson AG, Fuji RN. The successes and limitations of preclinical studies in predicting the pharmacodynamics and safety of cell-surface-targeted biological agents in patients. Br J Pharmacol

2012; 166(5): 1600-2.
[http://dx.doi.org/10.1111/j.1476-5381.2012.01916.x] [PMID: 22364106]

[90] Abu Fara D, Dadou SM, Rashid I, *et al.* A direct compression matrix made from xanthan gum and low molecular weight chitosan designed to improve compressibility in controlled release tablets. Pharmaceutics 2019; 11(11): 603-26.
[http://dx.doi.org/10.3390/pharmaceutics11110603] [PMID: 31726799]

[91] Reddy CM. Plasticity enhancement in pharmaceutical drugs by water of crystallization: unusual slip planes. IUCrJ 2019; 6(4): 505-6.
[http://dx.doi.org/10.1107/S205225251900890X] [PMID: 31316794]

[92] Mohan S. Compression physics of pharmaceutical powders: A review. Int J Pharm Sci Res 2012; 3: 1580-92.

[93] Marshall K. Compression and consolidation of powdered solids. In: Lachman L, Lieberman HA, Kanig JL, Eds. The Theory and Practice of Industrial Pharmacy 3rd ed. 1987; 66-99.

[94] Davies P. Oral solid dosage forms. In: Gibson M, Ed. Pharmaceutical Preformulation and Formulation Interpharm 2001; 379-456.

[95] Haruna F, Apeji YE, Oparaeche C, Oyi AR, Gamlen M. Compaction and tableting properties of composite particles of microcrystalline cellulose and crospovidone engineered for direct compression. Future Journal of Pharmaceutical Sciences 2020; 6(1): 35-43.
[http://dx.doi.org/10.1186/s43094-020-00055-9]

[96] Ansari JA. Drug interaction and pharmacist. J Young Pharm 2010; 2(3): 326-31.
[http://dx.doi.org/10.4103/0975-1483.66807] [PMID: 21042495]

[97] Snyder B, Polasek TM, Doogue MP. Drug interactions: principles and practice. Aust Prescr 2012; 35(3): 85-8.
[http://dx.doi.org/10.18773/austprescr.2012.037]

[98] Miranda V, Fede A, Nobuo M, *et al.* Adverse drug reactions and drug interactions as causes of hospital admission in oncology. J Pain Symptom Manage 2011; 42(3): 342-53.
[http://dx.doi.org/10.1016/j.jpainsymman.2010.11.014] [PMID: 21454043]

[99] Kotlinska-Lemieszek A, Klepstad P, Haugen DF. Clinically significant drug-Drug interactions involving medications used for symptom control in patients with advanced malignant disease: A systematic review. J Pain Symptom Manage 2019; 57(5): 989-98.
[http://dx.doi.org/10.1016/j.jpainsymman.2019.02.006] [PMID: 30776538]

[100] Farkas D, Shader RI, von Moltke LL, Greenblatt DJ. Mechanisms and consequences of drug-drug interactions. In: Gad SC, Ed. Preclinical Development Handbook: ADME and Biopharmaceutical Properties 2008; 879-917.

[101] Venkatakrishnan K, Shader R, Greenblatt D. Concepts and mechanisms of drug disposition and drug interactions. In: Ciraulo D, Shader R, Greenblatt D, Creelman W, Eds. Drug Interactions in Psychiatry 3rd ed. 2005; 1-46.

[102] Ochs HR, Steinhaus E, Locniskar A, Knüchel M, Greenblatt DJ. Desmethyldiazepam kinetics after intravenous, intramuscular, and oral administration of clorazepate dipotassium. Klin Wochenschr 1982; 60(8): 411-5.
[http://dx.doi.org/10.1007/BF01735933] [PMID: 6124654]

[103] Santoro MIRM, Singh AK. Pharmaceutical packaging. In: Yam KL, Ed. The Wiley encyclopedia of packaging technology 3rd ed. 2010; 949-57.

[104] Jatto E, Okhamafe AO. An overview of pharmaceutical validation and process controls in drug development. Trop J Pharm Res 2002; 1(2): 115-22.
[http://dx.doi.org/10.4314/tjpr.v1i2.14592]

[105] Dahiya S, Khar RK, Chhikara A. Opportunities, challenges and benefits of using HACCP as a quality

risk management tool in the pharmaceutical industry. Qual Assur J 2009; 12(2): 95-104. [http://dx.doi.org/10.1002/qaj.446]

[106] National Research Council. The prevention and treatment of missing data in clinical trials 2010.

[107] DerSimonian R, Charette LJ, McPeek B, Mosteller F. Reporting on methods in clinical trials. N Engl J Med 1982; 306(22): 1332-7. [http://dx.doi.org/10.1056/NEJM198206033062204] [PMID: 7070458]

[108] Gough S. Post-marketing surveillance: a UK/European perspective. Curr Med Res Opin 2005; 21(4): 565-70. [http://dx.doi.org/10.1185/030079905X41426] [PMID: 15899105]

[109] FDA. The drug development process. 2018. https://www.fda.gov/patients/drug-development-process/step-5-fda-post-market-drug-safety-monitoring

[110] Paluri RK, Li P, Anderson A, *et al.* First-in-human phase 1 clinical trials – A single-center experience in the era of modern oncotherapeutics. Sci Rep 2020; 10(1): 7935. [http://dx.doi.org/10.1038/s41598-020-64906-4] [PMID: 32404970]

[111] Fisher JA. Feeding and Bleeding. Sci Technol Human Values 2015; 40(2): 199-226. [http://dx.doi.org/10.1177/0162243914554838] [PMID: 25914430]

[112] Van den Eynde BJ, van Baren N, Baurain JF. Is there a clinical future for IDO1 inhibitors after the failure of epacadostat in melanoma? Annu Rev Cancer Biol 2020; 4(1): 241-56. [http://dx.doi.org/10.1146/annurev-cancerbio-030419-033635]

[113] Shamoo AE. The myth of equipoise in phase 1 clinical trials. Medscape J Med 2008; 10(11): 254-8. [PMID: 19099004]

[114] Umscheid CA, Margolis DJ, Grossman CE. Key concepts of clinical trials: a narrative review. Postgrad Med 2011; 123(5): 194-204. [http://dx.doi.org/10.3810/pgm.2011.09.2475] [PMID: 21904102]

[115] Meinert CL. Clinical trials handbook: Design and conduct 2013.

[116] Friedman LM, Furberg CD, Demets DL, *et al.* Fundamentals of clinical trials 5th ed., 2010. [http://dx.doi.org/10.1007/978-1-4419-1586-3]

[117] Gray R, Manola J, Saxman S, *et al.* Phase II clinical trial design: methods in translational research from the Genitourinary Committee at the Eastern Cooperative Oncology Group. Clin Cancer Res 2006; 12(7): 1966-9. [http://dx.doi.org/10.1158/1078-0432.CCR-05-1136] [PMID: 16609005]

[118] Van Norman GA. Phase II trials in drug development and adaptive trial design. JACC Basic Transl Sci 2019; 4(3): 428-37. [http://dx.doi.org/10.1016/j.jacbts.2019.02.005] [PMID: 31312766]

[119] Yan F, Thall PF, Lu KH, Gilbert MR, Yuan Y. Phase I–II clinical trial design: a state-of-the-art paradigm for dose finding. Ann Oncol 2018; 29(3): 694-9. [http://dx.doi.org/10.1093/annonc/mdx795] [PMID: 29267863]

[120] Fullerton T, Binneman B, David W, *et al.* A Phase 2 clinical trial of PF-05212377 (SAM-760) in subjects with mild to moderate Alzheimer's disease with existing neuropsychiatric symptoms on a stable daily dose of donepezil. Alzheimers Res Ther 2018; 10(1): 38. [http://dx.doi.org/10.1186/s13195-018-0368-9] [PMID: 29622037]

[121] Yuan J, Pang H, Tong T, Xi D, Guo W, Mesenbrink P. Seamless Phase IIa/IIb and enhanced dose-finding adaptive design. J Biopharm Stat 2016; 26(5): 912-23. [http://dx.doi.org/10.1080/10543406.2015.1094807] [PMID: 26390951]

[122] Mandrekar SJ, Sargent DJ. Randomized phase II trials: time for a new era in clinical trial design. J Thorac Oncol 2010; 5(7): 932-4. [http://dx.doi.org/10.1097/JTO.0b013e3181e2eadf] [PMID: 20581575]

[123] Fletcher RH, Fletcher SW. Clinical research in general medical journals: a 30-year perspective. N Engl J Med 1979; 301(4): 180-3.
[http://dx.doi.org/10.1056/NEJM197907263010403] [PMID: 377076]

[124] Riveros C, Dechartres A, Perrodeau E, Haneef R, Boutron I, Ravaud P. Timing and completeness of trial results posted at ClinicalTrials.gov and published in journals. PLoS Med 2013; 10(12): e1001566.
[http://dx.doi.org/10.1371/journal.pmed.1001566] [PMID: 24311990]

[125] Wong CH, Siah KW, Lo AW. Estimation of clinical trial success rates and related parameters. Biostatistics 2019; 20(2): 273-86.
[http://dx.doi.org/10.1093/biostatistics/kxx069] [PMID: 29394327]

[126] Thomas DW, Burns J, Audette J, *et al.* Clinical development success rates 2006-2015. BioMedTracker 1–28.

[127] Fennell DA, Kirkpatrick E, Cozens K, *et al.* CONFIRM: a double-blind, placebo-controlled phase III clinical trial investigating the effect of nivolumab in patients with relapsed mesothelioma: study protocol for a randomised controlled trial. Trials 2018; 19(1): 233.
[http://dx.doi.org/10.1186/s13063-018-2602-y] [PMID: 29669604]

[128] Rosenfeld PJ, Feuer WJ. Lessons from recent phase III trial failures: Don't design phase III trials based on retrospective subgroup analyses from phase II trials. Ophthalmology 2018; 125(10): 1488-91.
[http://dx.doi.org/10.1016/j.ophtha.2018.06.002] [PMID: 30243330]

[129] Pallmann P, Bedding AW, Choodari-Oskooei B, *et al.* Adaptive designs in clinical trials: why use them, and how to run and report them. BMC Med 2018; 16(1): 29.
[http://dx.doi.org/10.1186/s12916-018-1017-7] [PMID: 29490655]

[130] Lo AW, Siah KW, Wong CH. Estimating probabilities of success of vaccine and other antiinfective therapeutic development programs. Harvard Data Science Review 2020.
[http://dx.doi.org/10.3386/w27176]

[131] Arrowsmith J. Phase III and submission failures: 2007–2010. Nat Rev Drug Discov 2011; 10(2): 87.
[http://dx.doi.org/10.1038/nrd3375] [PMID: 21283095]

[132] Pazdur R. Endpoints for assessing drug activity in clinical trials. Oncologist 2008; 13(S2) (Suppl. 2): 19-21.
[http://dx.doi.org/10.1634/theoncologist.13-S2-19] [PMID: 18434634]

[133] FDA. 2019. New drug application. https://www.fda.gov/drugs/types-applications/new-dr-g-application-nda

[134] Liberti L, McAuslane JN, Walker S. Standardizing the benefit-risk assessment of new medicines. Pharmaceut Med 2011; 25(3): 139-46.
[http://dx.doi.org/10.1007/BF03256855]

[135] Holbein MEB. Understanding FDA regulatory requirements for investigational new drug applications for sponsor-investigators. J Investig Med 2009; 57(6): 688-94.
[http://dx.doi.org/10.2310/JIM.0b013e3181afdb26] [PMID: 19602987]

[136] Gogtay N, Ravi R, Thatte U. Regulatory requirements for clinical trials in India: What academicians need to know. Indian J Anaesth 2017; 61(3): 192-9.
[http://dx.doi.org/10.4103/ija.IJA_143_17] [PMID: 28405032]

[137] Wu E, Wang T, Lin T, *et al.* A comparative study of patients' attitudes toward clinical research in the United States and urban and rural China. Clin Transl Sci 2015; 8(2): 123-31.
[http://dx.doi.org/10.1111/cts.12254] [PMID: 25588611]

[138] Gungor B, Aylin M, Asena A, *et al.* Evaluation of legal legislation compliance and readability of clinical trial informed consent forms. Ther Innov Regul Sci 2019.
[http://dx.doi.org/10.1177/2168479019849520] [PMID: 32072599]

[139] Giezen TJ, Mantel-Teeuwisse AK, Leufkens HGM. Pharmacovigilance of Biopharmaceuticals. Drug

Saf 2009; 32(10): 811-7.
[http://dx.doi.org/10.2165/11316550-000000000-00000] [PMID: 19722725]

[140] https://www.cancer.org/treatment/treatments-and-side-effects/clinical-trials/what-you-need-to-know/phases-of-clinical-trials.html

[141] Suvarna V. Phase IV of drug development. Perspect Clin Res 2010; 1(2): 57-60.
[PMID: 21829783]

[142] Zhang X, Zhang Y, Ye X, Guo X, Zhang T, He J. Overview of phase IV clinical trials for postmarket drug safety surveillance: a status report from the ClinicalTrials.gov registry. BMJ Open 2016; 6(11): e010643.
[http://dx.doi.org/10.1136/bmjopen-2015-010643] [PMID: 27881517]

[143] Po ALW. Postmarketing surveillance. Elsevier Science 2007.

[144] Bell RL, O'Brian Smith E. Clinical trials in post-marketing surveillance of drugs. Control Clin Trials 1982; 3(1): 61-8.
[http://dx.doi.org/10.1016/0197-2456(82)90019-8] [PMID: 7067472]

[145] Chen BK, Yang YT. Post-marketing surveillance of prescription drug safety: past, present, and future. J Leg Med 2013; 34(2): 193-213.
[http://dx.doi.org/10.1080/01947648.2013.800797] [PMID: 23980746]

[146] Haque A, Daniel S, Maxwell T, Boerstoel M. Postmarketing surveillance studies—An industry perspective on changing global requirements and implications. Clin Ther 2017; 39(4): 675-85.
[http://dx.doi.org/10.1016/j.clinthera.2017.03.011] [PMID: 28392076]

[147] Hewitt C, Hahn S, Torgerson DJ, Watson J, Bland JM. Adequacy and reporting of allocation concealment: review of recent trials published in four general medical journals. BMJ 2005; 330(7499): 1057-8.
[http://dx.doi.org/10.1136/bmj.38413.576713.AE] [PMID: 15760970]

[148] Mehran R, Leon MB, Feigal DA, *et al.* Post-market approval surveillance: a call for a more integrated and comprehensive approach. Circulation 2004; 109(25): 3073-7.
[http://dx.doi.org/10.1161/01.CIR.0000134694.78653.B6] [PMID: 15226222]

[149] Raj N, Fernandes S, Charyulu NR, Dubey A, G S R, Hebbar S. Postmarket surveillance: a review on key aspects and measures on the effective functioning in the context of the United Kingdom and Canada. Ther Adv Drug Saf 2019; 10: 2042098619865413.
[http://dx.doi.org/10.1177/2042098619865413] [PMID: 31384423]

CHAPTER 4

Anti Bacterial Drugs: A Bright Past but a Challenging Future

Nitin Pal Kalia[1], Manoj Kumar[2], Souhaila Al Khodor[2] and **Sandeep Sharma[3,*]**

[1] *Department of Biological Sciences (Pharmacology & Toxicology), National Institute of Pharmaceutical Education and Research (NIPER), Hyderabad, Telangana, India*

[2] *Department of Research, Sidra Medicine, Doha, Qatar*

[3] *Department of Medical Laboratory Sciences, Lovely Professional University, Phagwara, Punjab, India*

Abstract: The discovery of penicillin opened the avenues for antibacterial drug discovery to address the global problem of deadly infectious diseases. However, despite the availability of potent antibiotics and effective vaccines, bacterial infections are still the major contributors to morbidity and mortality worldwide. The use of antibiotics is a two-edged sword; on the one hand, antibiotics have helped us combat deadly bacterial infections. On the other hand, overuse of antibiotics has led bacterial pathogens to develop drug resistance. The components of the bacterial cell-like cell wall, cell membrane, protein synthesis, and nucleic acid synthesis were targeted to develop effective drugs. Using these selective microbial targets, multiple potent antibiotic classes were developed in the last century, but emerging bacterial resistance and a decline in the number of new antibiotic approvals in recent years are pushing us back to the pre-antibiotic era. An increase in multidrug-resistant strains and the ineffectiveness of current drugs pose a challenge for researchers to develop new antibiotics with a novel mechanism to treat drug resistance. In the current chapter, we focus on the antibacterial drug used for the treatment of important human pathogens.

Keywords: Antibiotics, Anti-bacterial, Drug resistance, Drug discovery, Human pathogens.

INTRODUCTION

The golden era of antibiotic discovery began in the 1930s with the discovery of Penicillin by Alexzander Fleming, followed by the identification of several antibiotics. The discovery of antibiotics in the last century revolutionized the management and treatment of microbial infections caused by bacteria. The bacter-

* **Corresponding authors Sandeep Sharma:** Department of Medical Laboratory Sciences, Lovely Professional University, Phagwara, Punjab, India; E-mail: sandeep4380@gmail.com

ial infections that would normally have been fatal were now curable. Since then, antibiotics have revolutionized the medical field, saved lives and eased the suffering of billions of patients. Today, antibiotics are an indispensable part of our modern life, not only for the treatment of microbial infections but also for modern medical procedures of high-risk patients, such as procedure of organ transplants, cancer chemotherapy, prenatal care, intensive care *etc*. However, due to the misuse of antibiotics, these gains are going to be seriously paralyzed by the emergence of microbial-resistant strains. Unfortunately, the increasing bacterial resistance to antibiotics is not confined to a single country, but it poses a severe threat to human health globally. Moreover, the emergence of antimicrobial resistance did not come alone; it was also accompanied by a sharp decline in the approval of new antimicrobial drugs. The increase in antibiotic resistance is now an emerging challenge to global health agencies. Although all the health care agencies, including pharmaceutical industries, biotech companies, regulatory agencies, new ventures both at small and large scales, as well as academic laboratories, are working hard to revive the antibiotics discoveries programs, the discovery of new antibacterial drugs is continuously declining.

The consequences of emerging resistant bacteria are vast and going to impact all branches of medical practices. The discovery of new antibiotics is a two-edged sword one way, we are treating the infection, and another, we are imposing selective pressure on bacterial cells to develop resistance (mutants). Though different selective microbial targets such as cell wall, cell membrane, protein synthesis and nucleic acid were explored to overcome the emerging bacterial resistance; along with the antimicrobials discovery, bacteria also have evolved and adopted different mechanisms of antimicrobial resistance such as by impairing the selective antimicrobial agents permeability, enzymatically inactivating the antimicrobial structure, forming the antimicrobial resistant biofilm, inducing target mutations, expressing efflux pumps to decrease the effective concentration of antimicrobial agents and many others. This chapter focused on the current antibacterial drugs and the antibacterial pipeline under clinical development. We also highlighted the challenges of pharmaceutical companies and new approaches to antibacterial discoveries to combat the endless cycle of antibacterial resistance.

HISTORY OF ANTIMICROBIAL DRUGS

We normally correlate the beginning of the "antibiotic era" with the names of Paul Ehrlich and Alexander Fleming, but we have evidence of using some plant extracts or molds on wounds well before the discovery of bacteria as an infectious disease agent [1, 2]. During the ancient period, Egyptian healers used bread molds to treat surface infections well before the discovery of penicillin [3]. Later, Paul Ehrich came up with an idea of a "magic bullet" that can selectively target the infection-causing bacteria, not the host. This idea led him to discover a drug against syphilis, a disease almost incurable at that time [4]. During the same time, he coined the term "chemotherapy" in the early 1900s after identifying some chemicals and dyes to selectively color the bacterial cells [3, 4]. According to his theory, these chemicals were capable of inducing some substances to selectively kill bacteria without harming others [5].

The discovery of Paul Ehrlich and his systematic screening procedure for antimicrobials led to the discovery of many drugs. Following a similar screening procedure, Josef Klarer, Fritz Meitzsch, and Gerhard Domagk did the screening of a large number of chemical compounds against bacterial pathogens, which led to the discovery of the sulfa drug "sulfonamidochrysoidine" (KI-730, Prontosil).

Prontosil was later developed by a Bayer chemist, and the first human recipient of this drug was the daughter of Domagk, who was suffering from a *Streptococcus* infection and was completely cured after treatment [7]. Domagk received the Nobel Prize in medicine in 1939 for his discovery of prontosil [8]. Later, the sulfa drugs were further explored and led to the discovery of multiple synthetic antimicrobials, including quinolines and oxazolidinones [9].

Probably, many of us are aware of the serendipitous discovery on 3 September 1928 when Alexander Fleming noticed the growth of mold on his old plates of *Staphylococci* in his research laboratory at St. Mary's Hospital in London. He observed that the mold (Strain of *Penicillium notatum*) growth inhibited the nearby bacterial growth, which later led to the discovery of penicillin [10]. Even though Fleming tried for many years to scale up the purification of penicillin but failed, fortunately, an Oxford team named Howard Florey, Ernst Chain and Norman Heatley could optimize the large-scale production of penicillin [11], which was later used for treating the soldiers suffering from wound infections and pneumonia during World War II. Fleming was also among the first scientists who noticed the penicillin treatment failures if non-lethal quantities of the drug used, as a police officer suffering from scratch from a rose bush and treated with penicillin died due to an insufficient amount of penicillin [12]. Fleming, Florey and Chain received the Nobel Prize in physiology and medicine in 1945 for their incredible work. In 1945, Dorothy Hodgkin studied the crystallography of penicillin V and revealed its structure for the first time and was awarded the Nobel prize in chemistry in 1964.

Soon after the introduction of penicillin for clinical use, two major concerns were observed; 1) penicillin was found to be allergic to most of the patients, and 2) penicillin-resistant isolates were reported [13 - 15]. To overcome these concerns, erythromycin (macrolide) was introduced for allergic patients [16]. Following the same antimicrobial screening procedures, Selman Waksman also discovered various antibiotics in the 1940s, including actinomycin, streptomycin and neomycin [12]. A similar antimicrobial screening path was also followed by other researchers to discover several new antibiotics, which led to the golden era of antibacterial discoveries between the 1950s to 1970s. The discovery of different antibiotics and their reported resistance are summarized in Fig. (**1**).

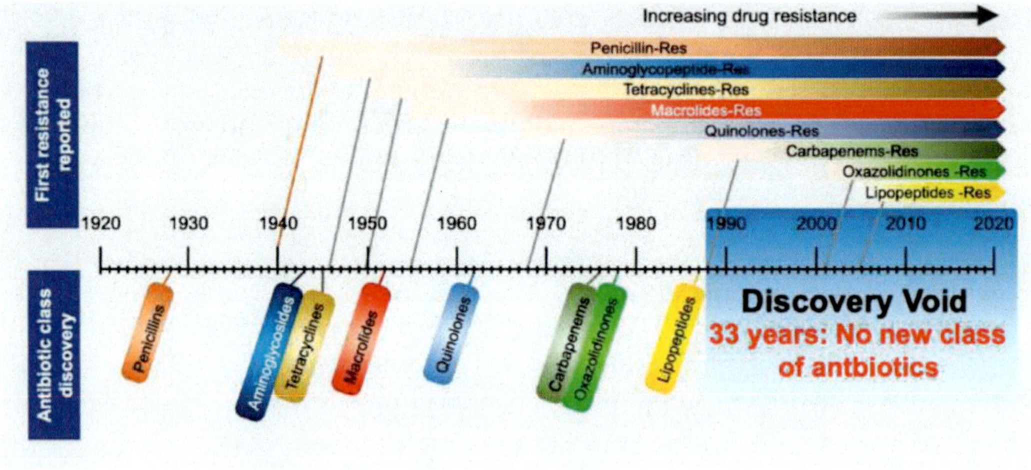

Fig. (1). Discovery of different antibiotics classes and emergence of first resistance. Timeline of the antibiotics discovery. Res: Resistance

A new class of antibiotic, tetracycline, was discovered in 1948 by Benjamin Duggar using soil, which was the only experimental antibiotic at that time. As a result of chemical modification, a new drug was developed, Rifamycin, by J.C Sheeman (the first person to make penicillin chemically), and this drug was used to treat tuberculosis. In 1962, the antibiotic Metronidazole was used to treat the bacterial infection named Gingivitis. During the course of antibiotic treatment, many of the bacterial strains developed resistance to the existing antibiotics using different mechanisms. This emergence of drug resistance allows researchers to discover new antibiotics that can treat these resistant bacterial strains in clinical settings [17].

Before antibiotic resistance, it was noticed that penicillin was allergic to most people. To reduce this allergic reaction, in 1940, the drug named Erythromycin (Macrolide) was an alternative drug for penicillin-allergic patients. In 1950, penicillin was the first drug that showed resistance to various strains by producing the enzyme penicillinase [12]. Thereafter colistin, the first polymyxin, was discovered in Japan by Y. Koyama in 1970 and was used as a replacement for toxic drugs. But bacteria managed to develop drug resistance to this drug as well. With time, *Staphylococcus aureus* also started showing resistance to methicillin and was first reported in 1961 in U.K. Vancomycin, the glycopeptide, was discovered and was known as the first drug which was effective against MRSA. The drug was first isolated from a soil sample, which was sent by Borneo. In Japan, vancomycin-intermediate *S. aureus* was reported, and in 1980, it was seen that some of the bacteria were getting resistant to Beta-lactamase and thereby becoming resistant to ampicillin. Thereafter, Merck, a well-known pharma company, discovered the "last resort" drug called Carbapenems, which was able to treat the infections caused by beta-lactam-resistant bacteria. During this period, carbapenems were considered the drug of last resort, and resistance to this drug left us with no other option [18].

In late 2001, linezolid, another drug that was the first new class of oxazolidinones for MRSA, was discovered as an alternative drug. In 2006, the European Union banned the use of antibiotics as growth promoters in the production of animals. On 9 July 2012, the Generating Antibiotic Incentives Now (GAIN) and the Food and Drug Administration Safety and Innovation Act (FDASIA) were signed into US law to handle the problem of antibiotic resistance [19 - 21]. Antibiotic resistance was recognized as "The World Economic Forum Global Risk "in 2013 and 2014. The first report on the global status of antibiotic resistance was published in 2014 by WHO. In July 2014, the UK prime minister announced that there is an urgent and global need to address the problem of antibiotic resistance therefore, he launched a commission: Review on Antimicrobial resistance. In 2015, the WHO released a global action plan and approved a new method (microchips) and drug (teixobactin) for resistant bacteria.

NOVEL ANTIBIOTICS FOR EFFECTIVE TREATMENT IN CLINICAL SETTINGS

According to the latest clinical trial data, there are few antibacterial drugs either with new mechanisms of action or better potency, under clinical development to combat the challenge of emerging multidrug-resistant bacteria. We have summarized the new antibacterial agents or combinations along with their target product profile (TPP) and mechanism of action in Table **1**. However, taken together, there is a lack of novel antibacterial agents for emerging Gram-negative pathogens. Indeed, the recent worldwide spread of multi-drug resistant pathogens (Table **2**), mainly ESBL-producing *Enterobacteriaceae* and carbapenem-resistant *Pseudomonas aeruginosa* isolates, illustrate that efforts to solve the antibacterial resistance problem cannot be combat with the current antibacterial pipeline, more refined approached and joined antibacterial discoveries programs are required. Drug-resistant concern is not only limited to critical priority pathogens; unregulated or misuse of antibiotics has increased the antibiotics resistant rate >50% among common pathogens [20]. Besides, high consumption of antibiotics is also known to correlate with an emerging high antibiotics resistance rate.

We all are worried about our future generations due to bacterial infections caused by drug resistant strains. If serious concerns are not shown now, we will not be able to resolve the issues later and ultimately, they may have to face problems like the non-availability of effective antibiotic drugs for treatment. Antibiotic resistance is a major concern of today's health care system. Though, many scientists worldwide are working towards a common goal *i.e.*, regulated use of antibiotics and the formulation of new antibiotics to replace the existing one which became less effective against resistant bacteria strains. The letter written by the President of Infectious Disease Society of America [IDSA] to the Assistant Secretary of Health and Human Services was about the importance of antibiotic therapies in almost all clinical treatments like organ transplants and cancer therapy; therefore, more focus should be on toward antibiotic stewardship, making new policies for R&D in antibiotic discovery and production.

The emergence of drug-resistant strains and the gap in antibiotic discovery is a major concerns. Normally it takes around 10-15 years to develop a new antibiotic, and on the other hand, bacteria start showing resistance to antibiotics immediately after exposure. For example, in the 2000s, Linezolid was approved for commercial use, and in 2002, the first linezolid-resistant isolate was reported [21], and by 2010 increasing linezolid resistance strains impacted its clinical uses [21]. Therefore, antibacterial discovery is an endless cycle, and we must continuously supply the new antibiotics to combat the resistance. In February, 2017 World Health Organization [WHO] published the list of pathogens for R&D of new antibiotic formulations. The type of resistance shown by the pathogens can be divided into three major categories (Table **1**). Even IDSA launched the "10×20' Initiative" to research and develop 10 novel antibiotics by 2020 [22, 23].

Table 1. List of antibiotics, class, mode of action and their indications.

Antibiotic	Class	Spectrum Against Organisms	Indications	Dose	Mechanism of Action
Ceftaroline [24, 25] [Teflaro/Zinforo]	Cephalosporin	ABSSSI: MRSA, MSSA, *S. pyogenes*, *S. agalactiae*, CABP: MSSA, *H. influenzae*, *S. pneumoniae*,	FDA: CABP and ABSSSI EMA: CABP and cSSSI	IV: 600 mg over 5 to 60 min every 12 hr	cell wall synthesis inhibition by inactivating penicillin-binding proteins [PBP]

(Table 1) cont.....

Antibiotic	Class	Spectrum Against Organisms	Indications	Dose	Mechanism of Action
Ceftobiprole [26, 27] [Zevtera/Mabelio]	Cephalosporin	MRSA, ampicillin-susceptible enterococci and penicillin-resistant pneumococci	EMA: HAP [excluding VAP] and CABP	IV: 500 mg over 2 hr every 8 hr	binding to PBPs, including the PBP-2a of MRSA, and blocking bacterial cell wall synthesis
Telavancin [28 - 29] [Vibativ]	Lipoglycopeptide	MRSA, vancomycin-intermediate S. aureus and penicillin-resistant S. pneumoniae	FDA: cSSSI, HAP [including VAP]	IV: 10 mg/kg over 60 min every 24 hr for 7–14 days [cSSSI] and 7–21 days [HAP/VAP]	Inhibition of cell wall synthesis of peptidoglycan chain and disrupting membrane barrier function by dissipating its potential
Dalbavancin [30 - 31] [Dalvance/ Xydalba]	Lipoglycopeptide	MRSA, S. pyogenes, S. agalactiae and E. faecalis strains susceptible to vancomycin	FDA: ABSSSI	IV: 1000 mg over 30 min loading dose followed by once weekly 500 mg over 30 min	binding to D-alanyl-D-alanine terminus of the stem pentapeptide in peptidoglycan, preventing cross-linking of cell wall synthesis
Oritavancin [32 - 33] [Orbactiv]	Glycopeptide	MSSA, MRSA, VRE and vancomycin-intermediate and vancomycin-resistant staphylococci	FDA: ABSSSI EMA: ABSSSI	IV: 1200 mg single dose over 3 hr	disrupts the bacterial membrane integrity leading to the bactericidal killing, inhibits bacterial RNA synthesis
Tedizolid [34, 35] Phosphate [Sivextro	Oxazolidinone	MRSA, vancomycin-intermediate Enterococcus spp.	FDA: ABSSSI EMA: ABSSSI	IV: 200 mg once daily over 1 h for 6 days PO: 200 mg once daily for 5 days	binding to the 23S ribosomal RNA [rRNA] of the 50S subunit of the bacterial ribosome and inhibits protein synthesis
Besifloxacin [36, 37] [Besivance]	Fluoroquinolone	MRSA, S. epidermidis, S. pneumoniae, and H. influenzae	FDA: bacterial conjunctivitis	Instill one drop in the affected eye[s] 3 times a day, for 7 days	inhibits DNA topoisomerases, including DNA gyrase and topoisomerase IV,
Delafloxacin [38, 39] [Baxdela]	Fluoroquinolone	S. aureus [including MRSA], S. pneumoniae, other fluoroquinolone resistant strains [Ineffective against Fluoroquinolone-resistant enterococci]	FDA: ABSSSI	IV: 300 mg over 1 hr every 12 hr PO: 450 mg tablet every 12 hr for 5 to 14 days	inhibits DNA gyrase and topoisomerase IV

(Table 1) cont.....

Antibiotic	Class	Spectrum Against Organisms	Indications	Dose	Mechanism of Action
Ozenoxacin [40, 41] [Ozaenex/Xepi]	Non-fluorinated quinolone	MRSA, MSSA, MRSE and S. pyogenes	FDA: impetigo	Topical: apply a thin layer to the affected area twice daily for 5 days	NA
Omadacycline [42, 43] [Nuzyra]	Tetracycline	MRSA, penicillin-resistant and multidrug-resistant S. pneumoniae, and vancomycin-resistant Enterococcus spp.	FDA: CABP, ABSSSI	Duration: 7–14 days Loading IV Day 1: 200 mg over 1 hr once daily or 100 mg over 30 min twice daily Maintenance: 100 mg over 30 min or 300 mg po once daily 6 Loading PO [ABSSSI] Day 1 and 2: 450 mg once daily Maintenance PO [ABSSSI] 300 mg once daily	binds to the 30S ribosomal subunit and inhibits protein synthesis
Ceftolozane/tazobactam [44]	cephalosporin combined with BLI	Gram-negative pathogens	complicated UTI [cUTI] or acute pyelo-nephritis [APN]	1.5 gm intravenous [IV] every 8 hr	Inhibits β-lactamase
Ceftazidim/avibactam [45]	cephalosporin combined with non–b-lactam BLI	Resistant Gram-negative bacilli producing extended spectrum β-lactamase (ESBL), klebsiella pneumoniae carbapenemases {KPCs} and / or AmpC β-lactamases and P. *aeruginosa*	complicated UTI [cUTI] or acute pyelo-nephritis [APN]	2000/500 mg IV every 8 hr	Inhibits β-lactamase

(Table 1) cont.....

Antibiotic	Class	Spectrum Against Organisms	Indications	Dose	Mechanism of Action
Meropenem/vaborbactam [46, 47]	Carbapenem	carbapenem-resistant Enterobacteriaceae Gram-negative bacilli-producing ESBL, KPC, and/or AmpC β-lactamases.	complicated UTI [cUTI] or acute pyelo-nephritis [APN]	2 gm/2 gm for 10 days	binds to several penicillin-binding proteins [PBPs], and inhibits of serine carbapenemases
Cefiderocol [48, 49]	Cephalosporin	Carbapenem resistant gram-negative bacteria	complicated UTI [cUTI] or acute pyelo-nephritis [APN]	2 gm IV every 8 hr	binds to PBP3 of Gram-negative bacteria and inhibits bacterial cell wall synthesis
Plazomicin [50 51]	Aminoglycoside	Escherichia co li, Klebsiella pne umoniae, Cit robacter spp., Enterobacter spp. and Serratia spp methicillin-resistant S. aureus [MRSA], methicillin -susceptible S.au reus, coagulase-negative staphylo cocci and Staphylococcus sa prophyticus	complicated UTI [cUTI] or acute pyelo-nephritis [APN]	15 mg/kg every 24 hr	inhibits bacterial protein synthesis and exhibits dose-dependent bactericidal activity
Finafloxacin [52]	Fluoroquinolone	Gram negative organism	complicated UTI [cUTI] or acute pyelo-nephritis [APN]	800 mg per day	inhibits the activity of bacterial type II topoisomerases
Eravacycline [46]	fluorocycline	Gram Negative organism	complicated UTI [cUTI] or acute pyelo-nephritis [APN]	1.5 mg/kg	

IV: Intravenous; PO: Per Oral administration; cUTI: Complicated Urinary Tract Infections; ABSSSI: Acute Bacterial Skin and Skin Structure Infections; MRSA: Methicillin-Resistant *Staphylococcus aureus*; MSSA: Methicillin-Sensitive *Staphylococcus aureus*; VRE: Vancomycin-Resistant Enterococci; CABP: Community-Acquired Bacterial Pneumonia; FDA: Food and Drug Administration; EMA: European Medicines Agency; HAP: Hospital-Acquired Pneumonia; VAP: Ventilator-Associated Pneumonia; cSSSI: Complicated Skin and Skin-Structure Infections; APN: Acute Pyelonephritis.

Table 2. List of antibiotic-resistant bacteria based on their priorities and hazard levels.

Bacterial pathogens and hazard level	Reported antibiotics resistant to
1. Hazard level: urgent: [Critical Priority] • *Acinetobacter baumannii* • *Enterobacteriaceae* • *Pseudomonas aeruginosa*	Carbapenem-resistant Carbapenem-resistant, ESBL-producing Carbapenem-resistant

(Table 2) cont.....

Bacterial pathogens and hazard level	Reported antibiotics resistant to
2. Hazard level: serious: [High Priority] • *Enterococcus faecium* • *Helicobacter pylori* • *Neisseria gonorrhoeae* • *Salmonellae* • *Staphylococcus aureus* •	Vancomycin-resistant Methicillin-resistant, vancomycin-intermediate and resistant Clarithromycin-resistant Fluoroquinolone-resistant Fluoroquinolone-resistant Cephalosporin-resistant, fluoroquinolone-resistant
3. Hazard level: concerning [Medium Priority] • *Haemophilus influenzae* • *Shigella* spp. • *Streptococcus pneumoniae*	Penicillin-non-susceptible Ampicillin-resistant Fluoroquinolone-resistant

Global priority list of antibiotic-resistant bacteria to guide research, discovery, and development of new antibiotics, World Health Organization, 2017. ESBL: Extended Spectrum Beta-Lactamases.

Most of these antibiotics are still under development and are supposed to be available in the near future to meet the emerging challenge of antibacterial resistance. But considering the lack of novel antibacterial agents with a new mechanism of action, and most of the antibiotics used presently are estimated to be useless within five to ten years, will this be the end of the antibiotics era, or are we going to enter the pre-antibiotics era.

ANTIBACTERIAL DRUGS: A BRIGHT PAST BUT A CHALLENGING FUTURE

Antibacterial Drugs Discoveries

In the last century, antibiotics developments have revolutionized our ability to manage bacterial infections and achieved remarkable advances on several fronts: a better understanding of bacterial pathogenesis and cure; a substantial improvement in the efficacy and therapeutic effects of antibiotics; a significant improvement in various medical interventions and their outcomes such as open-heart surgery and even transplantation medicine [52, 53]. The remarkable success of antibacterial drugs has led to substantial decreases in morbidity and mortality associated with bacterial infections worldwide [54]. However, excessive and inappropriate use of antibiotics has provided intense and selective pressure on microorganisms, leading to preferential survival and the emergence of antibiotic resistance mechanisms in bacteria [55, 56]. The increasing consequences of multidrug-resistant bacteria have posed severe threats to almost all branches of medical practice. Although, the pharmaceutical industries had worked hard to combat the increasing drug resistance by developing several antibacterial classes with different mechanisms of action [57 - 62]. However, resistance has already been observed for the majority of antibiotics classes (Fig. **1**) [63]. Moreover, a decline in the number of new antibiotic approvals and the growing emergence of multidrug-resistant bacteria has contributed to the situation where there are little or no treatment options for multidrug-resistant bacterial pathogens. Therefore, there is no escape from the emerging antibiotics resistance, and continuous antimicrobial research for antibiotics must be promoted to protect humanity. Though indirect, clear evidence suggests that pharmaceutical firms rejected their new initiatives for antibacterial discovery [64, 65].

Broad-target Activity Challenges

Over the past three decades, technological advancements and the recognition of microbial importance in maintaining human health [66, 67], have led to high expectations to develop new and specific approaches for the prevention and treatment of infection. The general consensus is that the gut microbiota is the target of broad-spectrum antibiotics, which, apart from eliminating the target pathogen, also influence the composition of symbiotic microbiota by killing or providing strong selective pressure to develop resistance [68]. Even early exposure to broad-spectrum antibiotics in early life was shown to play a pathogenic role in several human diseases [67, 69, 70]. Nevertheless, several pharmaceutical companies have based their antibacterial research and development on narrow-spectrum antibacterial agents and have started testing their pre-clinical R&D candidates for their selectivity and impact on the healthy gut microbiota [71 - 73]. However, the low success rate and poor financial returns from developing new narrow-spectrum antibacterial drugs, forced companies to withdraw from their new antibacterial research programs [74].

Medical Challenge

The consequences of emerging antibacterial-resistant pathogens are vast and pose a severe threat to almost all branches of medical procedures. Without access to effective antibiotics, practices such as prosthetic surgery, organ transplantation, management of immunosuppressed cancer patients, and care of elderly patients or neonates will be severely impacted. In addition, increasing health care costs associated with managing resistant bacteria and indirect costs such as days away from work or loss of output due to premature death are further jeopardizing our life.

Increasing Research and Development Cost

Preclinical evaluation of new antibacterial agents is an extensive process since several optimizations are required to bring a new chemical entity from the hit to the clinical candidate, which takes at least 6-8 years (Fig. **2**). This is in addition to the rising costs of preclinical and sclinical research estimated at about $400-800 million per new candidate [75], which has posed a considerable barrier to new antibacterial drug programs. Besides, antibiotics are also known as miracle bullets [76] and are typically used to treat disease for a short-course, thereby reducing their continuous need. In comparison, the treatment of non-communicable chronic diseases such as hypertension, diabetes, and arthritis require lifelong medication and is also very common in the elderly population; therefore, pharmaceutical firms have diverted their energies toward drug development to treat those conditions as they return more benefits to them.

Commercial Challenge for the Pharmaceutical Industry

Many patients use an ideal drug for a relatively long period of time from a commercial point of view. However, an effective antibacterial agent cured the patients within a few days. Therefore, from a business point of view, antibacterial agents do not have all the traits that an anti-cancer drug or other successful anti-diabetic drugs usually have. This is one of the main reasons, why Big Pharma is withdrawing from their anti-bacterial programs and the development of new antibiotics is turning out to be difficult.

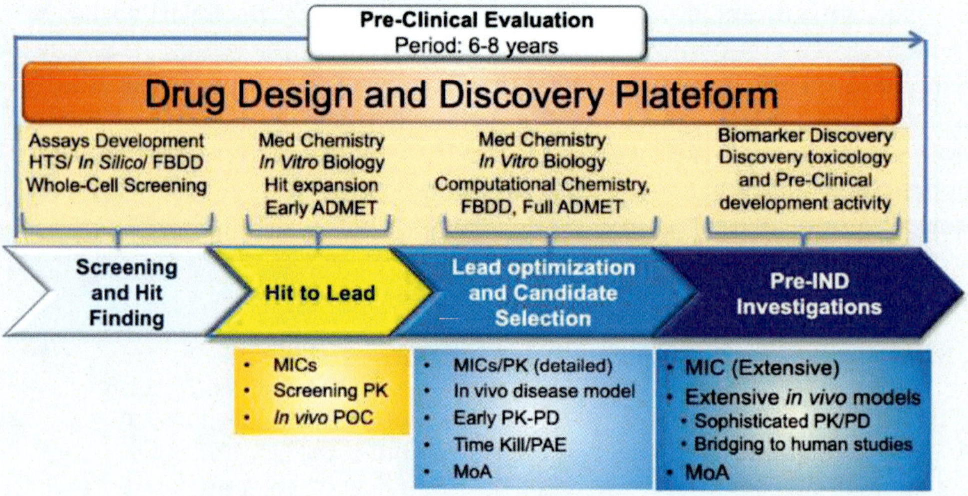

Fig. (2). Pre-clinical evaluation of new chemical entities from hit identification to lead optimization. HTS: High Throughput Screen; FBDD: Fragment-Based drug Discovery; ADMET: Absorption, Distribution, Metabolism, Elimination, Toxicity; Pre-IND: Pre-Investigational New Drug; PK: Pharmacokinetics; PK-PD: Pharmacokinetic-Pharmacodynamic; PAE: Post-Antibiotic Effect; MoA: Mechanism of Action; MIC: Minimum Inhibitory Concentration.

Challenges in Antibacterial Drug Discovery

Due to the emergence of resistance to almost all the available antibiotics, there is a pressing need to identify novel antibiotics with a unique mode of action. Aside from the resistance-mediated fall in clinical efficacy, available antibiotics have side effects, with issues in dosing regimens and their usage, particularly for children that limit their utility [77]. Antibiotic discovery faces various issues at economic and scientific levels. Despite strenuous attempts to utilize a wide range of modern antibiotic discovery platform strategies, including genomics, bioinformatics, systems biology and post-genomic approaches, success has been marginal. Therefore, it is not an easy task to find new classes of antibiotics using old strategies [78]. Also, failure to consider the importance of clinical use, manufacturing, regulatory practices, the feasibility of clinical study designs and reimbursement are not only inefficient but are the major causes behind the failure of Research led programs.

Bacterial cells have a cell wall (Gram-positive) and an additional outer membrane (Gram-negative) which plays a crucial role in protecting bacteria from chemical treatment. Except for the cell wall and outer membrane, almost all the targets are intracellular. Therefore, it is very important for the compound to penetrate these barriers and engage its biological targets. That is why whole cell-based high throughput assays are preferred over cell-free screening approaches in order to address the issue of permeability and efflux [79].

For more than two decades, different research groups have been focused on chasing new targets, and novel compound scaffolds and approaches have been considered inefficient in terms of output. Though it is always convenient to find compounds that kill bacteria, it is hard to find novel anti-bacterial classes eligible for development. Therefore, it is worth to change the strategies and focus on: {(i) innovative chemistry on and around known clinically effective drug scaffolds; (ii) alternative ways to inhibit the function of clinically validated targets; (iii) understanding resistance mechanisms and how they can be inhibited; (iv) establishing phenotypic screens to identify new molecules; (v) understanding the utility of animal models and the risks around reducing drug-efficacy hurdles; (vi) establishing the levels of *in vitro* resistance development that are unacceptable; (vii). Bioprospecting to identify novel drug scaffolds and making this an option in searching for new antibiotics or exploring existing ones for additional indications} [80, 81]. Improving bacterial culture techniques, molecular biology and metagenomics will make natural product drug discovery easier and more cost-effective, evading these limiting factors.

CONCLUSION AND FUTURE PERSPECTIVES

Failure to predict and fully understand the underlying mechanisms of bacterial resistance to antibiotics, the leading cause of nosocomial infections and mortalities, has limited our options for effective therapeutics. Moreover, despite the deployment of substantial capital and encouragement over the last three decades, discovering innovative targets or chemical scaffolds has failed, thus, different strategies are urgently required to rebuild sustainable discoveries programs and foundations for future antibacterial drugs.

To resolve this void of antibacterial research and discovery, we recommend that both academia, as well as pharmaceutical companies, play an important role. For example, academia should receive adequate funding to explore potential new bacterial targets and the resistance mechanism of bacterial pathogens, particularly selective permeabilities and the multi-drug efflux mechanism of Gram-negative bacteria. On the other hand, major pharmaceutical companies need to play a critical role in providing appropriate training for next-generation academic researchers on SAR (structure-activity relationship) optimization and making the antibacterial leads database publicly accessible to everyone.

Moreover, old natural product leads databases may also be helpful in developing new programs; however, an adequate strategy should be taken to ensure performance and not replicate past failures. In addition, an unprecedented level of investment should be provided by government agencies for (i) understanding the mechanisms of drug resistance and exploring the reverse-antibiotics programs [3], (ii) innovative and advanced animal models, not only for assessment of antibacterial efficacies but also for exploring the risks for resistance development and clinical efficacies [54, 66, 83] (iii) discovering alternative antibacterial targets such as virulence factors or multi-drug efflux inhibitor to enhance the efficacy of existing antibiotics (iv) streamlining the drug approval process [82] (v) repurposing drugs to fill the antibacterial discovery void [84] (vi) promising contract of government regulatory authorities with pharmaceutical industries to support academia in early pre-clinical programs and also support the clinical development to meet the unmet medical needs. Such methods are expected to help fill the vacuum of antibacterial exploration in the near future.

CONSENT FOR PUBLICATION

Not applicable.

CONFLICT OF INTEREST

The author declares no conflict of interest, financial or otherwise.

ACKNOWLEDGEMENTS

Declared none.

REFERENCES

[1] Forrest RD. Early history of wound treatment. J R Soc Med 1982; 75(3): 198-205. [http://dx.doi.org/10.1177/014107688207500310] [PMID: 7040656]

[2] Wainwright M. Moulds in ancient and more recent medicine. Mycologist 1989; 3(1): 21-3. [http://dx.doi.org/10.1016/S0269-915X(89)80010-2]

[3] Ehrlich P, Halta S. Die experimentelle chemotherapie der Spirillosen 1910. [http://dx.doi.org/10.1007/978-3-642-64926-4]

[4] Parascandola J. The theoretical basis of Paul Ehrlich's chemotherapy. J Hist Med Allied Sci 1981; XXXVI(1): 19-43. [http://dx.doi.org/10.1093/jhmas/XXXVI.1.19] [PMID: 7009720]

[5] Aminov RI. A brief history of the antibiotic era: lessons learned and challenges for the future. Front Microbiol 2010; 1: 134. [http://dx.doi.org/10.3389/fmicb.2010.00134] [PMID: 21687759]

[6] Carithers HA. The first use of an antibiotic in America. Am J Dis Child 1974; 128(2): 207-11. [PMID: 4605263]

[7] Henrik S, Sohlman R, Österling A, *et al.* The Prize in Physiology and Medicine: The Nobel Prizes in Wartime. In Nobel Foundation (ed.). Nobel: The Man and His Prizes. Stockholm: Klara Civiltryckeri. 1950; pp. 167–179.

[8] Aminov R. History of antimicrobial drug discovery: Major classes and health impact. Biochem Pharmacol 2017; 133: 4-19. [http://dx.doi.org/10.1016/j.bcp.2016.10.001] [PMID: 27720719]

[9] Fleming A. On the Antibacterial Action of Cultures of a Penicillium, with Special Reference to their Use in the Isolation of B. influenza. Br J Exp Pathol 1929; 10(3): 226-36.

[10] Gaynes R. The discovery of penicillin—new insights after more than 75 years of clinical use. Emerg Infect Dis 2017; 23(5): 849-53. [http://dx.doi.org/10.3201/eid2305.161556]

[11] Grossman TH. Tetracycline antibiotics and resistance. Cold Spring Harb Perspect Med 2016; 6(4): a025387. [http://dx.doi.org/10.1101/cshperspect.a025387] [PMID: 26989065]

[12] Levine DP. Vancomycin: a history. Clin Infect Dis 2006; 42 (Suppl. 1): S5-S12. [http://dx.doi.org/10.1086/491709] [PMID: 16323120]

[13] Castanon JIR. History of the use of antibiotic as growth promoters in European poultry feeds. Poult Sci 2007; 86(11): 2466-71. [http://dx.doi.org/10.3382/ps.2007-00249] [PMID: 17954599]

[14] Karageorgopoulos DE, Falagas ME. New antibiotics: optimal use in current clinical practice. Int J Antimicrob Agents 2009; 34 (Suppl. 4): S55-62. [http://dx.doi.org/10.1016/S0924-8579(09)70569-2] [PMID: 19931821]

[15] Stefani S, Bongiorno D, Mongelli G, Campanile F. Linezolid resistance in staphylococci. Pharmaceuticals (Basel) 2010; 3(7): 1988-2006. [http://dx.doi.org/10.3390/ph3071988] [PMID: 27713338]

[16] Govindaraj Vaithinathan A, Vanitha A. WHO global priority pathogens list on antibiotic resistance: an urgent need for action to integrate One Health data. Perspect Public Health 2018; 138(2): 87-8. [http://dx.doi.org/10.1177/1757913917743881] [PMID: 29465015]

[17] Boucher HW, Talbot GH, Benjamin DK Jr, *et al.* 10 x '20 Progress--development of new drugs active against gram-negative bacilli: an update from the Infectious Diseases Society of America. Clin Infect Dis 2013; 56(12): 1685-94. [http://dx.doi.org/10.1093/cid/cit152] [PMID: 23599308]

[18] Koulenti D, Xu E, Mok IYS, *et al.* Novel antibiotics for multidrug-resistant gram-positive microorganisms. Microorganisms 2019; 7(8): 270. [http://dx.doi.org/10.3390/microorganisms7080270] [PMID: 31426596]

[19] White BP, Barber KE, Stover KR. Ceftaroline for the treatment of methicillin-resistant *Staphylococcus aureus* bacteremia. Am J Health Syst Pharm 2017; 74(4): 201-8. [http://dx.doi.org/10.2146/ajhp160006] [PMID: 28179245]

[20] Awad SS, Rodriguez AH, Chuang YC, *et al.* A phase 3 randomized double-blind comparison of ceftobiprole medocaril *versus* ceftazidime plus linezolid for the treatment of hospital-acquired pneumonia. Clin Infect Dis 2014; 59(1): 51-61. [http://dx.doi.org/10.1093/cid/ciu219] [PMID: 24723282]

[21] Ye L, Zhang J, Xiao W, Liu S. Efficacy and mechanism of actions of natural antimicrobial drugs. Pharmacol Ther 2020; 216: 107671. [http://dx.doi.org/10.1016/j.pharmthera.2020.107671] [PMID: 32916205]

[22] Stryjewski ME, Graham DR, Wilson SE, *et al.* Telavancin *versus* vancomycin for the treatment of complicated skin and skin-structure infections caused by gram-positive organisms. Clin Infect Dis 2008; 46(11): 1683-93. [http://dx.doi.org/10.1086/587896] [PMID: 18444791]

[23] Boucher HW, Wilcox M, Talbot GH, Puttagunta S, Das AF, Dunne MW. Once-weekly dalbavancin *versus* daily conventional therapy for skin infection. N Engl J Med 2014; 370(23): 2169-79. [http://dx.doi.org/10.1056/NEJMoa1310480] [PMID: 24897082]

[24] Gales AC, Sader HS, Jones RN. Antimicrobial activity of dalbavancin tested against Gram-positive clinical isolates from Latin American medical centres. Clin Microbiol Infect 2005; 11(2): 95-100. [http://dx.doi.org/10.1111/j.1469-0691.2004.01051.x] [PMID: 15679482]

[25] Corey GR, Kabler H, Mehra P, *et al.* Single-dose oritavancin in the treatment of acute bacterial skin infections. N Engl J Med 2014; 370(23): 2180-90. [http://dx.doi.org/10.1056/NEJMoa1310422] [PMID: 24897083]

[26] Corey GR, Good S, Jiang H, *et al.* Single-dose oritavancin *versus* 7-10 days of vancomycin in the treatment of gram-positive acute bacterial skin and skin structure infections: the SOLO II noninferiority study. Clin Infect Dis 2015; 60(2): 254-62. [http://dx.doi.org/10.1093/cid/ciu778] [PMID: 25294250]

[27] Burdette SD, Trotman R. Tedizolid: the first once-daily oxazolidinone class antibiotic. Clin Infect Dis 2015; 61(8): 1315-21. [http://dx.doi.org/10.1093/cid/civ501] [PMID: 26105167]

[28] Prokocimer P, De Anda C, Fang E, Mehra P, Das A. Tedizolid phosphate *vs* linezolid for treatment of acute bacterial skin and skin structure infections: the ESTABLISH-1 randomized trial. JAMA 2013; 309(6): 559-69. [http://dx.doi.org/10.1001/jama.2013.241] [PMID: 23403680]

[29] Haas W, Pillar CM, Zurenko GE, Lee JC, Brunner LS, Morris TW. Besifloxacin, a novel fluoroquinolone, has broad-spectrum *in vitro* activity against aerobic and anaerobic bacteria. Antimicrob Agents Chemother 2009; 53(8): 3552-60. [http://dx.doi.org/10.1128/AAC.00418-09] [PMID: 19506065]

[30] Dutta NK, Pinn ML, Zhao M, Rudek MA, Karakousis PC. Thioridazine lacks bactericidal activity in an animal model of extracellular tuberculosis. J Antimicrob Chemother 2013; 68(6): 1327-30. [http://dx.doi.org/10.1093/jac/dkt037] [PMID: 23463208]

[31] Candel FJ, Peñuelas M. Delafloxacin: design, development and potential place in therapy. Drug Des Devel Ther 2017; 11: 881-91. [http://dx.doi.org/10.2147/DDDT.S106071] [PMID: 28356714]

[32] O'Riordan W, Mehra P, Manos P, Kingsley J, Lawrence L, Cammarata S. A randomized phase 2 study

comparing two doses of delafloxacin with tigecycline in adults with complicated skin and skin-structure infections. Int J Infect Dis 2015; 30: 67-73.
[http://dx.doi.org/10.1016/j.ijid.2014.10.009] [PMID: 25448332]

[33] Vila J, Hebert AA, Torrelo A, *et al.* Ozenoxacin: a review of preclinical and clinical efficacy. Expert Rev Anti Infect Ther 2019; 17(3): 159-68.
[http://dx.doi.org/10.1080/14787210.2019.1573671] [PMID: 30686133]

[34] Wren C, Bell E, Eiland LS. Ozenoxacin: A Novel Topical Quinolone for Impetigo. Ann Pharmacother 2018; 52(12): 1233-7.
[http://dx.doi.org/10.1177/1060028018786510] [PMID: 29962213]

[35] Villano S, Steenbergen J, Loh E. Omadacycline: development of a novel aminomethylcycline antibiotic for treating drug-resistant bacterial infections. Future Microbiol 2016; 11(11): 1421-34. [http://dx.doi.org/10.2217/fmb-2016-0100] [PMID: 27539442]

[36] Noel GJ, Draper MP, Hait H, Tanaka SK, Arbeit RD. A randomized, evaluator-blind, phase 2 study comparing the safety and efficacy of omadacycline to those of linezolid for treatment of complicated skin and skin structure infections. Antimicrob Agents Chemother 2012; 56(11): 5650-4.
[http://dx.doi.org/10.1128/AAC.00948-12] [PMID: 22908151]

[37] Wagenlehner FM, Umeh O, Steenbergen J, Yuan G, Darouiche RO. Ceftolozane-tazobactam compared with levofloxacin in the treatment of complicated urinary-tract infections, including pyelonephritis: a randomised, double-blind, phase 3 trial (ASPECT-cUTI). Lancet 2015; 385(9981): 1949-56.
[http://dx.doi.org/10.1016/S0140-6736(14)62220-0] [PMID: 25931244]

[38] Wagenlehner FM, Sobel JD, Newell P, *et al.* Ceftazidime-avibactam *versus* doripenem for the treatment of complicated urinary tract infections, including acute pyelonephritis: RECAPTURE, a phase 3 randomized trial program. Clin Infect Dis 2016; 63(6): 754-62.
[http://dx.doi.org/10.1093/cid/ciw378] [PMID: 27313268]

[39] Thaden JT, Pogue JM, Kaye KS. Role of newer and re-emerging older agents in the treatment of infections caused by carbapenem-resistant Enterobacteriaceae. Virulence 2017; 8(4): 403-16.
[http://dx.doi.org/10.1080/21505594.2016.1207834] [PMID: 27384881]

[40] Walsh TJ, Bhowmick T, Darouiche R, *et al.* Meropenem-Vaborbactam vs. Piperacillin-Tazobactam in TANGO I (a Phase 3 Randomized, Double-blind Trial): Outcomes by Baseline MIC in Adults with Complicated Urinary Tract Infections or Acute Pyelonephritis. In Open Forum Infectious Diseases (Vol 4, No suppl_1, pp S536-S536) US: Oxford University Press 2017.

[41] Ito A, Nishikawa T, Matsumoto S, *et al.* Siderophore cephalosporin cefiderocol utilizes ferric iron transporter systems for antibacterial activity against Pseudomonas aeruginosa. Antimicrob Agents Chemother 2016; 60(12): 7396-401.
[http://dx.doi.org/10.1128/AAC.01405-16] [PMID: 27736756]

[42] Falagas ME, Skalidis T, Vardakas KZ, Legakis NJ. Activity of cefiderocol (S-649266) against carbapenem-resistant Gram-negative bacteria collected from inpatients in Greek hospitals. J Antimicrob Chemother 2017; 72(6): 1704-8.
[http://dx.doi.org/10.1093/jac/dkx049] [PMID: 28369471]

[43] Zhanel GG, Lawson CD, Zelenitsky S, *et al.* Comparison of the next-generation aminoglycoside plazomicin to gentamicin, tobramycin and amikacin. Expert Rev Anti Infect Ther 2012; 10(4): 459-73. [http://dx.doi.org/10.1586/eri.12.25] [PMID: 22512755]

[44] Connolly LE, Riddle V, Cebrik D, Armstrong ES, Miller LG. A multicenter, randomized, double-blind, phase 2 study of the efficacy and safety of plazomicin compared with levofloxacin in the treatment of complicated urinary tract infection and acute pyelonephritis. Antimicrob Agents Chemother 2018; 62(4): e01989-17.
[http://dx.doi.org/10.1128/AAC.01989-17] [PMID: 29378708]

[45] Wagenlehner F, Nowicki M, Bentley C, *et al.* Explorative randomized phase II clinical study of the

efficacy and safety of finafloxacin *versus* ciprofloxacin for treatment of complicated urinary tract infections. Antimicrob Agents Chemother 2018; 62(4): e02317-17. [http://dx.doi.org/10.1128/AAC.02317-17] [PMID: 29339395]

[46] Bérdy J. Thoughts and facts about antibiotics: Where we are now and where we are heading. J Antibiot (Tokyo) 2012; 65(8): 385-95. [http://dx.doi.org/10.1038/ja.2012.27] [PMID: 22511224]

[47] Ventola CL. The antibiotic resistance crisis: part 1: causes and threats. P&T 2015; 40(4): 277-83. [PMID: 25859123]

[48] Fair RJ, Tor Y. Antibiotics and bacterial resistance in the 21st century. Persp Med Chem 2014; 6:PMC-S14459.

[49] Chanvatik S, Donnua S, Lekagul A, *et al.* Antibiotic use in mandarin production (Citrus reticulata Blanco) in major mandarin-producing areas in Thailand: A survey assessment. PLoS One 2019; 14(11): e0225172. [http://dx.doi.org/10.1371/journal.pone.0225172] [PMID: 31721802]

[50] Llor C, Bjerrum L. Antimicrobial resistance: risk associated with antibiotic overuse and initiatives to reduce the problem. Ther Adv Drug Saf 2014; 5(6): 229-41. [http://dx.doi.org/10.1177/2042098614554919] [PMID: 25436105]

[51] Butler MS, Blaskovich MAT, Cooper MA. Antibiotics in the clinical pipeline at the end of 2015. J Antibiot (Tokyo) 2017; 70(1): 3-24. [http://dx.doi.org/10.1038/ja.2016.72] [PMID: 27353164]

[52] Draenert R, Seybold U, Grützner E, Bogner JR. Novel antibiotics: Are we still in the pre–post-antibiotic era? Infection 2015; 43(2): 145-51. [http://dx.doi.org/10.1007/s15010-015-0749-y] [PMID: 25701222]

[53] Kumar M, Rao M, Purnapatre KP, *et al.* DS86760016, a leucyl-tRNA synthetase inhibitor with activity against Pseudomonas aeruginosa. Antimicrob Agents Chemother 2019; 63(4): e02122-18. [http://dx.doi.org/10.1128/AAC.02122-18] [PMID: 30670430]

[54] Barman TK, Kumar M, Mathur T, *et al. In vitro* and *in vivo* activities of a bi-aryl oxazolidinone, RBx 11760, against Gram-positive bacteria. Antimicrob Agents Chemother 2016; 60(12): 7134-45. [http://dx.doi.org/10.1128/AAC.00453-16] [PMID: 27645240]

[55] Barman TK, Kumar M, Mathur T, *et al. In vitro* and *in vivo* activities of DS-2969b, a novel GyrB inhibitor, and its water-soluble prodrug, DS11960558, against methicillin-resistant *Staphylococcus aureus.* Antimicrob Agents Chemother 2018; 62(6): e02556-17. [http://dx.doi.org/10.1128/AAC.02556-17] [PMID: 29610202]

[56] Kumar M, Mathur T, Barman TK, *et al. In vitro* and *in vivo* activities of the novel Ketolide RBx 14255 against Clostridium difficile. Antimicrob Agents Chemother 2012; 56(11): 5986-9. [http://dx.doi.org/10.1128/AAC.00015-12] [PMID: 22869573]

[57] Centers for Disease Control and Prevention Office of Infectious Disease Antibiotic Resistance Threats in the United States, 2013.

[58] Clemente JC, Ursell LK, Parfrey LW, Knight R. The impact of the gut microbiota on human health: an integrative view. Cell 2012; 148(6): 1258-70. [http://dx.doi.org/10.1016/j.cell.2012.01.035] [PMID: 22424233]

[59] Kumar M, Singh P, Murugesan S, *et al.* Microbiome as an immunological modifier. InBiomarkers for Immunotherapy of Cancer. 2020; 595-638.

[60] Gargiullo L, Del Chierico F, D'Argenio P, Putignani L. Gut microbiota modulation for multidrug-resistant organism decolonization: present and future perspectives. Front Microbiol 2019; 10: 1704. [http://dx.doi.org/10.3389/fmicb.2019.01704] [PMID: 31402904]

[61] Kumar M, Garand M, Al Khodor S. Integrating omics for a better understanding of Inflammatory

Bowel Disease: a step towards personalized medicine. J Transl Med 2019; 17(1): 419. [http://dx.doi.org/10.1186/s12967-019-02174-1] [PMID: 31836022]

[62] Akobeng AK, Singh P, Kumar M, Al Khodor S. Role of the gut microbiota in the pathogenesis of coeliac disease and potential therapeutic implications. Eur J Nutr 2020; 59(8): 3369-90. [http://dx.doi.org/10.1007/s00394-020-02324-y] [PMID: 32651763]

[63] Kumar M, Mathur T, Joshi V, Upadhyay DJ, Inoue S, Masuda N. Effect of DS-2969b, a novel GyrB inhibitor, on rat and monkey intestinal microbiota. Anaerobe 2018; 51: 120-3. [http://dx.doi.org/10.1016/j.anaerobe.2018.04.017] [PMID: 29758524]

[64] Louie TJ, Cannon K, Byrne B, *et al.* Fidaxomicin preserves the intestinal microbiome during and after treatment of Clostridium difficile infection (CDI) and reduces both toxin reexpression and recurrence of CDI. Clin Infect Dis 2012; 55 (Suppl. 2): S132-42. [http://dx.doi.org/10.1093/cid/cis338] [PMID: 22752862]

[65] Citron DM, Tyrrell KL, Dale SE, Chesnel L, Goldstein EJC. Impact of surotomycin on the gut microbiota of healthy volunteers in a phase 1 clinical trial. Antimicrob Agents Chemother 2016; 60(4): 2069-74. [http://dx.doi.org/10.1128/AAC.02531-15] [PMID: 26787687]

[66] Projan SJ. Why is big Pharma getting out of antibacterial drug discovery? Curr Opin Microbiol 2003; 6(5): 427-30. [http://dx.doi.org/10.1016/j.mib.2003.08.003] [PMID: 14572532]

[67] DiMasi JA, Hansen RW, Grabowski HG. The price of innovation: new estimates of drug development costs. J Health Econ 2003; 22(2): 151-85. [http://dx.doi.org/10.1016/S0167-6296(02)00126-1] [PMID: 12606142]

[68] Gradmann C. Magic bullets and moving targets: antibiotic resistance and experimental chemotherapy, 1900-1940. Dynamis 2011; 31(2): 305-21. [http://dx.doi.org/10.4321/S0211-95362011000200003] [PMID: 22332461]

[69] Hoffman PS. Antibacterial Discovery: 21st Century Challenges. Antibiotics (Basel) 2020; 9(5): 213. [http://dx.doi.org/10.3390/antibiotics9050213] [PMID: 32353943]

[70] Payne DJ, Gwynn MN, Holmes DJ, Pompliano DL. Drugs for bad bugs: confronting the challenges of antibacterial discovery. Nat Rev Drug Discov 2007; 6(1): 29-40. [http://dx.doi.org/10.1038/nrd2201] [PMID: 17159923]

[71] Singh SB. Confronting the challenges of discovery of novel antibacterial agents. Bioorg Med Chem Lett 2014; 24(16): 3683-9. [http://dx.doi.org/10.1016/j.bmcl.2014.06.053] [PMID: 25017034]

[72] Kumar M, Rao M, Mathur T, *et al.* Azithromycin exhibits activity against Pseudomonas aeruginosa in chronic rat lung infection model. Front Microbiol 2020. [PMID: 33967970]

[73] Jackson N, Czaplewski L, Piddock LJV. Discovery and development of new antibacterial drugs: learning from experience? J Antimicrob Chemother 2018; 73(6): 1452-9. [http://dx.doi.org/10.1093/jac/dky019] [PMID: 29438542]

[74] Hiramatsu K, Igarashi M, Morimoto Y, Baba T, Umekita M, Akamatsu Y. Curing bacteria of antibiotic resistance: reverse antibiotics, a novel class of antibiotics in nature. Int J Antimicrob Agents 2012; 39(6): 478-85. [http://dx.doi.org/10.1016/j.ijantimicag.2012.02.007] [PMID: 22534508]

[75] Barman T, Bhati A, Shukla G, *et al.* Non invasive real-time monitoring of bacterial infection & therapeutic effect of anti-microbials in five mouse models. Indian J Med Res 2011; 134(5): 688-95. [http://dx.doi.org/10.4103/0971-5916.90997] [PMID: 22199109]

[76] Marra A. Targeting virulence for antibacterial chemotherapy: identifying and characterising virulence factors for lead discovery. Drugs R D 2006; 7(1): 1-16.

[http://dx.doi.org/10.2165/00126839-200607010-00001] [PMID: 16620133]

[77] Kumar M, Khan FG, Sharma S, *et al.* Identification of Mycobacterium tuberculosis genes preferentially expressed during human infection. Microb Pathog 2011; 50(1): 31-8. [http://dx.doi.org/10.1016/j.micpath.2010.10.003] [PMID: 21035536]

[78] Garand M, Kumar M, Huang SSY, Al Khodor S. A literature-based approach for curating gene signatures in multifaceted diseases. J Transl Med 2020; 18(1): 279. [http://dx.doi.org/10.1186/s12967-020-02408-7] [PMID: 32650786]

[79] Kumar A, Khan IA, Koul S, *et al.* Novel structural analogues of piperine as inhibitors of the NorA efflux pump of *Staphylococcus aureus* . J Antimicrob Chemother 2008; 61(6): 1270-6. [http://dx.doi.org/10.1093/jac/dkn088] [PMID: 18334493]

[80] Sharma S, Kumar M, Sharma S, Nargotra A, Koul S, Khan IA. Piperine as an inhibitor of Rv1258c, a putative multidrug efflux pump of Mycobacterium tuberculosis. J Antimicrob Chemother 2010; 65(8): 1694-701. [http://dx.doi.org/10.1093/jac/dkq186] [PMID: 20525733]

[81] Hans H, Ehrlich P. Pioneer of Chemotherapy. Weizmann Institute of Science, Rehovat, Isarael. 1976 Mar 25; P-8. Wainwright M. Moulds in ancient and more recent medicine. Mycologist 1989; 3(1): 21-3.

[82] Domagk G. Chemotherapie der bakteriellen Infektionen. Angew Chem 1935; 48(42): 657-67. [http://dx.doi.org/10.1002/ange.19350484202]

[83] Henrik S, Sohlman R, Österling A, *et al.* The Prize in Physiology and Medicine: The Nobel Prizes in Wartime. In Nobel Foundation (ed.). Nobel: The Man and His Prizes. Stockholm: Klara Civiltryckeri. 1950; 167-79.

[84] Hutchings MI, Truman AW, Wilkinson B. Antibiotics: past, present and future. Curr Opin Microbiol 2019; 51: 72-80. [http://dx.doi.org/10.1016/j.mib.2019.10.008] [PMID: 31733401]

CHAPTER 5

Current Antifungal Drugs

Megha Kaushik[1], **Arvind Sharma**[2], **Shefali Gupta**[3] and **Pooja Gulati**[4,*]

[1] *Department of Microbiology, Maharshi Dayanand University, Rohtak, Haryana, India*

[2] *Department of Biochemistry and Molecular Biology, Howard Medical Sciences, University of Nevada, Reno, Nevada, USA*

[3] *Department of Industrial Microbiology, Sam Higginbottom University of Agriculture, Technology and Sciences (SHUATS), Allahabad, UP, India*

[4] *Department of Microbiology, Maharshi Dayanand University, Rohtak, Haryana, India*

Abstract: Fungal infections are one of the major causes of fatalities worldwide, causing an estimated 1.5 million deaths annually. Over the past few decades, the incidences of fungal infection have risen with the increase in the cases of immunocompromised patients. However, the impact of fungal diseases on public health is often underestimated. These infections are predominantly caused by the *Aspergillus*, *Candida,* and *Cryptococcus* species. Current therapeutic approaches to treat such fungal infections are limited to five distinct classes of antifungal drugs, *viz.* polyenes, echinocandins, azoles, pyrimidine analogs, and allylamines. Moreover, a few synthetic molecules are also used as fungicidal agents. Despite the current antifungal armamentarium, the burden of fungal infection is exacerbated by the emergence of drug resistance, host toxicity, and negative interactions with other drugs. The paucity of new antifungal drugs has further complicated the treatment of fungal infections. These limitations provide a rationale for developing novel antifungals preferably with new mechanisms and molecular targets. This chapter thus summarizes the currently used antifungal drugs, their effective combinations, and the challenges inherent to the development of new antifungal drugs. The chapter also addresses strategies to bolster the antifungal pipeline involving emerging new targets for better management of fungal infections.

Keywords: Antifungal agents, Antifungal drugs, Antifungal peptides, Azoles, Clinical trials, Echinocandins, Emerging targets, Ergosterol, Fungal diseases, Fungal infections, Fungal resistance, Mycosis, Multidrug resistance, Polyenes, Pyrimidine analogs.

[*] **Corresponding author Pooja Gulati:** Department of Microbiology, Maharshi Dayanand University,Rohtak, Haryana, India; E-mail: pooja.micro@mdurohtak.ac.in

M. Amin-ul Mannan & Gaurav Kumar (Eds.)

INTRODUCTION

Fungal diseases present a serious threat to human health, causing 1.5 million deaths annually [1]. The global burden caused by fungal diseases is underestimated due to the lack of diagnosis and public surveillance programs [2]. Even though fungi are ubiquitous in the environment and display an extensive diversity, only a few species are recognized as disease-causing pathogens in humans [3]. The majority of the fungal species are considered opportunistic pathogens, causing disease in immunosuppressed patients (like *Candida* and *Aspergillus*), while few can infect healthy humans (*Histoplasma* and *Paracoccidioides*). Lately, the clinical relevance of fungal diseases has substantially increased, owing to the rising population of immuno-compromised patients in large parts of the world. These high-risk individuals include those with multiple co-morbidities, HIV infection, organ transplantation, prolonged hospital stay, chemotherapy, and those treated with broad-spectrum antibiotics [4].

Fungal infections like superficial mycosis of skin and nails are the most common type of infections encountered in humans, followed by infections caused by *Candida* species [5]. While invasive fungal diseases are less frequent than superficial skin infections, they account for very high mortality rates (more than 50%, irrespective of provided antifungal treatment) [6]. The majority of the fungal-related mortalities (~90%) are associated with species belonging to *Candida*, *Aspergillus, Cryptococcus,* and *Pneumocystis*.

Although various antifungals with different targets are available, they have modest success in treating these life-threatening fungal infections. Factors like delay in disease diagnosis and fungal identification, narrow spectrum of activity of drugs, toxicities, and emergence of drug resistance have played major roles in defining the efficacy of these antifungal agents. The combinations of these existing drugs are therefore used to treat certain infections; however, their efficacy is still questionable. Initiatives by GAIN (Generating Antibiotic Incentives Now) and "Orphan Drug Act and Fast track" designation by US Food and Drug Administration (FDA) actively encourage the development of new antifungals by providing fast-track description and orphan drug market exclusivity [7]. Thus, the development of new inhibitors and new antifungals to tackle resistant fungal infections is the need of the hour.

In this book chapter, the current armamentarium of antifungal agents and their resistance mechanism are described. It will also provide an outline of the ongoing efforts to develop novel and promising antifungal agents under clinical trials. Further, the emerging targets for the development of antifungal compounds for better management of fungal infections are also discussed.

COMMON FUNGAL PATHOGENS

Candida is a ubiquitous fungus that lives as commensal on the mucosal membranes of humans. It is the most common fungal agent associated with invasive infections, especially in at-risk populations. *C. albicans*, *C. glabrata*, *C.krusei*, *C. lusitaniae*, *C. parapsilosis*, and *C. tropicalis* are some of the human colonizers and opportunistic pathogens of this genus [8]. *C. albicans* is the most important human pathogen causing illnesses ranging from mild mucosal infections (oral and vaginal candidiasis) to fatal systemic diseases. In oral infections, carriage of *Candida* is mediated *via* non-specific host barriers like saliva, while oropharyngeal candidiasis occurs mostly in patients undergoing chemotherapy or AIDS. In vaginal candidiasis, the epithelium becomes *Candida* sensitized, thereby causing hyperactivation of the innate immune system and generating a non-protective immune response [9]. Systemic infections occur when *C. albicans* enter the blood through medical implants or neutropenia [3]. Moreover, *Candida* non-*albicans* like *C. auris* have emerged as another bloodstream-infecting pathogen. CDC has laid concern over this pathogen owing to the increased incidences of multi-drug resistance and misidentification of the pathogen using standard laboratory methods (https://www.cdc.gov/fungal). The global dissemination of this emerging pathogen could be attributed to its ability to colonize skin and growth even at high temperatures.

Two closely related *Cryptococcus* species *viz. C. neoformans* and *C. gattii* are the main causative agent of *cryptococcosis*. *C. neoformans* is widely distributed, while *C. gattii* has limited geographical distribution. Recently, *C. gattii* infections have raised an alarming situation suggesting the evolution of virulent fungi affecting the healthy human community [10]. *C. neoformans* var. *grubii* is associated with approximately 95% of cryptococcal infections globally, affecting a majority of AIDS patients [11].

Aspergillus species are associated with invasive infections causing relentless pneumonia. It is known to infect patients with underlying lung diseases affecting 3 million people globally [12]. *A. fumigatus* is a common pathogen of aspergillosis, which grows well at human febrile temperatures but is well-adapted to higher temperatures as well. Transmission occurs through conidiophores producing conidia. This fatal pathogen is associated with an overall 50% mortality rate despite being treated swiftly.

Pneumocystis jirovecii is another common opportunistic fungal pathogen, associated with pneumonia, affecting individuals with weakened immunity, especially HIV/AIDS patients. Typically, patients with underlying diseases, like those suffering from cancer, transplant recipients, or under immunosuppressive medication, are more prone to pneumocystis infections. This infection is spread through an airborne route of transmission and accounts for 4 million pneumocystis infections annually. *Fusarium, penicillium,* and *mucor* are among other emerging opportunistic fungal pathogens. These fungal pathogens are also responsible for other health diseases, including fungal eye infections, sporotrichosis, and mycetoma. All these fungal infections are treated with the current arsenal of antifungal drugs. However, a rise in resistant species has been observed in the fungal kingdom, including drug resistant-resistant *Aspergillus* species and multidrug-resistant *C. auris* [13 - 15].

CLASSICAL ANTIFUNGAL DRUGS IN CLINICAL USE

Currently, five antifungal drug classes, namely, polyenes, azoles, echinocandins, allylamines, and pyrimidine analogs, are used to treat systemic fungal infections. The mode of action and mechanism of resistance for these main classes are depicted in Table **1**. These antifungal agents are directed against the fungal cell wall, ergosterol biosynthesis pathway, or fungal DNA/RNA. Important characteristics of major antifungal agents, along with their mechanism of resistance, are described in the section below.

Table 1. Major antifungal drug classes, their mode of action, and resistance mechanisms in them.

Class	Activity Spectrum	Antifungal Drugs	Target/Mode of Action	Resistance Mechanism
Azoles	Fungistatic	Fluconazole, Itraconazole, Miconazole, Voriconazole, Posaconazole, Clotrimazole Ketoconazole, Isavuconazole, *etc.*	Inhibition of lanosterol demethylase leads to loss of cell membrane integrity and inhibition of growth	1. Modifications of *ERG11* gene (encoding lanosterol demethylase) by: a) Point mutations b) Upregulation 2. Alterations in the ergosterol biosynthetic pathway 3. Overexpression of CDRs and MDR genes encoding efflux pumps 4. Metabolic bypass 5. Genome plasticity
Echinocandins	Fungicidal	Anidulafungin, Caspofungin, Micafungin	Inhibits enzyme β-(1,3) glucan synthase	Target alteration by point mutations in *FKS1* and *FKS2* genes.

(Table 1) cont.....

Class	Activity Spectrum	Antifungal Drugs	Target/Mode of Action	Resistance Mechanism
Polyenes	Fungicidal	Amphotericin B, Nystatin, Natamycin	Decreases the ergosterol content in the cell membrane	1. Increase in ergosterol synthesis 2. Point mutations in *ERG3* and *ERG6*
Allylamines	Fungicidal	Terbinafine, Naftifine	Ergosterol biosynthesis	CDR1 can use terbinafine as a substrate
Pyrimidine analogs	Fungicidal	5-fluorocytosine, 5-fluorouracil	Incorporation of fluorinated pyrimidine into DNA and RNA thereby inhibiting vital cell processes	1. Point mutations in *FCY1, FCY2, FUR1* 2. Increased synthesis of pyrimidines

Polyenes

Polyenes are the most common broad-spectrum antifungal compounds to be used for treating invasive fungal infections. Polyene macrolides are amphipathic molecules with a large hydrophobic lactone ring of 26–28 carbons having three to eight double bonds and a hydrophilic polyhydroxy chain (Fig. **1**) [16]. Although more than 100 different compounds classified as trienes (rapamycin), tetraenes (natamycin), heptanes (amphotericin B and candicidin), and a pseudoheptaene/tetraene (nystatin) have been isolated, but only amphotericin B (AmB), pimaricin, candicidin, and nystatin are the most widely used polyenes. Amongst them, AmB and nystatin are the first antifungals developed and are still the most commonly used. Polyenes show activity against both yeasts and filamentous fungi and are commonly used to treat invasive fungal infections like *Cryptococcus, Aspergillus,* and *Candida* infections [17].

All polyenes attach to sterols, especially ergosterol, present in the fungal cytoplasmic membrane, and form a complex. This complex has a hydrophobic outer core and a hydrophilic inner core due to the presence of a hydroxyl group [18, 19]. This configuration forms a pore making the cell membrane leaky, leading to the loss of vital cellular components, elimination of ions, and ultimately causing the death of the organism [20, 21]. Recently, another mode of action has also been reported by several researchers. It suggests the binding of polyenes to ergosterol initiates chromatin margination and condensation, DNA damage, and generation of reactive oxygen species due to the transfer of electrons in the cell membrane [22].

Polyenes specifically kill fungal cells because of the presence of high ergosterol content in fungal cell membranes as compared to mammalian cells. However, low selective toxicity toward mammalian cells occurs and leads to serious side effects [23]. The most common adverse effect of polyenes is chills, hypomagnesium, rigors, fever, and nephrotoxicity [24, 25]. Despite these, polyenes such as AmB and nystatin are regarded as a 'gold standard for treating severe invasive fungal infections.

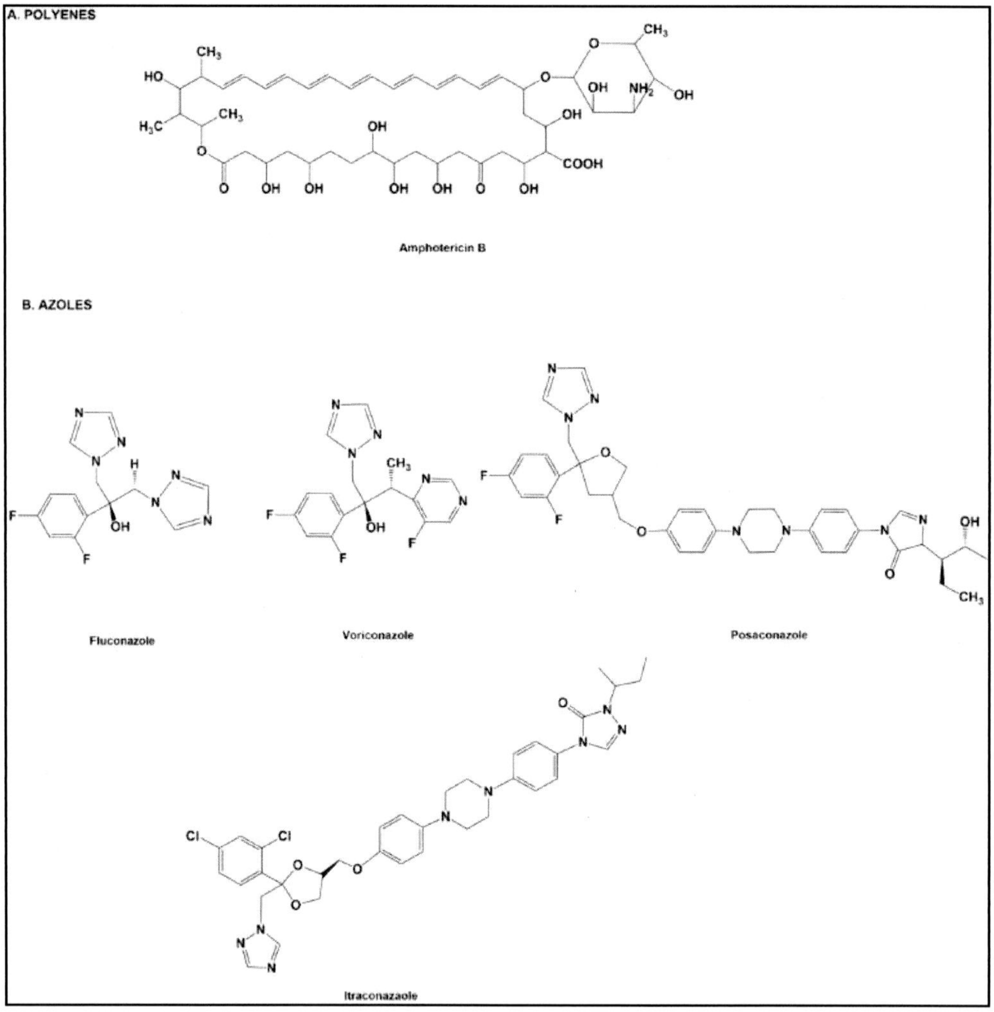

Fig. (1). Chemical structures of Polyenes **(A)** represented by AmB and Azoles **(B)** represented by fluconazole, voriconazole, posaconazole and itraconazole.

Azoles

Azoles are the second largest class of synthetic antimycotic compounds containing five-membered azole rings attached to other aromatic rings *via* a carbon-nitrogen bond [27] (Fig. **1**). The azoles in clinical use can be classified into either imidazole, which contains a basic azole ring with two nitrogen atoms (such as miconazole and ketoconazole) or triazoles containing a basic azole ring with three nitrogen atoms (such as fluconazole and itraconazole). Replacing the imidazole ring with triazole, azoles as antifungals are more advantageous, with increased specificity to fungal enzymes, improved solubility, and increased polarity, thereby increasing their efficacy [28]. Clotrimazole, miconazole, and ketoconazole, which are imidazole-based drugs, were among the first azole compounds to be used for treating fungal infections. Later, the next generation of triazole antifungals, such as voriconazole, posaconazole, and ravuconazole, with increased specificity, was developed.

Oral azoles, such as fluconazole, are widely used to treat cryptococcal meningitis [29]. It is the drug of choice in the intensive care unit (ICU), during organ transplantation, and at times in neutropenic patients [30, 31]. It is also used to treat adult neutropenic patients with invasive candidiasis. In children, fluconazole is used as the drug of choice in high-risk infants, which are susceptible to invasive candida infection [30, 31]. Itraconazole is effective against fluconazole-susceptible *Cryptococcus*, *Candida,* and *Coccidioides*, and shows little efficacy in treating aspergillosis [32]. Ketoconazole is employed for the treatment of infections caused by dimorphic fungi and *Candida.* Voriconazole is used for treating infections caused by *Aspergillus, Candida, Scedosporium,* and *Fusarium* species [33]. Miconazole is used against *C. albicans* [34]. The posaconazole is used as a drug of choice in invasive candidiasis, coccidioidomycosis, and other voriconazole-resistant invasive fungal diseases [35, 36].

The antifungal activity of the azole arises from its ability to inhibit the biosynthesis of ergosterol by inhibiting cytochrome P450-dependent enzyme, lanosterol demethylase (14α-sterol demethylase or P450DM) [37, 38]. The cytochrome P450, along with lanosterol demethylase removes the 14α-methyl group of lanosterol [39]. The azoles bind to the active site cytochrome P450 *via* Fe in the heme group [40]. The binding of the azole to cytochrome P450 inhibits the demethylation of lanosterol, leading to accumulation of 14α-methylated sterol and a reduction in ergosterol. The ergosterol depletion disrupts the membrane structure, inhibiting cellular growth [30, 31]. The strength of interaction between different types of azoles and cytochrome P450 determines their level of inhibition on different fungi.

Azoles also react with the mammalian cytochrome P450 but have a greater affinity for fungal P450DM due to different active site conformation [40] and thus cause various side effects. Thus, the oral and/or parenteral intake of azoles is restricted. Due to toxicity toward mammalian cells, most imidazoles have been restricted only for topical applications. Only triazoles *viz.* luconazole, itraconazole, isavuconazole, posaconazole and voriconazole have been licensed for treatment of invasive fungal disease [24].

Pyrimidines

Pyrimidines are a synthetic class of antimycotic drugs. 5- fluorocytosine (5-FC) is a fluorinated analog of pyrimidine cytosine [41] (Fig. **2**). In the 1950s, flucytosine was developed as a potential anti-cancer drug. Later it was discontinued as a potential antineoplastic drug because of a lack of specificity against tumor cells. Later in the 1960s, it was first shown to be active against many yeast infections [42]. Monotherapy with flucytosine is not recommended due to the rapid onset of resistance; therefore, it can only be used in combination with other known antifungals like AmB. Cryptococcal meningitis and other invasive fungal infections can easily be treated with this combination [43].

The fungistatic activity of 5-FC is mediated by cytosine permease that transports 5-FC inside the fungal cell [42]. Once inside the cell, it is converted to 5-fluorouracil (5-FU) *via* the enzyme cytosine deaminase. Subsequently, uracil phosphoribosyltransferase (UPRT) converts 5-fluorouracil into 5-fluorouridylic acid. The 5-fluorouridylic acid is later phosphorylated and can incorporate into RNA molecules, disrupting translation. The phosphorylated 5-FU is also a potent inhibitor of the enzyme thymidylate synthase. The thymidylate synthase is involved in nuclear division and DNA replication [44]. Hence, 5-FC interferes with the vital processes of replication, transcription, and translation in the fungal cell.

Allylamines

Allylamines are a class of ergosterol biosynthetic inhibitors that are structurally related to thiocarbamates (Fig. **2**). These include terbinafine, butenafine, and naftifine. Terbinafine is highly effective against dermatophytes. *In vitro* assay shows terbinafine to be highly active against *Fusarium* spp., *Aspergillus* spp., and other filamentous fungi, but exhibits varied efficacy for yeasts. When administered with azoles and AmB, it is found to be highly effective even against azole-resistant strains of *Aspergillus*, *C. neoformans,* and *C. albicans* [45]. *In vivo* studies have proved that terbinafine is not a very effective antifungal agent against systemic sporotrichosis, invasive aspergillosis, pulmonary *cryptococcosis* , and systemic candidiasis. Butenafine and naftifine are used as efficacious antimycotics for topical applications in dermatomycoses [46], whereas terbinafine is an oral-systemic agent.

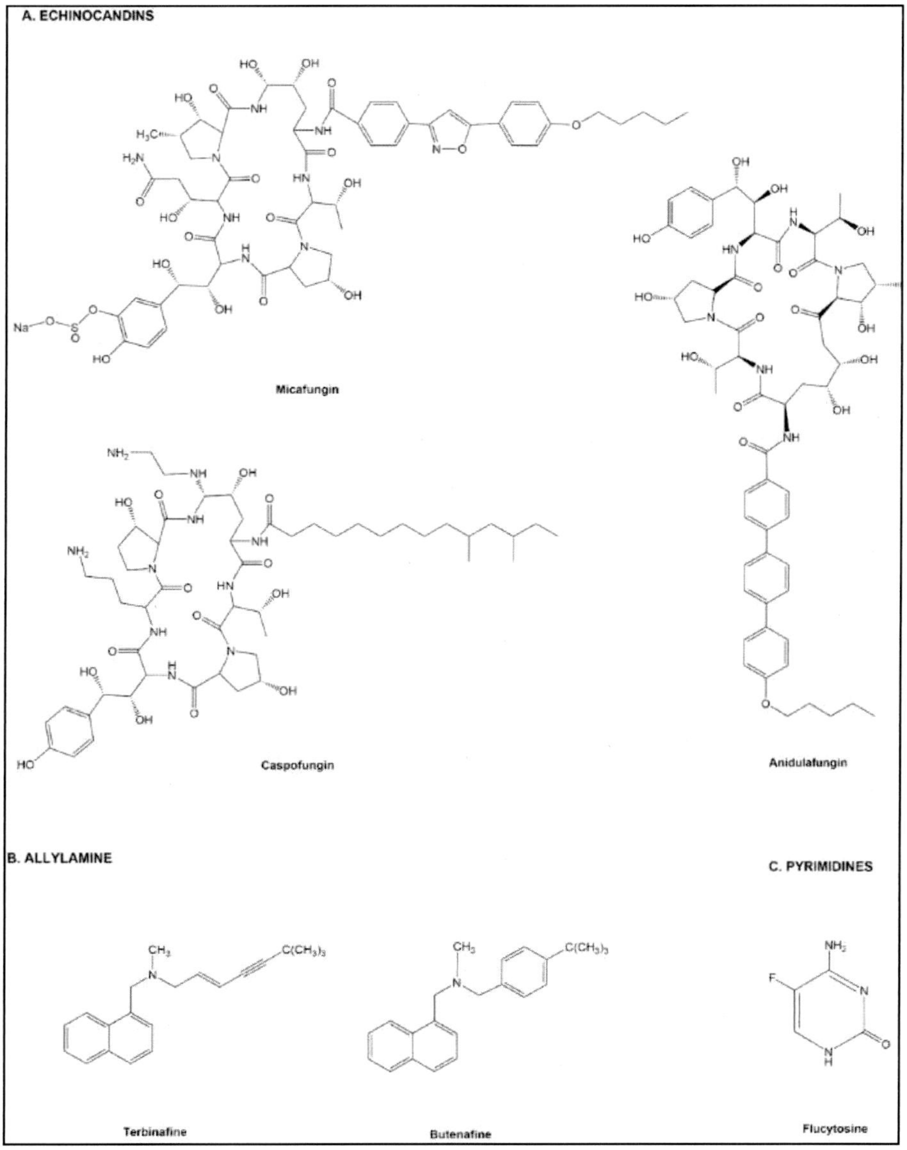

Fig. (2). Chemical structures of **(A)** Echinocandins represented by micafungin, caspofungin and anidulafungin **(B)** Allylamines represented by terbinafine and butenafine **(C)** Pyrimidine analogs represented by flucytosine.

The early steps of ergosterol biosynthesis are inhibited by allylamines. Allylamines are reversible, noncompetitive inhibitors of the enzyme squalene epoxidase. The squalene epoxidase, along with squalene cyclase, converts squalene to lanosterol, which later forms ergosterol in the fungal cell [47]. Hence, inhibition by allylamines leads to the accumulation of squalene rather than ergosterol in the cell [20]. Thereby, leading to increase membrane permeability, which leads to fungal cell death by disruption of cellular functions.

Echinocandins

The echinocandins are the recently developed antimycotic agents comprising anidulafungin, caspofungin, and micafungin. Echinocandins are cyclic lipopeptide molecules that are water-soluble and amphiphilic. The lipopeptide consists of a medium-sized cyclic peptide containing six amino acids, which is *N*-acylated with a branched or unbranched aliphatic fatty acid chain. The formation of aminal linkage between the d-amino group and the C-terminus of the dihydroxyomithine residue leads to the cyclization of peptide in a lipopeptide molecule (Fig. **2**). The various echinocandins class of antifungals differs from each other either due to variation in a fatty acid chain of amino acids in the hexapeptide [48]. Echinocandins are a drug of choice to cure invasive infections caused by azole or AmB-resistant strains of *Aspergillus* and *Candida* [49]. They have several additional merits like reduced emergence of drug-resistant isolates, fewer drug interactions, and an improved safety profile [50, 51].

1,3-β -glucan is an important component of the fungal cell wall that provides structural integrity. It is synthesized by the enzyme 1,3-β-D-glucan synthase. The echinocandins act as a non-competitive inhibitor of this enzyme [47, 52]. Hence, the fungal cell losses its structural integrity and undergo osmotic lysis [49, 52]. Echinocandins have high specificity for fungal cells because mammalian cells do not contain 1,3-β-glucan synthase, and therefore, they can mediate its antifungal activity without any side effects on human cells [53]. Caspofungin, an echinocandin, is as effective as AmB for treating mucosal and systemic candidiasis [54, 55]. It is also considered salvage therapy for invasive aspergillosis [54]. Although echinocandins are generally regarded as safe drugs, however, cardiac-associated side effects such as cardiac failure and arrhythmias have developed after caspofungin administration [56].

ANTIFUNGAL PEPTIDES

In addition to the five main classes, antimicrobial peptides with potent activity against fungi have also emerged as effective antifungal agents [57]. The origin of antifungal peptides is highly diverse, ranging from higher plants and even vertebrates to microorganisms. Regardless of their origin, they have some common features, such as overall positive charge, amphipathicity, and small size. The antifungal peptides include linear peptides, α-helical peptides, peptides with β-sheet configuration, peptides with a mixed β-sheet and α-helix configuration, transformed cyclic peptides, depsipeptides and lipopeptides [58]. The unicellular organisms generally produce small antifungal peptides (<1.5 kDa). These small antifungal peptides are cyclic structures consisting of a fatty acyl moiety and/or non-protein amino acid. The multicellular organisms produce either linear α-helical or cystine-stabilized defensin-like large antifungal peptides (>3 kDa). The peptides with antifungal activity are effective in the treatment of various invasive infections caused by *Candida* and *Aspergillus* [59]. With the increasing problem of drug resistance against classic antibiotics, the interest of researchers is diverting towards antifungal peptides with greater efficacy and safety [60].

Large numbers of antimicrobial peptides are obtained from plants. The plants express a large number of peptides in response to fungal infections. The peptides are either being constitutively produced or expressed upon fungal attack. These antifungal peptides have been classified into seventeen families. Many of the antifungal peptides are enzymes such as chitinases (PR-3), peroxidases (PR-9), (1 → 3)-β-glucanases (PR-2 -4, -8, and -11), oxalate oxidases (PR-16 and -17) and proteinases (PR-7) [61]. Other antifungal peptides of plant origin include small, disulfide-rich proteins such as plant defensins, thaumatin-like proteins, hevein- and knottin- type proteins, thionins, snakins, Ib-AMP's, and lipid transfer proteins (LTP) [62]. Vertebrates also secrete antifungal peptides as an adaptive immune response to fungal attacks [61]. The antifungal peptides in mammals are often expressed in the epithelial layers and neutrophils. The majority of antifungal peptides belong to histatins, cathelicidin, defensins, and lactoferricin groups. Humans and closely related primates express histatins, while cathelicidins have been identified only in mammals. These antifungal peptides from mammals are successfully used to treat infections caused by *C. albicans* [63]. Bacteria and fungi also secrete extracellular antifungal peptides to survive better in competing natural environments. A huge array of antimicrobial peptides has been isolated and identified from microorganisms. The antifungal peptides from microorganisms include aculeacins, bacilliomycin F, helioferins, cepacidins, leucinostatins, nikkomycin, polyoxin, syringomycin, echinocandin, and aureobasidin. These antifungal peptides with diverse origins show promising activity against *Candida, Aspergillus, Mucor,* and *Fusarium* strains [64].

Mode Of Action

The antifungal peptides initially target the cell membrane and cell wall of fungus and then attack their intracellular targets like protein synthesis machinery, mitochondrial membrane, and nucleotides. To get inside the cell, antifungal peptides interact with cell wall components, such as β-glucan, chitin, and mannoproteins [65]. The plant defensin *Na*D1 from tobacco (*Nicotiana alata*) targets a cell wall mannoprotein and induces leakage from the cell [66]. The cell wall protein Ssa1/2p of *C. albicans* is an ideal target for histidine-rich saliva peptides, the histatins [67]. The histatins bind to cell wall and then penetrates inside the cell through the cell membrane leading to intracellular effects [68].

β-glucans present in the cell wall polysaccharide network is the next unique target for the antifungal peptides [69]. Various fungi produce cyclic lipopeptides, which inhibit unique steps in the synthesis of β-glucan, which leads to a weaker cell wall and the death of fungal cells. These lipopeptides non-competitively inhibit enzymes involved in 1,3-β-glucan synthesis [70]. These peptides are not effective against fungi like *Cryptococcus, Scedosporium, Trichosporon,* and *Fusarium,* which lack 1,3-β-glucan in their cell walls [71, 72]. N-acetylglucosamine of chitin is an ideal target for several plant antifungal peptides having a lectin-like activity like α-thionins from barley and wheat [73], the peptides *Tu*-AMP1 and *Tu*-AMP2 from tulip [74], *Pp*-AMP1 and 2 from Japanese bamboo (*Phyllostachys pubescens*) [75] and *Ac*-AFP1 and 2 peptides from *Amaranthus caudatus* [76].

Once antifungal peptides cross the cell wall, their next target is the fungal cell membrane. The selectivity of the membrane-active antifungal peptides is dependent on the composition of the membrane, especially the charge on the membrane [77]. The glycerophospholipids (GPLs), sterols, lysolipids, and sphingolipids are the main lipids of eukaryotic cell membranes. The cationic antifungal peptides interact with glycerophospholipids disrupting the cell membrane integrity [78]. The human α-helical cationic peptide cathelicidin LL-37 shows antifungal activity against *Candida* species [62]. Defensin is another large class of cationic antifungals that shows varying activity against fungal pathogens. For example, human neutrophils (HNP1-3) produce α-defensins which exhibit weaker activity against yeast as compared to human β-defensins [79].

MOLECULAR MECHANISM OF RESISTANCE IN FUNGI

Fungi are among organisms that quickly adapt to their environment, making them resistant to cellular stress induced by antifungal compounds [80 - 82]. Resistance to most of the classes of antifungal compounds has already been reported in various pathogenic fungi [83] (Table **1**). The mechanism of antifungal resistance can be intrinsic or adaptive, depending on the fungal species and antifungal compound used against it. Furthermore, the resistance can develop to a specific drug or different classes of antifungal compounds. The molecular mechanism of resistance in common fungal pathogens is described in the following section (Fig. **3**).

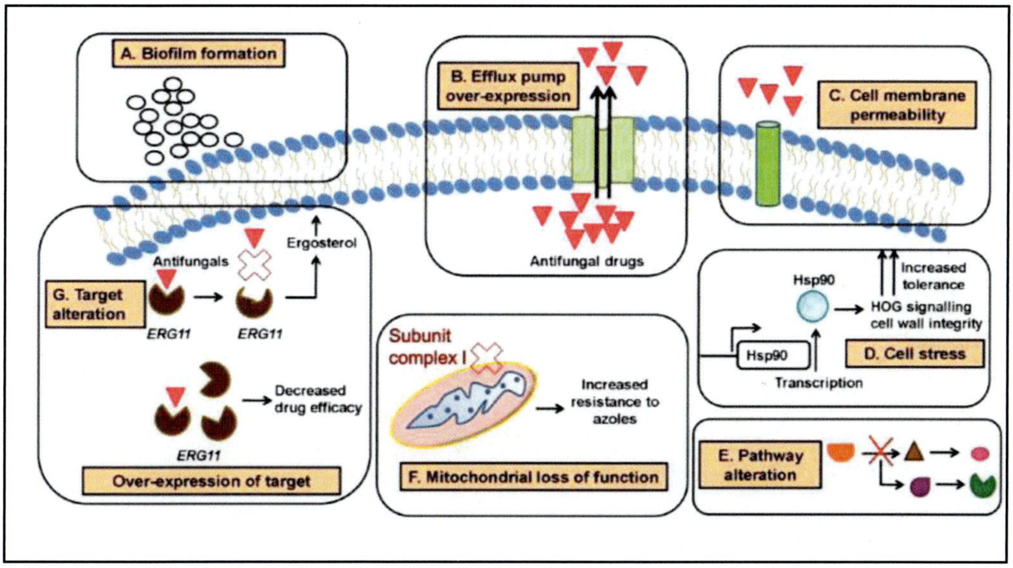

Fig. (3). Common resistance mechanisms against antifungal compounds. **A)** Biofilm formation forms a physical barrier against antifungal compounds. **B)** Overexpression of efflux pumps results in less accumulation of antifungal drugs. **C)** Exposure to antifungal drugs results in changes in cell membrane permeability, leading to restricted uptake of drugs. **D)** Increased production of Hsp90 induces cell-stress pathways resulting in increased tolerance. **E)** Prevention of cellular stress by inhibiting alternate pathways to reduce the accumulation of toxic intermediary products. **F)** Loss of mitochondrial complex I due to change in the amino acid sequence of subunit proteins results in increased resistance against azole. **G)** Alteration of drug targets results in resistance against antifungal compounds. Overexpression of the target for antifungal compounds results in decreased efficacy.

Alteration of the Target Site

The alteration of the target site by point mutation and/or overexpression is one of the most frequently found resistance mechanisms in microorganisms [83, 84]. A point mutation can potentially reduce the binding of antifungal drugs to target proteins. In contrast, the upregulation of the target gene can greatly reduce the efficacy of antifungal compounds by overproduction of the target protein [83, 84]. Resistance due to site alteration has been reported in *C. albicans*, *A. fumigatus*, *C. glabrata*, and *C. neoformans*, whereas overexpression of the target was found in *C. lusitaniae* [85].

This resistance mechanism is well accounted for by the drugs triggering membrane stress by an imbalance of sterol composition such as azole, echinocandins, and polyenes. The alteration in azole target lanosterol demethylase, encoded by the *ERG11* gene (Table **1**), leads to the depletion of ergosterol [84]. Azole and polyene resistance due to point mutation in the *ERG11* gene has been demonstrated in most pathogenic fungi [86 - 90]. The upregulation of the *ERG11* gene leading to its overexpression has also been described as a major resistance mechanism in clinical isolates of *C. albicans* and *A. fumigates* [83]. In the case of echinocandins, point mutation in the *FKS1* gene encoding (1,3) β-D-glucan synthase has been reported.

Overexpression of Efflux-pumps

Efflux pumps are transporter proteins found in the cell membrane and are responsible for maintaining cell viability by the extrusion of toxic products. Five major families of efflux transporters, namely "ABC (ATP binding cassette), MATE (multidrug and toxic efflux), MF (major facilitator), SMR (small multidrug resistance), and RND (resistance-nodulation-division)" are found in eukaryotes. Except for the ABC family which derives energy from the hydrolysis of ATP, the others obtain energy from the proton motive force [91]. ABC family derives energy from the hydrolysis of ATP to export toxic substances [91]. Overexpression of efflux pumps is another important resistance mechanism in microorganisms against toxic compounds. The increased expression causes the synthesis of more transporter proteins, which in turn leads to increased transport of antifungal compounds outside of the cell [92]. This upregulation of the efflux pump has been reported in *A. fumigatus* and several *Candida* spp. against the azole compounds [93 - 95]. In a recent study of *C. albicans* isolated from northern Indian states, it was found that increased expression of genes encoding ABC or MF efflux pumps (*CDR1*, *CDR2*, and *MDR1*) is among the common mechanisms of resistance [96].

In addition to efflux activity, these pumps also confer antifungal resistance by decreasing the influx of antifungal compounds, thereby reducing the intracellular drug accumulation [97]. This mechanism has been reported in *A. fumigatus*, *C. albicans,* and *C. neoformans* against azoles and pyrimidine [84]. This mechanism of resistance is not present in isolates exposed to echinocandin and polyenes, as these compounds exercise their fungicidal activity from outside of the cell [92].

Cellular Stress Pathways

Fungal pathogens relish a wide range of ecological settings, and their micro-niche, including the temperature, pH, and nutrient level, lay immense cellular stress on fungal cells. As a result, they have developed some innate stress-response mechanisms, which allow them to survive this cellular stress [83]. This stress response mechanism is also known to help them survive chemical stress induced by different antifungal agents and, in turn, helps in the development of resistance against these compounds. Heat shock proteins (Hsp90) and the target of rapamycin (TOR) protein kinase are two critical fighters in the arsenal of fungal cells, which enable them to tolerate the chemical stress induced by antifungal drugs. Chaperone proteins are highly conserved from bacteria to humans and help maintain cellular homeostasis by interacting with other proteins and helping them obtain their final functional forms [98]. Similarly, TOR kinase acts as a growth regulator and manages the cellular response to nutrient availability. In *Candida* and *Aspergillus*, Hsp90 is known to increase resistance to azoles and echinomycin by stabilizing key regulators of the stress response of cells to chemical stimuli [99]. Inhibition of Hsp90 is known to have an inverse effect on drug susceptibility in both *C. albicans* and *A. fumigatus* [100]. Similarly, the inhibition of the TOR function prevents the growth in the fungal cells and diminishes the *ERG3*-mediated azole resistance [84].

Alteration of Membrane Homeostasis

In fungi, the cell membrane serves as a protective barrier between the cell and its environment, like in other organisms. The fungal cell membrane ergosterol is mainly responsible for membrane homeostasis in fungi [83]. Azoles and AmB target ergosterol content resulting in disruption of membrane homeostasis [101]. Along with ergosterol, sphingolipids and sterols are also important constituents of fungal membranes working together in complexes [102]. It has been shown that any disturbance in ergosterol-sphingolipid interaction can result in increased or decreased drug susceptibility in common pathogenic fungi. In *C. albicans*, a mutation in *FEN1* or *FEN12* genes, which encodes an enzyme for the synthesis of a precursor of sphingolipids, resulted in increased resistance to azoles [102]. In another study, exposure to echinocandins inhibits the synthesis of β-D-glucan, thereby generating membrane stress, which further stimulates the formation of a second cell wall polysaccharide-chitin, rejuvenating the cells and weakening the effects of the drug [103]. Similarly, in *A. fumigatus,* exposure to caspofungin triggers changes in the lipid composition of the plasma membrane surrounding β-(1,3)-D-glucan synthase, which leads to echinocandin resistance [104]. In *C. glabrata,* resistance to AmB was found when grown in lactate or oleate due to alterations in the architecture of the cell wall with a lower content of β-glucan [105].

Biofilm Formation

Biofilm formation is an important adaptation strategy employed by microorganisms to ensure their survival in a stressful environment. A biofilm refers to a network of cells attached to each other and/or to a surface in an extracellular matrix (ECM). This ECM surrounding fungal biofilms acts as a physical barrier and thus prevents the diffusion of antifungal compounds into the fungal cells, thereby reducing drug efficacy [106]. Although widely studied in bacteria, biofilm formation by pathogenic fungi has only recently gained the attention of the scientific community [107]. Many filamentous yeasts and dimorphic fungi prefer this community life because it can easily overcome host defense reactions, protect against antifungal compounds, and increase host surface adhesion [108].

Biofilm formation has been reported in *C. albicans*, *T. asahii*, *A. fumigatus,* and *C. neoformans* and was found to be associated with elevated levels of resistance against antifungals belonging to different classes [109 - 112]. The mechanism by which fungal biofilms promote antifungal resistance is not well understood. In addition to acting as a physical barrier, drug efflux pumps formed during the early phase of biofilm formation also contribute to resistance, thus making it a potentially multifaceted resistance mechanism [106, 107]. Additionally, the ECM seems to play an essential role against antifungal drugs due to its ability to capture and store antifungal agents. This has been demonstrated in the case of fluconazole and AmB in *C. albicans* [113 - 115].

Genome Plasticity

The ability of fungal species to adapt to their ever-changing and diverse environmental conditions can be attributed to their remarkable genome plasticity. Genome plasticity contributes to long-term evolutionary changes, and this genetic exchange between distantly related strains is considered to be a vital process for ecological adaptations in many species [116]. In addition, genome changes, such as aneuploidy, loss of heterozygosity (LOH), and chromosomal rearrangements, can also lead to drug resistance [83]. Aneuploidy refers to the presence of an abnormal number of chromosomes in cells compared to wild-type populations. Aneuploidy in fungi either occurs spontaneously or as a result of stress generated by genetic or environmental factors. Although microorganisms poorly tolerate aneuploidy, however, it provides benefits of rapidly developing genetic diversity without permanent commitment to mutant genotype.

The role of aneuploidy in developing azole resistance in both clinical and laboratory setups is well established [84]. In *C. albicans*, i(5L) isochromosome is formed due to a duplication of the left arm at chromosome 5. Addition and deletion of i(5L) isochromosome directly corresponds to the gain and loss of azole resistance. The elevated azole resistance in i(5L) isolates is due to the enhanced expression of two genes *viz.ERG11* and *TAC1*, which are present at the same locus [117]. Also, the formation of i(5L) is typically followed by a loss of heterozygosity event, which further contributes to the increased resistance for azole. This increased resistance is due to hyperactive alleles of *ERG11*, *TAC1,* and *MRR1* (that encodes for the target of azole drugs and the transcription factors which upregulate drug efflux pumps) [118].

Metabolic Bypass

This lesser-known mechanism of antifungal resistance is the bypass of the target enzyme in the metabolic pathway. It was first noted in *C. albicans* against the azole compounds [119]. In this mechanism, the fungal cells adopt a defensive strategy of inactivating some pathway enzymes to prevent the accumulation of toxic metabolites [83]. In *C. albicans*, 14α-methylated sterol intermediates were not converted to toxic sterols intermediate when sterol Δ5,6-desaturase *ERG3* was non-functional, further preventing membrane stress. In *C. albicans* isolates resistant to AmB, a similar loss of function was found in *ERG3* and *ERG11* genes [120]. This mechanism has also been found in *A. fumigatus*, suggesting that its distribution in the fungal world may be higher than we currently think [120, 121].

Resistance By Mitochondrial Alteration

Mitochondria have recently been identified as an important contributor to virulence and antifungal drug resistance, especially in human fungal pathogens [122, 123]. A recent study has established the involvement of mitochondrial complex I subunits in the development of resistance to azoles in *A. fumigatus* [123]. A mutation in the 29.9 KDa subunit protein resulting in a change in E180D amino acid was associated with resistance to azoles in clinical isolates. This disruption leads to the biochemical inhibition of enzymes or loss of ability to change state, causing resistance to azoles. The underlying mechanism of cell death is an abnormal hypoxic response, which seems to be restored by the dysregulation of mitochondrial complex I. Another study supports the presence of mitochondrial dysfunction and triazole resistance in *A. fumigatus*. Mutations were found in mitochondrial superoxide dismutase (MnSOD), Mdm31 (inner membrane protein, and Tom70 (outer membrane protein), which resulted in antifungal resistance [124]. Lately, a study on *A. fumigatus* also reported that mitochondrial dysfunction was related to multidrug resistance caused by defects in heme A biosynthesis [125]. It triggers calcium signaling, thereby up-regulating a series of calcified neuron-dependent response element genes leading to antifungal drug resistance [125].

EMERGING TARGETS FOR NEW ANTIFUNGAL DRUGS

Nowadays, the rapid emergence of drug resistance and the lack of new antifungals are the main impediment to the control of mycosis. The advancements and extensive studies on the fungal lifecycle have led to the identification of new and potential targets for antifungal activity. The components of the fungal architecture, particularly the fungal cell wall, cell membrane, and metabolic pathways, could act as a promising antifungal target.

Fungal Cell Wall Proteins

The fungal cell is enclosed in a cell wall, unlike the mammalian cell. It is made up of chitin, beta-glucan, and mannoproteins. Cell surface proteins are one of the important components of the fungal cell wall. These are characterized by the modification/addition of glycosylphosphatidylinositol (GPI) anchor at their C terminus [126]. This post-translational modification results in the transportation of the protein from the endoplasmic reticulum to the cell membrane and cell wall. The GPI anchor synthesis genes like *G17* and *Ecm33* have been identified as crucial for the maintenance of virulence and morphogenesis [127, 128]. It was observed that the repression of GPI anchor genes resulted in the cell wall and morphological defects, resulting in fungal cell death due to autophagy and necrosis [129, 130]. Recently, studies have identified molecules inhibiting GPI anchor synthesis. One such molecule is gemination (ER stress inducer) which specifically inhibits Gwt1, an acyltransferase involved in GPI anchor biosynthesis. Gepinacin-treated *Saccharomyces cerevisiae* cells showed altered localization of Emp24, which is responsible for the trafficking of GPI anchor proteins from ER to Golgi bodies [131]. The sublethal concentration of gepinacin in *C. albicans* causes thinning out of the outer mannoprotein coat and further exposure of β-glucans to the cell surface. This, in turn, resulted in enhanced host immunogenicity against *C. albicans* [131, 132]. G365 and G384 are among other inhibitors of Gwt1, which result in growth suppression and altered virulence and enhanced immune response in fungal isolates of *C. albicans* and *A. fumigatus* [133].

Fungal Membrane Sphingolipids

Sphingolipids are ubiquitous membrane components essential for cellular metabolism, signal transduction, and virulence in fungal species. Studies have shown that targeting sphingolipid biosynthesis negatively affects fungal virulence. In *C. albicans*, deletions of genes responsible for sphingolipid biosynthesis are associated with loss of virulence [134]. Glucosylceramides (GlcCer), one of the fungal-specific sphingolipids, have emerged as a potent target for antifungal studies. In a mouse study, the *C. neoformans* strains devoid of GlcCer synthase I were observed to be avirulent [135]. These studies showed that molecules targeting the inhibition or deletion of sphingolipid biosynthesis enzymes may serve as potential antifungal agents. Two compounds have been identified as potent inhibitors of fungal sphingolipid GlcCer *viz.* N′-(3-brom- -4-hydroxybenzylidene)-2-methylbenzohydrazide (BHBM) and 3-bromo-N′-(3-bromo- 4-hydroxybenzylidene) benzohydrazide (D0). These two compounds have shown antifungal activity against *C. neoformans* and invasive candidiasis [136].

Metabolic Pathways

Fungal metabolic pathways also provide important targets for the identification of potential antifungal agents. The metabolic glyoxylate cycle aids in fungal survival in nutrient-deficit host niches. An enzyme of this pathway, isocitrate lyase, can serve as a target of some potent inhibitors. It has been regarded as important in the invasion of fungal isolates of *C. albicans* and *A. fumigatus* but not for *C. neoformans*. Compounds like apigenin (API), caffeic acid (CAFF), mohangamide A and mohangamide B and rosmarinic acid (ROS) have been reported as potent inhibitors to isocitrate lyase in the *C. albicans* model [137, 138].

Heme biosynthesis is another metabolic pathway that has emerged as a potential target for antifungal studies. Molecules like hemofungin and azaoxoaporphine alkaloid sampagine have been identified as potent inhibitors of the heme biosynthesis pathway. Hemofungin displayed antifungal activities against *A. fumigatus* and insect models of fungal infections, while sampagine showed strong inhibitory activity on fungal biofilms [139, 140].

Lately, drugs disrupting the essential micronutrients, zinc, and iron homeostasis have been identified as appealing antifungal targets. Compounds like artemisinin and pyrvinium pamoate have been shown to possess antifungal properties [141]. These compounds interfere with the metal uptake and transcriptional regulation in the fungal pathogen, *C. albicans*. One of the anticancer protein kinase inhibitors, AR12, has been repurposed for its antifungal activity. AR12 is an antitumor celecoxib-derivative with a wide range of activity against molds, dimorphic fungi, and yeasts, including azole and echinocandins-resistant *Candida* [142]. AR12 is reported to inhibit fungal acetyl-CoA synthetase activity (involved in acetate metabolism) along with the downregulation of host chaperone proteins, thus increasing host immune response [143].

Signal Transduction Pathways

Mitogen-activated protein kinase (MAP kinase) pathway is one of the stress signaling pathways which allows the adaptation of the fungal cell to environmental changes. Several MAP kinase pathways like Mkc1, Cek1, Cek2, and HOG pathways have been recognized with potential therapeutic implications. These pathways play important roles in virulence and act as promising targets for antifungal drugs. The high osmolarity glycerol (HOG) pathway, responsible for cellular stress response, has emerged as one of the most promising pathways for antifungal studies. Two compounds, namely, 4- and 5-substituted 1,2,3-triazoles, have been reported to inhibit Hog1-type MAPK cascade both *in vitro* and *in vivo* in *S. cerevisiae* [144]. Another compound, ambruticin S (cyclopropyl-pyran acid), is found to be a potent fungicidal agent against coccidioidal infections [145]. Recently, coccidioidal murine models were treated with KOSN-2079 and KOSN-2089 (analogs to ambruticin S) [146]. These agents also exhibit antifungal activities, suggesting the ability of inhibitors of the HOG pathway as potent antifungal agents.

Inhibitors to other important pathways, like the PDK-1 pathway and calcium signaling pathway have also displayed antifungal activities [147]. Further investigations into the calcium signaling pathway have identified compounds like Crz1, fingolimod hydrochloride (FTY720), and AFPNN5353 which act as anti-infective agents against infections caused by *C. albicans*, *C. neoformans,* and *A. fumigatus* [148 - 150]. Inhibitors to farnesyltransferases (catalyzes post-translational lipidation) have also been studied in various fungal models and reported as efficient antifungal agents. One of the inhibitors, manumycin A, originally an anti-cancer agent is reported to inhibit *C. neoformans* by hindering the localization of Ras1 to the cell membrane [151]. Tipifarnib and ethylenediamine inhibitor 2 are among other farnesyltransferase inhibitors with potent antifungal activities against *C. neoformans*.

Biofilm formation has also been explored as a target for antifungal activities. Bcr1 is a transcription factor known to regulate the expression of adhesins like Als1, Als3, Hwp1, and a novel mutation Ece1. Mutations or deletions in Bcr 1/or in one of the adhesins have resulted in incomplete biofilm formation *in vitro*, suggesting that targeting the biofilm adhesion proteins or transcription factors could be a promising antifungal agent [152].

ANTIFUNGAL AGENTS IN CLINICAL TRIALS

In the last decade, new formulations or new drugs with different targets or the use of combined drug therapy have strengthened the antifungal pipeline in combating fungal infections. Several new antifungal agents have been developed and are under different phases of clinical trial studies. They have been reported to show high efficacy in combating invasive fungal infections and are regarded as safe. These agents target fungal cell membranes, cell walls, or pathways for the synthesis of genetic materials. Some important antifungal agents in clinical trials are discussed below.

Novel Azoles/tetrazoles

Viet Pharmaceuticals has developed new metalloenzyme inhibitors of lanosterol 14 α-demethylase with higher specificity to fungal cytochrome P51 (CYP51) enzyme than related human CYP450 enzyme. The improved target specificity was achieved by replacing the imidazole or triazole moiety of azoles (which binds to CYP450 with high affinity) with tetrazole moiety [153]. VT1161, VT1129, and VT1598 are such tetrazoles presently under clinical trials [154].

VT1161 is a novel ergosterol synthesis inhibitor directed against *Candida* infections [155]. It was found to have more efficacy than fluconazole in preventing the dissemination of infection, owing to its longer half-life and high plasma levels. VT1161 has also demonstrated high oral absorption in murine pharmacokinetic models [156]. VT1161 has successfully entered into phase 2 clinical trials for the treatment of vaginal candidiasis. A phase 2 randomized and double-blind study showed the drug to be efficacious compared to fluconazole for the treatment of vulvovaginal candidiasis (VVC) [157]. It has also demonstrated positive results in various phase 2b clinical trials for the treatment of onychomycosis [158]. Lately, VT1661 has been granted the "Qualified Infectious Disease Product (QIDP)" and "Fast Track" designations for the treatment of RVVC.

VT1129, another promising oral tetrazole, has depicted high efficacy *in vitro* against *C. neoformans*, *C. gattii* and *Candida* species [159]. When administrated orally, this tetrazole reduced the fungal burden and increased the survival time in the murine model of cryptococcal meningitis [160]. Recently, VT1129 has been granted "fast track", "orphan drug," and "QIDP" designations by the FDA for treating cryptococcal meningitis [155]. VT1598 has the broadest spectrum activity amongst the tetrazoles, being highly potent against most fungal pathogens [161].

Ibrexafungerp (formerly Known As Scy078)

Ibrexafungerp (formerly known as SCY078) is a novel, first-in-class antifungal agent developed by Scynexis, Inc. (Jersey City, USA) [162]. Ibrexafungerp (a triterpenoid) is an enfumafungin derivative, functionally similar but structurally distinct to echinocandins. They are also glucan synthase inhibitors and can be administered orally. Unlike echinocandins, the activity of ibrexafungerp is not compromised by mutations in the target protein, fks [163]. This triterpenoid has recently been granted QIDP designation by FDA.

Ibrexafungerp showed excellent *in vitro* activity against *Candida* species (including MDR *C. glabrata* and *C. auris*), *Paecilomyces variotti*, *Aspergillus* species, and pan-resistant *S. prolificans* [164 - 166]. Its activity is also reported against molds and echinocandins-resistant yeast [165]. Ibrexafungerp has completed a few human clinical trial studies as an effective antifungal agent. It was found to have higher cure rates than fluconazole in VVC patients (NCT02679456). Phase 3 studies have also displayed ibrexafungerp as a safe and efficient oral antifungal option to treat patients with acute VVC (NCT03734991, VANISH303). Lately, ibrexafungerp has shown promising activity against *C. auris* species [167, 168]. Interim results of an ongoing phase 3 clinical study have also shown successful treatment of patients with failed conventional antifungal therapy by administration of ibrexafungerp *per os* (NCT03363841). Other ongoing clinical trials of this drug have been listed in Table **2**.

Rezafungin (Formerly Known As CD101)

Rezafungin is a structural analog of anidulafungin (an echinocandin), with modified choline moiety of the cyclic echinocandins core, which resulted in its increased stability and solubility. This antifungal agent was developed by 'Cidara Therapeutics Inc., CA, USA' and inhibits β-1,3 glucan synthase. Rezafungin exhibits a prolonged half-life and high stability in aqueous solution, plasma, and at high temperatures [169]. Lately, this antifungal agent has received 'QIDP' and 'fast track' designations. Rezafungin is effective against several pathogens, including *Candida* spp.; *Aspergillus* spp and *Pneumocystis* spp [170 - 172]. Intravenous administration of rezafungin with less frequency, *i.e.*, once weekly, has been approved as an effective dose in treating various fungal infections. Preclinical studies further validated rezafungin as a chemically and metabolically stable antifungal with no hepatotoxicity, which was observed in subjects treated with anidulafungin [173].

Several clinical trial studies have been conducted to evaluate the efficacy of rezafungin. A phase 1 study has reported the pharmacokinetic profile of rezafungin and declared it a safe antifungal treatment option (NCT02516904; NCT02551549). A two-part phase 2 clinical trial study was carried out to compare the loading doses of rezafungin and caspofungin followed by an optional step down with oral fluconazole for treating candidiasis (NCT02734862, STRIVE).

The most effective dosage was exhibited by rezafungin as; 400mg for the 1st first week, followed by 200mg once a week (for two weeks). This regimen is further explored in an ongoing phase 3 invasive candidiasis clinical trial (NCT03667690, ReSTORE). The other ongoing clinical trials have been summarized in Table **2**.

Fosmanogepix (APX001)

Fosmanogepix (formerly known as APX001, developed by Amplyx, California) displays activity against several fungi *viz.Candida, Fusarium, Aspergillus, S. prolificans,* and *P. lilacinus* [174]. Fosmanogepix is a glycosyl-phosphatidy-linositol (GPI) inhibitor, which acts primarily as an inhibitor of fungal enzyme GPI-anchor wall transfer protein 1, Gwt1 [131]. This antifungal agent is a pro-drug of E1210, which inhibits inositol acyltransferase (an enzyme involved in the synthesis of GPI anchor protein). This GPI anchor protein is responsible for colonization and infection by serving as an agent for adherence and invasion of micro-organisms to host mucosal and epithelial surfaces, thus maintaining the integrity of fungal cells [155]. Lately, fosmanogepix has received "QIDP" and "orphan drug" designations. Additionally, fosmanogepix acts as a promising and effective therapy for infections like coccidioidomycosis and pulmonary aspergillosis [175, 176]. Clinical trials have been carried out to evaluate the efficacy of fosmanogepix (Table **2**). Phase 1 and phase 2 trials for both intravenous and oral formulations have identified it as a safe and well-tolerated drug.

Mat2203

MAT2203, developed by "Matinas BioPharma Holdings, Inc, NJ, USA", is the first polyene cochleate provisioned for the oral administration of AmB. Cochleates are lipid-based drug vehicles with lipid bilayer sheets rolled up in spirals devoid of any aqueous space and can be used to deliver hydrophobic drugs such as AmB within the cell cytoplasm either by phagocytosis or direct delivery through the fusion of membranes [155]. Nanocochleates have also been developed with increased stability, reduced susceptibility to oxidation, and resistance to enzyme degradation. Although, loss of stability at temperatures >4°C, precipitation during storage, and high cost involved in manufacture are a few of its limitations [177]. Recently, this antifungal agent has received QIDP and fast-track designation by the FDA for the treatment of fungal infections. MAT2203 has undergone various phase 1 & 2 clinical trials proving to be a potential candidate for oral administration (Table **2**). Single-dose oral administration of MAT2203 (200-800mg) was found to be safe and well-tolerated, with no adverse effects or abnormalities reported [178]. A phase 2 study has demonstrated the safety and efficacy of MAT2203 in a randomized clinical trial study where two dose regimens (200mg & 400mg) were compared to a single dose regimen of fluconazole (150mg) for the treatment of VVC (NCT02971007).

Ceragenins

Ceragenins, also known as cationic steroid antimicrobials (CSAs), are small, synthetic compounds containing a sterol backbone with amino acids and other chemical groups attached. These are amphiphilic derivatives of bile acids that mimic the properties of endogenous antimicrobial peptides. They act as antifungal agents by targeting their membrane and creating morphological changes to the membrane structure [179]. They also inhibit adhesion and biofilm formation, and even disruption of established biofilms as well [180]. Ceragenins have reported better *in vitro* antifungal and anti-biofilm activity than human cathelicidin antimicrobial peptide, LL-37 [181]. They have also displayed potent activity against MDR *C. auris* in planktonic and biofilm forms [182]. It is also reported as a potential candidate for topical or mucosal application to treat fungal infections. Hacioglu and colleagues have recommended the use of carrageenin for the prophylaxis of vulvovaginal candidiasis (caused by fluconazole-resistant *Candida*) [183].

Table (2). Details of the ongoing clinical trials on new antifungal drugs.

Antifungal Drug	Clinical Trial	The Clinical Phase of the Study	Characteristics
Ibrexafungerp	NCT03987620 (Vanish 306)	Phase 3	Oral ibrexafungerp *vs.* placebo to treat patients with acute VVC
	NCT04029116 (CANDLE)	Phase 3	Oral ibrexafungerp for the treatment of patients with RVVC
	NCT03672292 (SCYNERGIA)	Phase 2	Co-administration of ibrexafungerp and voriconazole compared to voriconazole alone for the treatment of invasive pulmonary aspergillosis
	NCT03059992 (FURI)	Phase 3	In patients with invasive or severe fungal diseases which were intolerant or refractory to standard antifungal treatments
	NCT03363841 (CARES)	Phase 3	For the cure of *C. auris* infections

(Table 2) cont.....

Antifungal Drug	Clinical Trial	The Clinical Phase of the Study	Characteristics
Rezafungin	NCT03667690	Phase 3	Intravenous administration of rezafungin in comparison to caspofungin (followed by optional oral fluconazole) to treat candidemia
	NCT04368559	Phase 3	A weekly dose of the drug will be compared to standard antifungal agents (including trimethoprim-sulfamethoxazole and fluconazole or posaconazole) for the prevention of invasive fungal diseases
Fosmanogepix	NCT04240886 (AEGIS)	Phase 2	For treating infections caused by *Aspergillus* sp. and rare molds
	NCT04148287	Phase 2	In treating candidemia and/or invasive candidiasis caused by *C. auris*
MAT2203	NCT02629419	Phase 2a	MAT2203 at (200mg, 400mg, 800mg) for the treatment of mucocutaneous candidiasis
	NCT04031833 (EnACT)	Phase 1/2	Treatment of cryptococcal meningitis Phase 1: to test the efficacy and tolerability of encochleated AmB (MAT2203). Phase 2: evaluate the toxicity and fungicidal activity of MAT2203 with flucytosine

* The program names associated with their respective clinical trials are mentioned in the parenthesis

Others

Several other drugs are in the pre-clinical or early phase of clinical studies targeted for the treatment or prophylaxis of fungal infections (Table **3**). These include drugs targeting chitin synthases (nikkomycins), ergosterol synthesis (SUBA-itraconazole), pyrimidine synthesis (olorofim), and histone deacetylases (MGCD290).

Table (3). Antifungal drugs in the early stages of clinical study.

Antifungal Drug	Mechanism of action	Clinical Phase of the study	Status
Nikkomycins	Inhibitor of chitin synthases	Phase 1 (NCT00834184)	Completed
		Phase 1/2 (NCT00614666)	Terminated
SUBA-itraconazole	Interferes with ergosterol synthesis	Phase 2 (NCT00791219)	Completed
		Phase 2/3 (NCT03572049)	Recruiting

(Table 3) cont.....

Antifungal Drug	Mechanism of action	Clinical Phase of the study	Status
Olorofim (F901318)	Blocks the fungal dihydroorotate dehydrogenase (DHODH) pathway	Phase 1 (NCT02142153)	Completed
		Phase 1 (NCT02342574)	Completed
		Phase 1 (NCT02394483)	Completed
		Phase 1 (NCT02737371)	Completed
		Phase 2 (NCT03583164)	Recruiting
MGCD290	Fungal histone deactylase inhibitor	Phase 2 (NCT01497223)	Completed

ANTIFUNGAL VACCINES

The development of effective vaccines is regarded as a milestone in the triumph over a disease or infection. Significant efforts have progressed in the production of an efficient antifungal vaccine. Research focusing on various animal models and human trials testing the efficacy of vaccines against fungal diseases is underway. Some human trials have shown promising results in the early phases of clinical trials but require further investigation to be transformed into a proficient therapeutic agent for the complete eradication of fungal diseases. None of the vaccines developed so far have achieved a 'bench to bedside' transition.

A phase 1 study was carried out to assess a recombinant vaccine, NDV-3, as a safe, tolerable, and immunogenic agent for the prevention of candidiasis (NCT01273922). This cross-kingdom vaccine claimed by NovaDigm Therapeutics, Inc., carried a recombinant N-terminus of *C. Albicans* agglutinin-like sequence 3 protein (rAls3p-N) with alum as an adjuvant. The study assessed the immune response of this intramuscular administered vaccine within 6 months of low dosage followed by 3 months of the ten-fold higher dosage. This vaccine was shown to induce strong humoral and cell-mediated immune responses indicating better survival and reduced fungal burden. Recently, a Phase 1b/2a multi-center, randomized, placebo-controlled clinical study determined the efficacy of the NDV-3A vaccine against RVVC infections. The study showed NDV-3A vaccine was safe and highly immunogenic in reducing the symptoms or prevention of RVVC [184].

A virosome-based vaccine (PEV7) developed by Pevion Biotech used *C. albicans* N terminus truncated, recombinant secretory aspartyl proteinase (SAP2). It displayed satisfactory results in the phase 1 study for the prevention of RVVC (NCT01067131). A clinical trial against *Candida* infections has also reported an oral ribosomal vaccine with moderate protection in A phase 2 clinical trial against *Candida* infections has also reported an oral ribosomal vaccine with moderate protection [185]. However, no follow-up study has been reported yet.

Vaccines directed against coccidioidomycosis have also entered human clinical trials. Different vaccine approaches like formalin-killed spherules (FKS) or antigen 2/proline rich antigen (Ag2/PRA) were tested in both animal and human trials but were significantly less protective for humans [186]. Thus, additional efforts are required to obtain highly effective vaccines against *Coccidioides* infections. Apart from these human trials, antifungal vaccine strategies have been studied in animal models. These are designed as cross-kingdom, subunit, heat-killed or recombinant vaccines for the prevention of fungal infections. A few of these vaccines against important fungal pathogens have been listed in Table **4**.

Table 4. Vaccines designed against important fungal pathogens

Candidate	Vaccine Target	Vaccine Characteristics	References
β-glucan laminarin	*Candida* species and *Aspergillus* species	• Cross-kingdom vaccine • Brown algae-derived β-glucan laminarin diphtheria toxoid CRM197 conjugate • Complete Freund's used as an adjuvant	[187]
Cell wall glucanase	*Candida* species and *Aspergillus* species	• Cross-kingdom vaccine • *A. fumigatus* derived cell wall glucanase (Crf1) • Murine cytosine guanine dinucleotide (CpG) oligodeoxynucleotide is used as an adjuvant	[188]
Cytosolic and cell wall peptide	*Candida albicans*	• Subunit vaccine • Peptide designated as Fba consisting of 14 a.a. sequence from the N-terminus of fructose bisphosphoaldolase • Intraperitoneal administration of Fba pulsed dendritic cells as priming and first booster dose • Subcutaneous administration of Fba emulsified in Complete Freund adjuvant	[189]
Cell surface protein	*Candida* species	• Recombinant vaccine • Recombinant N-terminus of *C. albicans* cell surface protein Hyr1 (rHyr1p-N) • Alum used as an adjuvant	[190]
Heat killed yeast	*Coccidioides*	• Live-attenuated vaccine • Heat-killed *Saccharomyces cerevisiae* • Showed significant Th1-dependent protection against glucan and mannan	[191]
Heat shock proteins	*Paracoccidioides*	• Recombinant DNA vaccine • *Mycobacterium leprae* derived Hsp65 DNA fragment inserted into a pVAX1vector • Induction of Th1 cytokine production	[192]

CONCLUSION

Despite the existing current antifungal agents, morbidity and mortality associated with fungal infections have escalated in the past few years. Over the past 3-4 decades, fungal infections have exponentially increased, largely due to the increased number of the immunocompromised population susceptible to fungal pathogens. With a limited understanding of the impact of fungal diseases, it is critical to raise the general awareness among the masses of these underappreciated fatal pathogens. Various factors need to be addressed to combat fungal infections. These include (i) unraveling the mechanism of resistance against current antifungals (ii) implementation of more rapid diagnostic approaches for quick and accurate identification (iii) development of safe, effective, and novel antifungal agents. Pharmacodynamic modeling and focused clinical studies should be encouraged to strengthen the development and validation of antifungal agents. Repurposing old drugs and using combination therapy have also been evaluated as promising alternatives to counter fungal infections. Domains of structural biology and structural genomics should be explored as they might provide insights into the complex fungal architecture. Combining the classical approaches with advanced technologies like genomics-era and next-generation sequencing technologies will provide impetus to the current developmental strategies. Further, appropriate drug stewardship and proper patient management should also be prioritized to tackle fungal diseases. These strategies will complement the ongoing initiatives of global agencies like the CDC and WHO to thwart antifungal resistance and fungal diseases. Rigorous efforts are required to bolster the antifungal pipeline for the management of life-threatening fungal infections, however, several drugs have reached clinical trials and exhibited better efficacy in comparison to classical antifungals. These new antifungal agents are equipped with improved bioavailability and activity against drug-resistant isolates, reduced toxicity, and drug interactions, thus are anticipated as promising antifungal therapeutics. The development of effective antifungal agents might seem elusive, but the progression of novel antifungal agents and promising vaccine candidates to clinical trials will strongly augment the better management of fungal diseases.

CONSENT FOR PUBLICATION

Not applicable.

CONFLICT OF INTEREST

The author declares no conflict of interest, financial or otherwise.

ACKNOWLEDGMENT

The authors acknowledge the Council of Scientific & Industrial Research (Grant No. 09/382(0212)/19-EMR-I) for providing a fellowship to MK and 'The Department of Science and Technology, India for the FIST grant (1196 SR/FST/LS-I/2017/4)' for the development of infrastructural facilities.

REFERENCES

[1] Rodrigues ML, Nosanchuk JD. Fungal diseases as neglected pathogens: A wake-up call to public health officials. PLoS Negl Trop Dis 2020; 14(2): e0007964.
[http://dx.doi.org/10.1371/journal.pntd.0007964] [PMID: 32078635]

[2] Vallabhaneni S, Mody RK, Walker T, Chiller T. The global burden of fungal diseases. Infect Dis Clin North Am 2016; 30(1): 1-11.
[http://dx.doi.org/10.1016/j.idc.2015.10.004] [PMID: 26739604]

[3] Hall RA, Noverr MC. Fungal interactions with the human host: exploring the spectrum of symbiosis. Curr Opin Microbiol 2017; 40: 58-64.
[http://dx.doi.org/10.1016/j.mib.2017.10.020] [PMID: 29132066]

[4] Richardson M, Lass-Flörl C. Changing epidemiology of systemic fungal infections. Clin Microbiol Infect 2008; 14 (Suppl. 4): 5-24.
[http://dx.doi.org/10.1111/j.1469-0691.2008.01978.x] [PMID: 18430126]

[5] Havlickova B, Czaika VA, Friedrich M. Epidemiological trends in skin mycoses worldwide. Mycoses 2008; 51 (Suppl. 4): 2-15.
[http://dx.doi.org/10.1111/j.1439-0507.2008.01606.x] [PMID: 18783559]

[6] Brown GD, Denning DW, Gow NAR, Levitz SM, Netea MG, White TC. Hidden killers: human fungal infections. Sci Transl Med 2012; 4(165): 165rv13.
[http://dx.doi.org/10.1126/scitranslmed.3004404] [PMID: 23253612]

[7] Tillotson J, Tillotson GS. The regulatory pathway for antifungal drugs: A US Perspective. Clin Infect Dis 2015; 61 (Suppl. 6): S678-83.
[http://dx.doi.org/10.1093/cid/civ819] [PMID: 26567287]

[8] Köhler JR, Casadevall A, Perfect J. The spectrum of fungi that infects humans. Cold Spring Harb Perspect Med 2015; 5(1): a019273.
[http://dx.doi.org/10.1101/cshperspect.a019273] [PMID: 25367975]

[9] Fidel PL Jr, Barousse M, Espinosa T, *et al.* An intravaginal live *Candida* challenge in humans leads to new hypotheses for the immunopathogenesis of vulvovaginal candidiasis. Infect Immun 2004; 72(5): 2939-46.
[http://dx.doi.org/10.1128/IAI.72.5.2939-2946.2004] [PMID: 15102806]

[10] Dixit A, Carroll SF, Qureshi ST. *Cryptococcus gattii*: An emerging cause of fungal disease in North America. Interdiscip Perspect Infect Dis 2009; 2009: 1-13.
[http://dx.doi.org/10.1155/2009/840452] [PMID: 19503836]

[11] Simwami SP, Khayhan K, Henk DA, *et al.* Low diversity *Cryptococcus neoformans* variety *grubii* multilocus sequence types from Thailand are consistent with an ancestral African origin. PLoS Pathog 2011; 7(4): e1001343.
[http://dx.doi.org/10.1371/journal.ppat.1001343] [PMID: 21573144]

[12] Bongomin F, Gago S, Oladele R, Denning D. Global and multi-national prevalence of fungal diseases-estimate precision. J Fungi (Basel) 2017; 3(4): 57.
[http://dx.doi.org/10.3390/jof3040057] [PMID: 29371573]

[13] Arendrup MC, Perlin DS. Echinocandin resistance. Curr Opin Infect Dis 2014; 27(6): 484-92.

[http://dx.doi.org/10.1097/QCO.0000000000000111] [PMID: 25304391]

[14] Chowdhary A, Sharma C, Meis JF. Azole-resistant Aspergillosis: epidemiology, molecular mechanisms, and treatment. J Infect Dis 2017; 216 (Suppl. 3): S436-44. [http://dx.doi.org/10.1093/infdis/jix210] [PMID: 28911045]

[15] Forsberg K, Woodworth K, Walters M, *et al. Candida auris* : The recent emergence of a multidrug-resistant fungal pathogen. Med Mycol 2019; 57(1): 1-12. [http://dx.doi.org/10.1093/mmy/myy054] [PMID: 30085270]

[16] Lakhani P, Patil A, Majumdar S. Challenges in the polyene- and azole-based pharmacotherapy of ocular fungal infections. J Ocul Pharmacol Ther 2019; 35(1): 6-22. [http://dx.doi.org/10.1089/jop.2018.0089] [PMID: 30481082]

[17] Pitman SK, Drew RH, Perfect JR. Addressing current medical needs in invasive fungal infection prevention and treatment with new antifungal agents, strategies and formulations. Expert Opin Emerg Drugs 2011; 16(3): 559-86. [http://dx.doi.org/10.1517/14728214.2011.607811] [PMID: 21846302]

[18] Thevissen K, François IEJA, Takemoto JY, Ferket KK, Meert EM, Cammue BPA. DmAMP1, an antifungal plant defensin from dahlia (Dahlia merckii), interacts with sphingolipids from Saccharomyces cerevisiae. FEMS Microbiol Lett 2003; 226(1): 169-73. [http://dx.doi.org/10.1016/S0378-1097(03)00590-1] [PMID: 13129623]

[19] Thevissen K, Warnecke DC, François IEJA, *et al.* Defensins from insects and plants interact with fungal glucosylceramides. J Biol Chem 2004; 279(6): 3900-5. [http://dx.doi.org/10.1074/jbc.M311165200] [PMID: 14604982]

[20] Kerridge D. The plasma membrane of *Candida albicans* and its role in the action of anti-fungal drugs . In Thirteenth Symposium of th e Society for General Microbiology . The Eukaryotic Microbial Cell (Gooday , G . W. , Lloyd , D . & Trinci , A . P . J. , Eds) , pp . 103-26 . Cambridge University Press, Cambridge.

[21] Anderson TM, Clay MC, Cioffi AG, *et al.* Amphotericin forms an extramembranous and fungicidal sterol sponge. Nat Chem Biol 2014; 10(5): 400-6. [http://dx.doi.org/10.1038/nchembio.1496] [PMID: 24681535]

[22] Phillips AJ, Sudbery I, Ramsdale M. Apoptosis induced by environmental stresses and amphotericin B in *Candida albicans*. Proc Natl Acad Sci USA 2003; 100(24): 14327-32. [http://dx.doi.org/10.1073/pnas.2332326100] [PMID: 14623979]

[23] Odds FC, Brown AJP, Gow NAR. Antifungal agents: mechanisms of action. Trends Microbiol 2003; 11(6): 272-9. [http://dx.doi.org/10.1016/S0966-842X(03)00117-3] [PMID: 12823944]

[24] Mathew BP, Nath M. Recent approaches to antifungal therapy for invasive mycoses. ChemMedChem 2009; 4(3): 310-23. [http://dx.doi.org/10.1002/cmdc.200800353] [PMID: 19170067]

[25] Kamiński DM, Czernel G, Murphy B, Runge B, Magnussen OM, Gagoś M. Effect of cholesterol and ergosterol on the antibiotic amphotericin B interactions with dipalmitoylphosphatidylcholine monolayers: X-ray reflectivity study. Biochim Biophys Acta Biomembr 2014; 1838(11): 2947-53. [http://dx.doi.org/10.1016/j.bbamem.2014.08.004] [PMID: 25128151]

[26] Sheehan DJ, Hitchcock CA, Sibley CM. Current and emerging azole antifungal agents. Clin Microbiol Rev 1999; 12(1): 40-79. [http://dx.doi.org/10.1128/CMR.12.1.40] [PMID: 9880474]

[27] Saag MS, Dismukes WE. Azole antifungal agents: emphasis on new triazoles. Antimicrob Agents Chemother 1988; 32(1): 1-8. [http://dx.doi.org/10.1128/AAC.32.1.1] [PMID: 2831809]

[28] Ehrenfreund-Kleinman T, Azzam T, Falk R, Polacheck I, Golenser J, Domb AJ. Synthesis and

characterization of novel water soluble amphotericin B–arabinogalactan conjugates. Biomaterials 2002; 23(5): 1327-35.
[http://dx.doi.org/10.1016/S0142-9612(01)00251-4] [PMID: 11804288]

[29] Roemer T, Krysan DJ. Antifungal drug development: challenges, unmet clinical needs, and new approaches. Cold Spring Harb Perspect Med 2014; 4(5): a019703.
[http://dx.doi.org/10.1101/cshperspect.a019703] [PMID: 24789878]

[30] Charlier C, Hart E, Lefort A, *et al.* Fluconazole for the management of invasive candidiasis: where do we stand after 15 years? J Antimicrob Chemother 2006; 57(3): 384-410.
[http://dx.doi.org/10.1093/jac/dki473] [PMID: 16449304]

[31] Wade KC, Benjamin DK Jr, Kaufman DA, *et al.* Fluconazole dosing for the prevention or treatment of invasive candidiasis in young infants. Pediatr Infect Dis J 2009; 28(8): 717-23.
[http://dx.doi.org/10.1097/INF.0b013e31819f1f50] [PMID: 19593252]

[32] Lass-Flörl C. Triazole antifungal agents in invasive fungal infections: a comparative review. Drugs 2011; 71(18): 2405-19.
[http://dx.doi.org/10.2165/11596540-000000000-00000] [PMID: 22141384]

[33] Pascual A, Calandra T, Bolay S, Buclin T, Bille J, Marchetti O. Voriconazole therapeutic drug monitoring in patients with invasive mycoses improves efficacy and safety outcomes. Clin Infect Dis 2008; 46(2): 201-11.
[http://dx.doi.org/10.1086/524669] [PMID: 18171251]

[34] Kobayashi D, Kondo K, Uehara N, *et al.* Endogenous reactive oxygen species is an important mediator of miconazole antifungal effect. Antimicrob Agents Chemother 2002; 46(10): 3113-7.
[http://dx.doi.org/10.1128/AAC.46.10.3113-3117.2002] [PMID: 12234832]

[35] González GM, Tijerina R, Najvar LK, *et al.* In vitro and in vivo activities of posaconazole against *Coccidioides immitis.* Antimicrob Agents Chemother 2002; 46(5): 1352-6.
[http://dx.doi.org/10.1128/AAC.46.5.1352-1356.2002] [PMID: 11959568]

[36] Torres HA, Hachem RY, Chemaly RF, Kontoyiannis DP, Raad II. Posaconazole: a broad-spectrum triazole antifungal. Lancet Infect Dis 2005; 5(12): 775-85.
[http://dx.doi.org/10.1016/S1473-3099(05)70297-8] [PMID: 16310149]

[37] Georgopapadakou NH, Walsh TJ. Antifungal agents: chemotherapeutic targets and immunologic strategies. Antimicrob Agents Chemother 1996; 40(2): 279-91.
[http://dx.doi.org/10.1128/AAC.40.2.279] [PMID: 8834867]

[38] White TC, Marr KA, Bowden RA. Clinical, cellular, and molecular factors that contribute to antifungal drug resistance. Clin Microbiol Rev 1998; 11(2): 382-402.
[http://dx.doi.org/10.1128/CMR.11.2.382] [PMID: 9564569]

[39] Petrikkos G, Skiada A. Recent advances in antifungal chemotherapy. Int J Antimicrob Agents 2007; 30(2): 108-17.
[http://dx.doi.org/10.1016/j.ijantimicag.2007.03.009] [PMID: 17524625]

[40] Ullmann AJ, Sanz MA, Tramarin A, *et al.* Prospective study of amphotericin B formulations in immunocompromised patients in 4 European countries. Clin Infect Dis 2006; 43(4): e29-38.
[http://dx.doi.org/10.1086/505969] [PMID: 16838223]

[41] Robbins N, Wright GD, Cowen LE. Antifungal drugs: The current armamentarium and development of new agents. Microbiol Spectr 2016; 4:10.
[http://dx.doi.org/10.1128/microbiolspec.FUNK-0002-2016]

[42] Kerridge D. Mode of action of clinically important antifungal drugs. Adv Microb Physiol 1986; 27: 1-72.
[http://dx.doi.org/10.1016/S0065-2911(08)60303-3] [PMID: 3532715]

[43] Day JN, Chau TTH, Wolbers M, *et al.* Combination antifungal therapy for cryptococcal meningitis. N Engl J Med 2013; 368(14): 1291-302.

[http://dx.doi.org/10.1056/NEJMoa1110404] [PMID: 23550668]

[44] Diasio RB, Bennett JE, Myers CE. Mode of action of 5-fluorocytosine. Biochem Pharmacol 1978; 27(5): 703-7.
[http://dx.doi.org/10.1016/0006-2952(78)90507-5] [PMID: 350227]

[45] Liu W, May GS, Lionakis MS, Lewis RE, Kontoyiannis DP. Extra copies of the *Aspergillus fumigatus* squalene epoxidase gene confer resistance to terbinafine: genetic approach to studying gene dose-dependent resistance to antifungals in *A. fumigatus.* Antimicrob Agents Chemother 2004; 48(7): 2490-6.
[http://dx.doi.org/10.1128/AAC.48.7.2490-2496.2004] [PMID: 15215099]

[46] Stütz A. Synthesis and structure-activity correlations within allylamine antimycotics. Ann N Y Acad Sci 1988; 544(1 Antifungal Dr): 46-62.
[http://dx.doi.org/10.1111/j.1749-6632.1988.tb40388.x] [PMID: 3214088]

[47] Francois I, Aerts A, Cammue B, Thevissen K. Currently used antimycotics: spectrum, mode of action and resistance occurrence. Curr Drug Targets 2005; 6(8): 895-907.
[http://dx.doi.org/10.2174/138945005774912744] [PMID: 16375673]

[48] Debono M, Turner WW, LaGrandeur L, *et al.* Semisynthetic chemical modification of the antifungal lipopeptide echinocandin B (ECB): structure-activity studies of the lipophilic and geometric parameters of polyarylated acyl analogs of ECB. J Med Chem 1995; 38(17): 3271-81.
[http://dx.doi.org/10.1021/jm00017a012] [PMID: 7650681]

[49] Nami S, Aghebati-Maleki A, Morovati H, Aghebati-Maleki L. Current antifungal drugs and immunotherapeutic approaches as promising strategies to treatment of fungal diseases. Biomed Pharmacother 2019; 110: 857-68.
[http://dx.doi.org/10.1016/j.biopha.2018.12.009] [PMID: 30557835]

[50] Kofla G, Ruhnke M. Pharmacology and metabolism of anidulafungin, caspofungin and micafungin in the treatment of invasive candidosis - review of the literature. Eur J Med Res 2011; 16(4): 159-66.
[http://dx.doi.org/10.1186/2047-783X-16-4-159] [PMID: 21486730]

[51] Walker SS, Xu Y, Triantafyllou I, *et al.* Discovery of a novel class of orally active antifungal beta-1,--D-glucan synthase inhibitors. Antimicrob Agents Chemother 2011; 55(11): 5099-106.
[http://dx.doi.org/10.1128/AAC.00432-11] [PMID: 21844320]

[52] Wingard JR, Leather H. A new era of antifungal therapy. Biol Blood Marrow Transplant 2004; 10(2): 73-90.
[http://dx.doi.org/10.1016/j.bbmt.2003.09.014] [PMID: 14750074]

[53] Mroczyńska M, Brillowska-Dąbrowska A. Review on current status of echinocandins use. Antibiotics (Basel) 2020; 9(5): 227.
[http://dx.doi.org/10.3390/antibiotics9050227] [PMID: 32370108]

[54] Denning DW. Echinocandin antifungal drugs. Lancet 2003; 362(9390): 1142-51.
[http://dx.doi.org/10.1016/S0140-6736(03)14472-8] [PMID: 14550704]

[55] Saravolatz LD, Deresinski SC, Stevens DA. Caspofungin. Clin Infect Dis 2003; 36(11): 1445-57.
[http://dx.doi.org/10.1086/375080] [PMID: 12766841]

[56] Stover KR, Farley JM, Kyle PB, Cleary JD. Cardiac toxicity of some echinocandin antifungals. Expert Opin Drug Saf 2014; 13(1): 5-14.
[http://dx.doi.org/10.1517/14740338.2013.829036] [PMID: 24047086]

[57] Fernández de Ullivarri M, Arbulu S, Garcia-Gutierrez E, Cotter PD. Antifungal peptides as therapeutic agents. Front Cell Infect Microbiol 2020; 10: 105.
[http://dx.doi.org/10.3389/fcimb.2020.00105] [PMID: 32257965]

[58] Tam J, Wang S, Wong K, Tan W. Antimicrobial peptides from plants. Pharmaceuticals (Basel) 2015; 8(4): 711-57.
[http://dx.doi.org/10.3390/ph8040711] [PMID: 26580629]

[59] Scarsini M, Tomasinsig L, Arzese A, D'Este F, Oro D, Skerlavaj B. Antifungal activity of cathelicidin peptides against planktonic and biofilm cultures of *Candida* species isolated from vaginal infections. Peptides 2015; 71: 211-21.
[http://dx.doi.org/10.1016/j.peptides.2015.07.023] [PMID: 26238597]

[60] Bradshaw JP. Cationic antimicrobial peptides : issues for potential clinical use. BioDrugs 2003; 17(4): 233-40.
[http://dx.doi.org/10.2165/00063030-200317040-00002] [PMID: 12899640]

[61] Edmond MB, Wallace SE, McClish DK, Pfaller MA, Jones RN, Wenzel RP. Nosocomial bloodstream infections in United States hospitals: a three-year analysis. Clin Infect Dis 1999; 29(2): 239-44.
[http://dx.doi.org/10.1086/520192] [PMID: 10476719]

[62] Rautenbach M, Troskie AM, Vosloo JA. Antifungal peptides: To be or not to be membrane active. Biochimie 2016; 130: 132-45.
[http://dx.doi.org/10.1016/j.biochi.2016.05.013] [PMID: 27234616]

[63] Troskie AM, de Beer A, Vosloo JA, Jacobs K, Rautenbach M. Inhibition of agronomically relevant fungal phytopathogens by tyrocidines, cyclic antimicrobial peptides isolated from *Bacillus aneurinolyticus.* Microbiology (Reading) 2014; 160(9): 2089-101.
[http://dx.doi.org/10.1099/mic.0.078840-0] [PMID: 24996824]

[64] Mygind PH, Fischer RL, Schnorr KM, *et al.* Plectasin is a peptide antibiotic with therapeutic potential from a saprophytic fungus. Nature 2005; 437(7061): 975-80.
[http://dx.doi.org/10.1038/nature04051] [PMID: 16222292]

[65] Cabib E, Arroyo J. How carbohydrates sculpt cells: chemical control of morphogenesis in the yeast cell wall. Nat Rev Microbiol 2013; 11(9): 648-55.
[http://dx.doi.org/10.1038/nrmicro3090] [PMID: 23949603]

[66] van der Weerden NL, Hancock REW, Anderson MA. Permeabilization of fungal hyphae by the plant defensin NaD1 occurs through a cell wall-dependent process. J Biol Chem 2010; 285(48): 37513-20.
[http://dx.doi.org/10.1074/jbc.M110.134882] [PMID: 20861017]

[67] Oppenheim FG, Xu T, McMillian FM, *et al.* Histatins, a novel family of histidine-rich proteins in human parotid secretion. Isolation, characterization, primary structure, and fungistatic effects on *Candida albicans*. J Biol Chem 1988; 263(16): 7472-7.
[http://dx.doi.org/10.1016/S0021-9258(18)68522-9] [PMID: 3286634]

[68] Li XS, Reddy MS, Baev D, Edgerton M. *Candida albicans* Ssa1/2p is the cell envelope binding protein for human salivary histatin 5. J Biol Chem 2003; 278(31): 28553-61.
[http://dx.doi.org/10.1074/jbc.M300680200] [PMID: 12761219]

[69] Bowman SM, Free SJ. The structure and synthesis of the fungal cell wall. BioEssays 2006; 28(8): 799-808.
[http://dx.doi.org/10.1002/bies.20441] [PMID: 16927300]

[70] Beaulieu D, Tang J, Zeckner DJ, Parr TR Jr. Correlation of cilofungin in vivo efficacy with its activity against *Aspergillus fumigatus* (1,3)-beta-D-glucan synthase. FEMS Microbiol Lett 1993; 108(2): 133-7.
[PMID: 8486238]

[71] Espinel-Ingroff A. In vitro antifungal activities of anidulafungin and micafungin, licensed agents and the investigational triazole posaconazole as determined by NCCLS methods for 12,052 fungal isolates: review of the literature. Rev Iberoam Micol 2003; 20(4): 121-36.
[PMID: 15456349]

[72] Bormann A, Morrison VA. Review of the pharmacology and clinical studies of micafungin. Drug Des Devel Ther 2009; 3: 295-302.
[http://dx.doi.org/10.2147/DDDT.S3242] [PMID: 20054447]

[73] Oita S, Ohnishi-Kameyama M, Nagata T. Binding of barley and wheat α-thionins to polysaccharides.

Biosci Biotechnol Biochem 2000; 64(5): 958-64.
[http://dx.doi.org/10.1271/bbb.64.958] [PMID: 10879464]

[74] Fujimura M, Ideguchi M, Minami Y, Watanabe K, Tadera K. Purification, characterization, and sequencing of novel antimicrobial peptides, Tu-AMP 1 and Tu-AMP 2, from bulbs of tulip (*Tulipa gesneriana* L.). Biosci Biotechnol Biochem 2004; 68(3): 571-7.
[http://dx.doi.org/10.1271/bbb.68.571] [PMID: 15056889]

[75] Fujimura M, Ideguchi M, Minami Y, Watanabe K, Tadera K. Amino acid sequence and antimicrobial activity of chitin-binding peptides, Pp-AMP 1 and Pp-AMP 2, from Japanese bamboo shoots (*Phyllostachys pubescens*). Biosci Biotechnol Biochem 2005; 69(3): 642-5.
[http://dx.doi.org/10.1271/bbb.69.642] [PMID: 15784998]

[76] Broekaert WF, Mariën W, Terras FRG, *et al.* Antimicrobial peptides from *Amaranthus caudatus* seeds with sequence homology to the cysteine/glycine-rich domain of chitin-binding proteins. Biochemistry 1992; 31(17): 4308-14.
[http://dx.doi.org/10.1021/bi00132a023] [PMID: 1567877]

[77] Nguyen LT, Haney EF, Vogel HJ. The expanding scope of antimicrobial peptide structures and their modes of action. Trends Biotechnol 2011; 29(9): 464-72.
[http://dx.doi.org/10.1016/j.tibtech.2011.05.001] [PMID: 21680034]

[78] Yeaman MR, Yount NY. Mechanisms of antimicrobial peptide action and resistance. Pharmacol Rev 2003; 55(1): 27-55.
[http://dx.doi.org/10.1124/pr.55.1.2] [PMID: 12615953]

[79] Feng Z, Jiang B, Chandra J, Ghannoum M, Nelson S, Weinberg A. Human beta-defensins: differential activity against candidal species and regulation by *Candida albicans*. J Dent Res 2005; 84(5): 445-50.
[http://dx.doi.org/10.1177/154405910508400509] [PMID: 15840781]

[80] Berkow EL, Lockhart SR. Activity of novel antifungal compound APX001A against a large collection of *Candida auris*. J Antimicrob Chemother 2018; 73(11): 3060-2.
[http://dx.doi.org/10.1093/jac/dky302] [PMID: 30085167]

[81] Ksiezopolska E, Gabaldón T. Evolutionary emergence of drug resistance in *Candida* opportunistic pathogens. Genes (Basel) 2018; 9(9): 461.
[http://dx.doi.org/10.3390/genes9090461] [PMID: 30235884]

[82] du Pré S, Beckmann N, Almeida MC, *et al.* Effect of the novel antifungal drug F901318 (Olorofim) on growth and viability of *Aspergillus fumigatus*. Antimicrob Agents Chemother 2018; 62(8): e00231-18.
[http://dx.doi.org/10.1128/AAC.00231-18] [PMID: 29891595]

[83] Hokken MWJ, Zwaan BJ, Melchers WJG, Verweij PE. Facilitators of adaptation and antifungal resistance mechanisms in clinically relevant fungi. Fungal Genet Biol 2019; 132: 103254.
[http://dx.doi.org/10.1016/j.fgb.2019.103254] [PMID: 31326470]

[84] Robbins N, Caplan T, Cowen LE. Molecular evolution of antifungal drug resistance. Annu Rev Microbiol 2017; 71(1): 753-75.
[http://dx.doi.org/10.1146/annurev-micro-030117-020345] [PMID: 28886681]

[85] Prasad R, Banerjee A, Shah AH. Resistance to antifungal therapies. Essays Biochem 2017; 61(1): 157-66.
[http://dx.doi.org/10.1042/EBC20160067] [PMID: 28258238]

[86] Garcia-Effron G, Dilger A, Alcazar-Fuoli L, Park S, Mellado E, Perlin DS. Rapid detection of triazole antifungal resistance in *Aspergillus fumigatus*. J Clin Microbiol 2008; 46(4): 1200-6.
[http://dx.doi.org/10.1128/JCM.02330-07] [PMID: 18234874]

[87] Morio F, Loge C, Besse B, Hennequin C, Le Pape P. Screening for amino acid substitutions in the *Candida albicans* Erg11 protein of azole-susceptible and azole-resistant clinical isolates: new substitutions and a review of the literature. Diagn Microbiol Infect Dis 2010; 66(4): 373-84.
[http://dx.doi.org/10.1016/j.diagmicrobio.2009.11.006] [PMID: 20226328]

[88] Rodero L, Mellado E, Rodriguez AC, *et al.* G484S amino acid substitution in lanosterol 14-α demethylase (ERG11) is related to fluconazole resistance in a recurrent *Cryptococcus neoformans* clinical isolate. Antimicrob Agents Chemother 2003; 47(11): 3653-6. [http://dx.doi.org/10.1128/AAC.47.11.3653-3656.2003] [PMID: 14576140]

[89] Sanglard D, Coste A, Ferrari S. Antifungal drug resistance mechanisms in fungal pathogens from the perspective of transcriptional gene regulation. FEMS Yeast Res 2009; 9(7): 1029-50. [http://dx.doi.org/10.1111/j.1567-1364.2009.00578.x] [PMID: 19799636]

[90] Wang H, Kong F, Sorrell TC, *et al.* Rapid detection of ERG11 gene mutations in clinical *Candida albicans* isolates with reduced susceptibility to fluconazole by rolling circle amplification and DNA sequencing. BMC Microbiol 2009; 9(1): 167. [http://dx.doi.org/10.1186/1471-2180-9-167] [PMID: 19682357]

[91] Webber MA, Piddock LJV. The importance of efflux pumps in bacterial antibiotic resistance. J Antimicrob Chemother 2003; 51(1): 9-11. [http://dx.doi.org/10.1093/jac/dkg050] [PMID: 12493781]

[92] Cannon RD, Lamping E, Holmes AR, *et al.* Efflux-mediated antifungal drug resistance. Clin Microbiol Rev 2009; 22(2): 291-321. [http://dx.doi.org/10.1128/CMR.00051-08] [PMID: 19366916]

[93] Choi MJ, Won EJ, Shin JH, *et al.* Resistance mechanisms and clinical features of fluconazole-nonsusceptible *Candida tropicalis* isolates compared with fluconazole-less-susceptible isolates. Antimicrob Agents Chemother 2016; 60(6): 3653-61. [http://dx.doi.org/10.1128/AAC.02652-15] [PMID: 27044550]

[94] Moran GP, Sanglard D, Donnelly SM, Shanley DB, Sullivan DJ, Coleman DC. Identification and expression of multidrug transporters responsible for fluconazole resistance in *Candida dubliniensis.* Antimicrob Agents Chemother 1998; 42(7): 1819-30. [http://dx.doi.org/10.1128/AAC.42.7.1819] [PMID: 9661028]

[95] Sanglard D, Ischer F, Monod M, Bille J. Cloning of *Candida albicans* genes conferring resistance to azole antifungal agents: characterization of CDR2, a new multidrug ABC transporter gene. Microbiology (Reading) 1997; 143(2): 405-16. [http://dx.doi.org/10.1099/00221287-143-2-405] [PMID: 9043118]

[96] Kumar A, Nair R, Kumar M, *et al.* Assessment of antifungal resistance and associated molecular mechanism in *Candida albicans* isolates from different cohorts of patients in North Indian state of Haryana. Folia Microbiol (Praha) 2020; 65(4): 747-54. [http://dx.doi.org/10.1007/s12223-020-00785-6] [PMID: 32219719]

[97] Gsaller F, Furukawa T, Carr PD, Rash B, Jöchl C, Bertuzzi M, *et al.* Mechanistic basis of pH-dependent 5-flucytosine resistance in *Aspergillus fumigatus.* Antimicrob Agents Chemother 2018. 62e02593-17

[98] Schopf FH, Biebl MM, Buchner J. The HSP90 chaperone machinery. Nat Rev Mol Cell Biol 2017; 18(6): 345-60. [http://dx.doi.org/10.1038/nrm.2017.20] [PMID: 28429788]

[99] Cowen LE, Lindquist S. Hsp90 potentiates the rapid evolution of new traits: drug resistance in diverse fungi. Science 2005; 309(5744): 2185-9. [http://dx.doi.org/10.1126/science.1118370] [PMID: 16195452]

[100] Cowen LE, Singh SD, Köhler JR, *et al.* Harnessing Hsp90 function as a powerful, broadly effective therapeutic strategy for fungal infectious disease. Proc Natl Acad Sci USA 2009; 106(8): 2818-23. [http://dx.doi.org/10.1073/pnas.0813394106] [PMID: 19196973]

[101] Georgopapadakou NH, Dix BA, Smith SA, Freudenberger J, Funke PT. Effect of antifungal agents on lipid biosynthesis and membrane integrity in *Candida albicans.* Antimicrob Agents Chemother 1987; 31(1): 46-51.

[http://dx.doi.org/10.1128/AAC.31.1.46] [PMID: 3551826]

[102] Guan XL, Souza CM, Pichler H, *et al.* Functional interactions between sphingolipids and sterols in biological membranes regulating cell physiology. Mol Biol Cell 2009; 20(7): 2083-95. [http://dx.doi.org/10.1091/mbc.e08-11-1126] [PMID: 19225153]

[103] Walker LA, Munro CA, de Bruijn I, Lenardon MD, McKinnon A, Gow NAR. Stimulation of chitin synthesis rescues *Candida albicans* from echinocandins. PLoS Pathog 2008; 4(4): e1000040. [http://dx.doi.org/10.1371/journal.ppat.1000040] [PMID: 18389063]

[104] Satish S, Jiménez-Ortigosa C, Zhao Y, *et al.* Stress-induced changes in the lipid microenvironment of β-(1,3)-d-glucan synthase cause clinically important echinocandin resistance in *Aspergillus fumigatus*. MBio 2019; 10(3): e00779-19. [http://dx.doi.org/10.1128/mBio.00779-19] [PMID: 31164462]

[105] Chew SY, Ho KL, Cheah YK, Sandai D, Brown AJP, Than LTL. Physiologically relevant alternative carbon sources modulate biofilm formation, cell wall architecture, and the stress and antifungal resistance of *Candida glabrata.* Int J Mol Sci 2019; 20(13): 3172. [http://dx.doi.org/10.3390/ijms20133172] [PMID: 31261727]

[106] Ramage G, Rajendran R, Sherry L, Williams C. Fungal biofilm resistance. Int J Microbiol 2012; 2012: 1-14. [http://dx.doi.org/10.1155/2012/528521] [PMID: 22518145]

[107] Costa-Orlandi C, Sardi J, Pitangui N, *et al.* Fungal biofilms and polymicrobial diseases. J Fungi (Basel) 2017; 3(2): 22. [http://dx.doi.org/10.3390/jof3020022] [PMID: 29371540]

[108] Fanning S, Mitchell AP. Fungal Biofilms. PLoS Pathog 2012; 8(4): e1002585. [http://dx.doi.org/10.1371/journal.ppat.1002585] [PMID: 22496639]

[109] Di Bonaventura G, Pompilio A, Picciani C, Iezzi M, D'Antonio D, Piccolomini R. Biofilm formation by the emerging fungal pathogen *Trichosporon asahii*: development, architecture, and antifungal resistance. Antimicrob Agents Chemother 2006; 50(10): 3269-76. [http://dx.doi.org/10.1128/AAC.00556-06] [PMID: 17005804]

[110] Müller FMC, Seidler M, Beauvais A. *Aspergillus fumigatus* biofilms in the clinical setting. Med Mycol 2011; 49(S1) (Suppl. 1): S96-S100. [http://dx.doi.org/10.3109/13693786.2010.502190] [PMID: 21254964]

[111] Kaur S, Singh S. Biofilm formation by *Aspergillus fumigatus*. Med Mycol 2014; 52(1): 2-9. [PMID: 23962172]

[112] Martinez LR, Casadevall A. Biofilm formation by *Cryptococcus neoformans.* Microbiol Spectr 2015; 3(3): 3.3.05. [http://dx.doi.org/10.1128/microbiolspec.MB-0006-2014] [PMID: 26185073]

[113] Vediyappan G, Rossignol T, d'Enfert C. Interaction of *Candida albicans* biofilms with antifungals: transcriptional response and binding of antifungals to beta-glucans. Antimicrob Agents Chemother 2010; 54(5): 2096-111. [http://dx.doi.org/10.1128/AAC.01638-09] [PMID: 20194705]

[114] Bonhomme J, d'Enfert C. *Candida albicans* biofilms: building a heterogeneous, drug-tolerant environment. Curr Opin Microbiol 2013; 16(4): 398-403. [http://dx.doi.org/10.1016/j.mib.2013.03.007] [PMID: 23566895]

[115] Mitchell KF, Zarnowski R, Sanchez H, *et al.* Community participation in biofilm matrix assembly and function. Proc Natl Acad Sci USA 2015; 112(13): 4092-7. [http://dx.doi.org/10.1073/pnas.1421437112] [PMID: 25770218]

[116] Naranjo Ortiz MA, Gabaldón T. Fungal evolution: cellular, genomic and metabolic complexity 2020. [http://dx.doi.org/10.1111/brv.12605]

[117] Selmecki A, Gerami-Nejad M, Paulson C, Forche A, Berman J. An isochromosome confers drug resistance in vivo by amplification of two genes, ERG11 and TAC1. Mol Microbiol 2008; 68(3): 624-41.
[http://dx.doi.org/10.1111/j.1365-2958.2008.06176.x] [PMID: 18363649]

[118] Selmecki A, Forche A, Berman J. Genomic plasticity of the human fungal pathogen *Candida albicans*. Eukaryot Cell 2010; 9(7): 991-1008.
[http://dx.doi.org/10.1128/EC.00060-10] [PMID: 20495058]

[119] Kelly SL, Lamb DC, Kelly DE, *et al.* Resistance to fluconazole and cross-resistance to amphotericin B in *Candida albicans* from AIDS patients caused by defective sterol $\Delta^{5,6}$ -desaturation. FEBS Lett 1997; 400(1): 80-2.
[http://dx.doi.org/10.1016/S0014-5793(96)01360-9] [PMID: 9000517]

[120] Vincent BM, Lancaster AK, Scherz-Shouval R, Whitesell L, Lindquist S. Fitness trade-offs restrict the evolution of resistance to amphotericin B. PLoS Biol 2013; 11(10): e1001692.
[http://dx.doi.org/10.1371/journal.pbio.1001692] [PMID: 24204207]

[121] Vale-Silva LA, Coste AT, Ischer F, *et al.* Azole resistance by loss of function of the sterol Δ^5,^{6}-desaturase gene (ERG3) in *Candida albicans* does not necessarily decrease virulence. Antimicrob Agents Chemother 2012; 56(4): 1960-8.
[http://dx.doi.org/10.1128/AAC.05720-11] [PMID: 22252807]

[122] Sanglard D, Ischer F, Bille J. Role of ATP-binding-cassette transporter genes in high-frequency acquisition of resistance to azole antifungals in *Candida glabrata.* Antimicrob Agents Chemother 2001; 45(4): 1174-83.
[http://dx.doi.org/10.1128/AAC.45.4.1174-1183.2001] [PMID: 11257032]

[123] Bromley M, Johns A, Davies E, *et al.* Mitochondrial Complex I Is a Global Regulator of Secondary Metabolism, Virulence and Azole Sensitivity in Fungi. PLoS One 2016; 11(7): e0158724.
[http://dx.doi.org/10.1371/journal.pone.0158724] [PMID: 27438017]

[124] Zhang J, van den Heuvel J, Debets AJM, *et al.* Evolution of cross-resistance to medical triazoles in *Aspergillus fumigatus* through selection pressure of environmental fungicides. Proc Biol Sci 2017; 284(1863): 20170635.
[http://dx.doi.org/10.1098/rspb.2017.0635] [PMID: 28931745]

[125] Li Y, Zhang Y, Zhang C, *et al.* Mitochondrial dysfunctions trigger the calcium signaling-dependent fungal multidrug resistance. Proc Natl Acad Sci USA 2020; 117(3): 1711-21.
[http://dx.doi.org/10.1073/pnas.1911560116] [PMID: 31811023]

[126] Yadav U, Khan MA. Targeting the GPI biosynthetic pathway. Pathog Glob Health 2018; 112(3): 115-22.
[http://dx.doi.org/10.1080/20477724.2018.1442764] [PMID: 29484956]

[127] Richard M, Ibata-Ombetta S, Dromer F, Bordon-Pallier F, Jouault T, Gaillardin C. Complete glycosylphosphatidylinositol anchors are required in *Candida albicans* for full morphogenesis, virulence and resistance to macrophages. Mol Microbiol 2002; 44(3): 841-53.
[http://dx.doi.org/10.1046/j.1365-2958.2002.02926.x] [PMID: 11994163]

[128] Romano J, Nimrod G, Ben-Tal N, *et al.* Disruption of the *Aspergillus fumigatus* ECM33 homologue results in rapid conidial germination, antifungal resistance and hypervirulence. Microbiology (Reading) 2006; 152(7): 1919-28.
[http://dx.doi.org/10.1099/mic.0.28936-0] [PMID: 16804168]

[129] Li H, Zhou H, Luo Y, Ouyang H, Hu H, Jin C. Glycosylphosphatidylinositol (GPI) anchor is required in *Aspergillus fumigatus* for morphogenesis and virulence. Mol Microbiol 2007; 64(4): 1014-27.
[http://dx.doi.org/10.1111/j.1365-2958.2007.05709.x] [PMID: 17501924]

[130] Yan J, Du T, Zhao W, *et al.* Transcriptome and biochemical analysis reveals that suppression of GPI-anchor synthesis leads to autophagy and possible necroptosis in *Aspergillus fumigatus*. PLoS One

2013; 8(3): e59013.
[http://dx.doi.org/10.1371/journal.pone.0059013] [PMID: 23527074]

[131] McLellan CA, Whitesell L, King OD, Lancaster AK, Mazitschek R, Lindquist S. Inhibiting GPI anchor biosynthesis in fungi stresses the endoplasmic reticulum and enhances immunogenicity. ACS Chem Biol 2012; 7(9): 1520-8.
[http://dx.doi.org/10.1021/cb300235m] [PMID: 22724584]

[132] Shen H, Chen SM, Liu W, *et al.* Abolishing cell wall glycosylphosphatidylinositol-anchored proteins in *Candida albicans* enhances recognition by host dectin-1. Infect Immun 2015; 83(7): 2694-704.
[http://dx.doi.org/10.1128/IAI.00097-15] [PMID: 25895969]

[133] Mann PA, McLellan CA, Koseoglu S, *et al.* Chemical genomics-based antifungal drug discovery: targeting Glycosylphosphatidylinositol (GPI) precursor biosynthesis. ACS Infect Dis 2015; 1(1): 59-72.
[http://dx.doi.org/10.1021/id5000212] [PMID: 26878058]

[134] Noble SM, French S, Kohn LA, Chen V, Johnson AD. Systematic screens of a *Candida albicans* homozygous deletion library decouple morphogenetic switching and pathogenicity. Nat Genet 2010; 42(7): 590-8.
[http://dx.doi.org/10.1038/ng.605] [PMID: 20543849]

[135] Rittershaus PC, Kechichian TB, Allegood JC, *et al.* Glucosylceramide synthase is an essential regulator of pathogenicity of *Cryptococcus neoformans.* J Clin Invest 2006; 116(6): 1651-9.
[http://dx.doi.org/10.1172/JCI27890] [PMID: 16741577]

[136] Mor V, Rella A, Farnoud AM, *et al.* Identification of a new class of antifungals targeting the synthesis of fungal sphingolipids. MBio 2015; 6(3): e00647-15.
[http://dx.doi.org/10.1128/mBio.00647-15] [PMID: 26106079]

[137] Cheah HL, Lim V, Sandai D. Inhibitors of the glyoxylate cycle enzyme ICL1 in *Candida albicans* for potential use as antifungal agents. PLoS One 2014; 9(4): e95951.
[http://dx.doi.org/10.1371/journal.pone.0095951] [PMID: 24781056]

[138] Bae M, Kim H, Moon K, *et al.* Mohangamides A and B, new dilactone-tethered pseudo-dimeric peptides inhibiting *Candida albicans* isocitrate lyase. Org Lett 2015; 17(3): 712-5.
[http://dx.doi.org/10.1021/ol5037248] [PMID: 25622093]

[139] Jiang Z, Liu N, Hu D, *et al.* The discovery of novel antifungal scaffolds by structural simplification of the natural product sampangine. Chem Commun (Camb) 2015; 51(78): 14648-51.
[http://dx.doi.org/10.1039/C5CC05699C] [PMID: 26289663]

[140] Ben Yaakov D, Rivkin A, Mircus G, *et al.* Identification and characterization of haemofungin, a novel antifungal compound that inhibits the final step of haem biosynthesis. J Antimicrob Chemother 2016; 71(4): 946-52.
[http://dx.doi.org/10.1093/jac/dkv446] [PMID: 26747101]

[141] Simm C, May RC. Zinc and Iron Homeostasis: Target-Based Drug Screening as New Route for Antifungal Drug Development. Front Cell Infect Microbiol 2019; 9: 181.
[http://dx.doi.org/10.3389/fcimb.2019.00181] [PMID: 31192169]

[142] Koselny K, Green J, DiDone L, *et al.* The celecoxib derivative AR-12 has broad-spectrum antifungal activity in vitro and improves the activity of fluconazole in a murine model of *cryptococcosis* . Antimicrob Agents Chemother 2016; 60(12): 7115-27.
[http://dx.doi.org/10.1128/AAC.01061-16] [PMID: 27645246]

[143] Perfect JR. The antifungal pipeline: a reality check. Nat Rev Drug Discov 2017; 16(9): 603-16.
[http://dx.doi.org/10.1038/nrd.2017.46] [PMID: 28496146]

[144] Dinér P, Veide Vilg J, Kjellén J, *et al.* Design, synthesis, and characterization of a highly effective Hog1 inhibitor: a powerful tool for analyzing MAP kinase signaling in yeast. PLoS One 2011; 6(5): e20012.

[http://dx.doi.org/10.1371/journal.pone.0020012] [PMID: 21655328]

[145] Vetcher L, Menzella HG, Kudo T, Motoyama T, Katz L. The antifungal polyketide ambruticin targets the HOG pathway. Antimicrob Agents Chemother 2007; 51(10): 3734-6. [http://dx.doi.org/10.1128/AAC.00369-07] [PMID: 17698623]

[146] Shubitz LF, Galgiani JN, Tian ZQ, Zhong Z, Timmermans P, Katz L. Efficacy of ambruticin analogs in a murine model of coccidioidomycosis. Antimicrob Agents Chemother 2006; 50(10): 3467-9. [http://dx.doi.org/10.1128/AAC.00670-06] [PMID: 17005834]

[147] McCarthy MW, Kontoyiannis DP, Cornely OA, Perfect JR, Walsh TJ. Novel agents and drug targets to meet the challenges of resistant fungi. J Infect Dis 2017; 216 (Suppl. 3): S474-83. [http://dx.doi.org/10.1093/infdis/jix130] [PMID: 28911042]

[148] Welsch CA, Roth LWA, Goetschy JF, Movva NR. Genetic, biochemical, and transcriptional responses of *Saccharomyces cerevisiae* to the novel immunomodulator FTY720 largely mimic those of the natural sphingolipid phytosphingosine. J Biol Chem 2004; 279(35): 36720-31. [http://dx.doi.org/10.1074/jbc.M406179200] [PMID: 15190065]

[149] Steinbach WJ, Reedy JL, Cramer RA Jr, Perfect JR, Heitman J. Harnessing calcineurin as a novel anti-infective agent against invasive fungal infections. Nat Rev Microbiol 2007; 5(6): 418-30. [http://dx.doi.org/10.1038/nrmicro1680] [PMID: 17505522]

[150] Binder U, Bencina M, Eigentler A, Meyer V, Marx F. The *Aspergillus giganteus* antifungal protein AFPNN5353activates the cell wall integrity pathway and perturbs calcium homeostasis. BMC Microbiol 2011; 11(1): 209. [http://dx.doi.org/10.1186/1471-2180-11-209] [PMID: 21943024]

[151] Hast MA, Nichols CB, Armstrong SM, *et al.* Structures of *Cryptococcus neoformans* protein farnesyltransferase reveal strategies for developing inhibitors that target fungal pathogens. J Biol Chem 2011; 286(40): 35149-62. [http://dx.doi.org/10.1074/jbc.M111.250506] [PMID: 21816822]

[152] Nobile CJ, Andes DR, Nett JE, *et al.* Critical role of Bcr1-dependent adhesins in *C. albicans* biofilm formation *in vitro* and *in vivo.* PLoS Pathog 2006; 2(7): e63. [http://dx.doi.org/10.1371/journal.ppat.0020063] [PMID: 16839200]

[153] Hoekstra WJ, Garvey EP, Moore WR, Rafferty SW, Yates CM, Schotzinger RJ. Design and optimization of highly-selective fungal CYP51 inhibitors. Bioorg Med Chem Lett 2014; 24(15): 3455-8. [http://dx.doi.org/10.1016/j.bmcl.2014.05.068] [PMID: 24948565]

[154] Wiederhold NP. The antifungal arsenal: alternative drugs and future targets. Int J Antimicrob Agents 2018; 51(3): 333-9. [http://dx.doi.org/10.1016/j.ijantimicag.2017.09.002] [PMID: 28890395]

[155] Gonzalez-Lara MF, Sifuentes-Osornio J, Ostrosky-Zeichner L. Drugs in clinical development for fungal infections. Drugs 2017; 77(14): 1505-18. [http://dx.doi.org/10.1007/s40265-017-0805-2] [PMID: 28840541]

[156] Garvey EP, Hoekstra WJ, Schotzinger RJ, Sobel JD, Lilly EA, Fidel PL Jr. Efficacy of the clinical agent VT-1161 against fluconazole-sensitive and -resistant *Candida albicans* in a murine model of vaginal candidiasis. Antimicrob Agents Chemother 2015; 59(9): 5567-73. [http://dx.doi.org/10.1128/AAC.00185-15] [PMID: 26124165]

[157] Brand S, Dehenhardt T, Nyirjesy P, Augenbraun M, Schotzinger RJ. Efficacy and safety of oral VT-1161, a novel inhibitor of fungal CYP51, in a randomized phase 2 study in patients with acute vulvovaginal candidiasis: six-month follow-up. Portland: Infectious Diseases Society for Obstetrics and Gynecology Annual Meeting; 2015.

[158] Elewski B, Kempers S, Bhatia N, Blauvelt A, Curelop S, Brand S, *et al.* Efficacy and safety of VT-1161 in a randomized, double-blinded, placebo-controlled phase 2 study of four oral dosing regimens

in the treatment of patients with moderate-to-severe distal-lateral subungual onychomycosis (DLSO). 4th International Summit on Nail Diseases; 2017; Athens, Greece. 2017.

[159] Lockhart SR, Fothergill AW, Iqbal N, *et al.* The investigational fungal cyp51 inhibitor VT-1129 demonstrates potent *in vitro* activity against *Cryptococcus neoformans* and *Cryptococcus gattii.* Antimicrob Agents Chemother 2016; 60(4): 2528-31. [http://dx.doi.org/10.1128/AAC.02770-15] [PMID: 26787697]

[160] Wiederhold NP, Najvar LK, Garvey EP, *et al.* The fungal Cyp51 inhibitor VT-1129 is efficacious in an experimental model of cryptococcal meningitis. Antimicrob Agents Chemother 2018; 62(9): e01071-18. [http://dx.doi.org/10.1128/AAC.01071-18] [PMID: 29987152]

[161] Hargrove TY, Garvey EP, Hoekstra WJ, *et al.* Crystal structure of the new investigational drug candidate VT-1598 in complex with *Aspergillus fumigatus* sterol 14alpha-demethylase provides insights into its broad-spectrum antifungal activity. Antimicrob Agents Chemother 2017; 61(7): e00570-17. [http://dx.doi.org/10.1128/AAC.00570-17] [PMID: 28461309]

[162] Peláez F, Cabello A, Platas G, *et al.* The discovery of enfumafungin, a novel antifungal compound produced by an endophytic *Hormonema* species biological activity and taxonomy of the producing organisms. Syst Appl Microbiol 2000; 23(3): 333-43. [http://dx.doi.org/10.1016/S0723-2020(00)80062-4] [PMID: 11108011]

[163] Jiménez-Ortigosa C, Paderu P, Motyl MR, Perlin DS. Enfumafungin derivative MK-3118 shows increased *in vitro* potency against clinical echinocandin-resistant *Candida* Species and *Aspergillus* species isolates. Antimicrob Agents Chemother 2014; 58(2): 1248-51. [http://dx.doi.org/10.1128/AAC.02145-13] [PMID: 24323472]

[164] Lamoth F, Alexander BD. Antifungal activities of SCY-078 (MK-3118) and standard antifungal agents against clinical non-*Aspergillus* mold isolates. Antimicrob Agents Chemother 2015; 59(7): 4308-11. [http://dx.doi.org/10.1128/AAC.00234-15] [PMID: 25896696]

[165] Pfaller MA, Messer SA, Rhomberg PR, Borroto-Esoda K, Castanheira M. Differential activity of the oral glucan synthase inhibitor SCY078 against wild-type and echinocandin-resistant strains of *Candida* species. Antimicrob Agents Chemother 2017; 61(8): e00161-17. [http://dx.doi.org/10.1128/AAC.00161-17] [PMID: 28533234]

[166] Ghannoum M, Long L, Larkin EL, *et al.* Evaluation of the antifungal activity of the novel oral glucan synthase inhibitor SCY-078, singly and in combination, for the treatment of invasive aspergillosis. Antimicrob Agents Chemother 2018; 62: e00244-18. [http://dx.doi.org/10.1128/AAC.00244-18]

[167] Berkow EL, Angulo D, Lockhart SR. *In vitro* activity of a novel glucan synthase inhibitor, SCY-078, against clinical isolates of *Candida auris.* Antimicrob Agents Chemother 2017; 61(7): e00435-17. [http://dx.doi.org/10.1128/AAC.00435-17] [PMID: 28483955]

[168] Arendrup MC, Jørgensen KM, Hare RK, Chowdhary A. *In vitro* activity of ibrexafungerp (SCY-078) against *Candida auris* isolates as determined by EUCAST methodology and comparison with activity against *C. albicans* and *C. glabrata* and with the activities of six comparator agents. Antimicrob Agents Chemother 2020; 64(3): e02136-19. [http://dx.doi.org/10.1128/AAC.02136-19] [PMID: 31844005]

[169] Pfaller MA, Carvalhaes C, Messer SA, Rhomberg PR, Castanheira M. Activity of a long-acting echinocandin, rezafungin, and comparator antifungal agents tested against contemporary invasive fungal isolates (SENTRY Program, 2016 to 2018). Antimicrob Agents Chemother 2020; 64(4): e00099-20. [http://dx.doi.org/10.1128/AAC.00099-20] [PMID: 32015043]

[170] Pfaller MA, Messer SA, Rhomberg PR, Castanheira M. CD101, a long-acting echinocandin, and

comparator antifungal agents tested against a global collection of invasive fungal isolates in the SENTRY 2015 Antifungal Surveillance Program. Int J Antimicrob Agents 2017; 50(3): 352-8. [http://dx.doi.org/10.1016/j.ijantimicag.2017.03.028] [PMID: 28689871]

[171] Wiederhold NP, Locke JB, Daruwala P, Bartizal K. Rezafungin (CD101) demonstrates potent *in vitro* activity against *Aspergillus*, including azole-resistant *Aspergillus fumigatus* isolates and cryptic species. J Antimicrob Chemother 2018; 73(11): 3063-7. [http://dx.doi.org/10.1093/jac/dky280] [PMID: 30032217]

[172] Cushion MT, Ashbaugh A, Ong V. Rezafungin prevention of *pneumocystis* pneumonia and *pneumocystis* reactivation using different doses and durations of prophylaxis in a mouse model. Blood 2019; 134 (Suppl. 1): 3266. [http://dx.doi.org/10.1182/blood-2019-127392]

[173] Ong V, Hough G, Schlosser M, *et al.* Preclinical evaluation of the stability, safety, and efficacy of CD101, a novel echinocandin. Antimicrob Agents Chemother 2016; 60(11): 6872-9. [http://dx.doi.org/10.1128/AAC.00701-16] [PMID: 27620474]

[174] Miyazaki M, Horii T, Hata K, *et al. In vitro* activity of E1210, a novel antifungal, against clinically important yeasts and molds. Antimicrob Agents Chemother 2011; 55(10): 4652-8. [http://dx.doi.org/10.1128/AAC.00291-11] [PMID: 21825291]

[175] Viriyakosol S, Kapoor M, Okamoto S, *et al.* APX001 and other Gwt1 inhibitor prodrugs are effective in experimental *Coccidioides immitis* pneumonia. Antimicrob Agents Chemother 2019; 63(2): e01715-18. [http://dx.doi.org/10.1128/AAC.01715-18] [PMID: 30455238]

[176] Gebremariam T, Alkhazraji S, Alqarihi A, *et al.* APX001 is effective in the treatment of murine invasive pulmonary aspergillosis. Antimicrob Agents Chemother 2019; 63(2): e01713-18. [http://dx.doi.org/10.1128/AAC.01713-18] [PMID: 30455236]

[177] Shende P, Khair R, Gaud RS. Nanostructured cochleates: a multi-layered platform for cellular transportation of therapeutics. Drug Dev Ind Pharm 2019; 45(6): 869-81. [http://dx.doi.org/10.1080/03639045.2019.1583757] [PMID: 30767577]

[178] Biederdorf FA, Breithaupt I, Mannino R, Blum DE. Oral administration of amphotericin B (CAmB)) in humans: a phase I study of tolerability and pharmacokinetics. Matinas Biopharma 2016. http://content.equisolve.net/_b471c5f8d5c5228ccb2fffeb89c7d38b/matinasbiopharma/db/128/515/pdf/CAmB-Focus-Phase-1-Poster.pdf

[179] Rapala-Kozik M, Bochenska O, Zawrotniak M, *et al.* Inactivation of the antifungal and immunomodulatory properties of human cathelicidin LL-37 by aspartic proteases produced by the pathogenic yeast *Candida albicans*. Infect Immun 2015; 83(6): 2518-30. [http://dx.doi.org/10.1128/IAI.00023-15] [PMID: 25847962]

[180] Wnorowska U, Fiedoruk K, Piktel E, *et al.* Nanoantibiotics containing membrane-active human cathelicidin LL-37 or synthetic ceragenins attached to the surface of magnetic nanoparticles as novel and innovative therapeutic tools: current status and potential future applications. J Nanobiotechnology 2020; 18(1): 3. [http://dx.doi.org/10.1186/s12951-019-0566-z] [PMID: 31898542]

[181] Durnaś B, Wnorowska U, Pogoda K, *et al.* Candidacidal activity of selected ceragenins and human cathelicidin LL-37 in experimental settings mimicking infection sites. PLoS One 2016; 11(6): e0157242. [http://dx.doi.org/10.1371/journal.pone.0157242] [PMID: 27315208]

[182] Hashemi MM, Rovig J, Holden BS, *et al.* Ceragenins are active against drug-resistant *Candida auris* clinical isolates in planktonic and biofilm forms. J Antimicrob Chemother 2018; 73(6): 1537-45. [http://dx.doi.org/10.1093/jac/dky085] [PMID: 29635279]

[183] Hacioglu M, Guzel CB, Savage PB, Tan ASB. Antifungal susceptibilities, *in vitro* production of

virulence factors and activities of ceragenins against *Candida* spp. isolated from vulvovaginal candidiasis. Med Mycol 2019; 57(3): 291-9.
[http://dx.doi.org/10.1093/mmy/myy023] [PMID: 29846682]

[184] Edwards JE Jr, Schwartz MM, Schmidt CS, *et al.* A fungal immunotherapeutic vaccine (NDV-3A) for treatment of recurrent vulvovaginal candidiasis-A phase 2 randomized, double-blind, placebo-controlled trial. Clin Infect Dis 2018; 66(12): 1928-36.
[http://dx.doi.org/10.1093/cid/ciy185] [PMID: 29697768]

[185] Levy DA, Bohbot JM, Catalan F, Normier G, Pinel AM, Dussourd d'Hinterland L. Phase II study of D.651, an oral vaccine designed to prevent recurrences of vulvovaginal candidiasis. Vaccine 1989; 7(4): 337-40.
[http://dx.doi.org/10.1016/0264-410X(89)90197-7] [PMID: 2683457]

[186] Cox RA, Magee DM. Coccidioidomycosis: host response and vaccine development. Clin Microbiol Rev 2004; 17(4): 804-39.
[http://dx.doi.org/10.1128/CMR.17.4.804-839.2004] [PMID: 15489350]

[187] Bromuro C, Romano M, Chiani P, *et al.* Beta-glucan-CRM197 conjugates as candidates antifungal vaccines. Vaccine 2010; 28(14): 2615-23.
[http://dx.doi.org/10.1016/j.vaccine.2010.01.012] [PMID: 20096763]

[188] Stuehler C, Khanna N, Bozza S, *et al.* Cross-protective TH1 immunity against *Aspergillus fumigatus* and *Candida albicans*. Blood 2011; 117(22): 5881-91.
[http://dx.doi.org/10.1182/blood-2010-12-325084] [PMID: 21441461]

[189] Cutler JE, Corti M, Lambert P, Ferris M, Xin H. Horizontal transmission of *Candida albicans* and evidence of a vaccine response in mice colonized with the fungus. PLoS One 2011; 6(7): e22030.
[http://dx.doi.org/10.1371/journal.pone.0022030] [PMID: 21818288]

[190] Luo G, Ibrahim AS, French SW, Edwards JE Jr, Fu Y. Active and passive immunization with rHyr1p-N protects mice against hematogenously disseminated candidiasis. PLoS One 2011; 6(10): e25909.
[http://dx.doi.org/10.1371/journal.pone.0025909] [PMID: 22028796]

[191] Capilla J, Clemons KV, Liu M, Levine HB, Stevens DA. *Saccharomyces cerevisiae* as a vaccine against coccidioidomycosis. Vaccine 2009; 27(27): 3662-8.
[http://dx.doi.org/10.1016/j.vaccine.2009.03.030] [PMID: 19464548]

[192] Ribeiro AM, Bocca AL, Amaral AC, *et al.* HSP65 DNA as therapeutic strategy to treat experimental paracoccidioidomycosis. Vaccine 2010; 28(6): 1528-34.
[http://dx.doi.org/10.1016/j.vaccine.2009.11.062] [PMID: 20045500]

CHAPTER 6

Antiviral Drugs - in Past and Current Pandemics

Arijit Nandi[1,†], **Anwesha Das**[2,†], **Wei-Yu Lin**[3] and **Payal Chakraborty**[4,*]

[1] *Dr. B.C. Roy College of Pharmacy and Allied Health Sciences, Durgapur-713206, West Bengal, India*

[2] *Drug Design and Medicinal Chemistry Laboratory, Department of Pharmaceutical Chemistry, School of Pharmaceutical Education and Research, Jamia Hamdard, New Delhi 110062, India*

[3] *Department of Medicinal and Applied Chemistry, Kaohsiung Medical University, Kaohsiung, Taiwan*

[4] *Department of Pharmaceutical Technology, JIS University, 81, Nilgunj Road, Kolkata – 700109, W.B. India*

Abstract: Antiviral drug discovery and its developmental processes happen to be the need of the hour. The break-out burden of complications and mortality caused by viruses like Influenza, Ebola, MERS, SARS and presently, the subtype SARS-CoV-2 are randomly growing in an exponential arc. Under such critical circumstances, there occurs an urgent paradigm shift in the research domain where antimalarial drugs like Hydoxychloroquine were given as a prophylactic treatment to improve the condition of the patients affected by the SARS-CoV-2-COVID-19 disease. For the use of emergency purposes in this global pandemic, a ground breaking development has taken place in vaccine therapy with mRNA-based technologies by pharma giants like Pfizer-BioNTech, Moderna Inc. and AstraZeneca Plc. All three newly launched successful mRNA vaccines, like Comirnaty, mRNA-1273 and AZD1222, in their late-phase clinical trials showed an effective rate up to ~ 95%. Many alternative approaches use translational medicines and artificial intelligence tools to mitigate clinical morbidities within a given timeframe. Hence in this particular book chapter, we tried to highlight the pros and cons of all the possible antiviral drug interventions and strategies that have been implemented from the past till the present to combat several epidemics and global pandemics. At present, the occurrence of the COVID-19 pandemic imposed a greater threat and unprecedented challenge in the antiviral drug discovery platform that needs to be focused on in detail.

Keywords: Antigen, Antiviral drug interactions, Artificial Intelligence, Baltimore Classification, Convalescent Plasma, COVID-19, Immunity, Immunoglobulins, Interferons, Monoclonal antibody, Mutation, mRNAs, Receptor, Self-amplifying RNA, Severe Acute Respiratory Syndrome (SARS), Small interfering RNA, Spike protein, Translational medicine, Vaccines, Virus life cycle.

* **Corresponding author Payal Chakraborty**: Department of Pharmaceutical Technology, JIS University, 81, Nilgunj Road, Kolkata – 700109, W.B. India E-mail: payal.chakraborty@jisuniversity.ac.in

M. Amin-ul Mannan & Gaurav Kumar (Eds.)

INTRODUCTION

The severe acute respiratory syndrome coronavirus-2, formally known as SARS-CoV-2, was identified with unique protein characteristics, *i.e.*, spike protein on the surface of the virus and central nucleocapsid protein, which holds the viral genome. These features are particularly linked to high fatality rate as well as their ability to move at ease from animal to human hosts. The emergence of the current global pandemic of COVID-19 caused by the novel SARS-CoV-2 has resulted in an estimated 2,553,081 deaths worldwide to date. It imposed an unprecedented challenge to uncover the viral specificity and identify effective drugs for their prevention, cure and treatment. Due to this emergency, the development and procurement of safe and effective therapeutics were under accelerated clinical trials. Thus, in this chapter, we have elucidated in detail the possible interventions and strategies in the pipeline to cope with the current crisis that waits for fast-track screening and approvals by the regulatory authorities on a larger scale.

Different Types of Viruses

Viruses are sub-microscopic infectious agents which are mainly classified based on their morphology and genetic characteristics [1]. With the aid of two methods, *i.e.*, electron microscopy and X-ray crystallography, morphological symmetry of the viral capsid can be viewed and based upon such findings, it can be classified into helical and icosahedral symmetry types. In the helical symmetry, the viral nucleic acid acquires a helical form inside the helical protein coat, having a molecular weight of 18000 Dalton (Da) (exceptionally up to ~ 2 million Da in the case of Tobacco Mosaic Virus) which is made up of 2130 units of single protein such as Mumps virus and Measles virus. Whereas in the case of icosahedral symmetry, the molecular weight ranges between 15000-130000 Da of the structural protein capsid. Each icosahedron consists of 12 combinations of capsomere (*i.e.*, groups of structural proteins which form the viral capsid). Every capsomere contains 5 structural proteins located at the vertex of the icosahedral capsid, with an additional varying number of capsomeres containing 6 structural proteins at the facet and edges of the capsid, *e.g.*, Enterovirus and Rubella virus [2]. Further, based on their genetic characteristics, viruses were broadly classified as DNA viruses and RNA viruses. All viruses were specifically grouped into 7 families based on their type of genome and their method of replication by the American biologist David Baltimore in the year 1971, which is popularly known as the "Baltimore classification" (Table **1**) [1, 3].

Life Cycle of a Virus

Viruses are obligate intracellular parasites; they need a host that can either be a prokaryote or eukaryote to replicate. In a prokaryote, the life cycle of a typical bacteriophage takes place mainly *via* 3 mechanisms: lytic cycle, lysogenic cycle and transduction. In the lytic cycle, the virus destroys the host cell to reproduce new bacteriophages. In the lysogenic cycle, instead of killing the host, the viral genome integrates with the bacterial chromosome and reproduces new viruses (Fig. **1**). At the end of the lytic cycle or during the lysogenic cycle, in some cases, viruses can also inject the former host's DNA into the newly infected host by a process known as transduction. In this process, the virus often packs the host genome within its capsid during its separation from the host, which causes an alteration in their genetic characteristics and DNA recombination might occur within the newly attacked host [4].

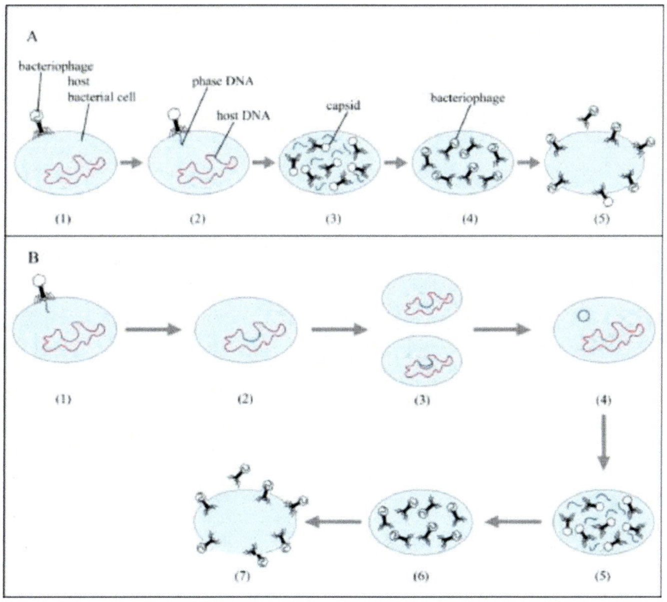

Fig. (1). Lytic **(A)**: (1) Attachment of the virus to the bacterial cell surface, (2) Penetration of the viral DNA into the bacterial cell wall, (3) Biosynthesis of all viral components and degradation of the bacterial chromosome, (4) Assembly of all the viral particles to form a matured virus, (5) Lysis of bacterial cell wall and release of the virus from the host bacteria, and Lysogenic **(B)** Cycle of Bacteriophage: (1) Attachment and Penetration of the phage DNA into the bacterial cell wall, (2) Integration of the phage DNA with the bacterial chromosome, (3) Reproduction of the bacterial host which also contains the genetic characters of phage, (4) Excision of the phage DNA from the host chromosome and entering into the lytic cycle, (5) Biosynthesis of all viral components and degradation of the bacterial chromosome, (6) Assembly of all the viral particles to form a matured virus, (7) Lysis of bacterial cell wall and release of the virus from the host bacteria.

Table 1. Baltimore classification of viruses.

Classes of Viruses	Examples
Double-stranded DNA viruses.	T4-like bacteriophage viruses, P1-like bacteriophage viruses
Single-stranded DNA viruses	Circoviruses, Gyroviruses
Double-stranded RNA (dsRNA) viruses	Chrysoviruses, Systoviruses
Single-stranded RNA viruses whose RNA is identical in base sequence to mRNA	Arteriviruses, Coronaviruses
Single-stranded RNA viruses whose RNA is complementary in base sequence to mRNA	Bornaviruses, Ebola-like viruses
RNA reverse transcribing viruses	Retroviruses, Hemiviruses
DNA reverse transcribing viruses	Orthohepadnaviruses, Badnaviruses

Host–Virus Interactions

When a virus infects a host, it mainly interacts by three mechanisms: apoptosis, cell-signalling and immune response (Fig. **2**). The most common form of host-virus interactions is mediated by apoptosis, which leads to acute cytolytic infections by the massive destruction of the host cells. Cytolytic infections mainly occur in 3 possible ways, *i.e.*, when the viral antigens get triggered by (i) host cell lysis, (ii) before cell death, (iii) before the complete multiplication of the virus [5]. Another common type is cell-signalling, where virus-specific antigens and several viral progenies are produced in the host cell, causing an either persistent or latent viral infection. Sometimes in response to the host-virus interaction, due to underlying cellular mechanisms, the virus might also get cleared or prevented from entering the host cell by any other alternative infectious processes [6]. Lastly, through a wide range of innate or adaptive immune responses cascade of host-virus interaction occurs. Several rapid innate immune responses are triggered by a receptor named Pattern Recognition Receptor, where virus-infected host cells are directly killed by the natural killer (NK) cells or disposed of within the macrophages by the process of phagocytosis. Even the rate of viral multiplication declines by producing interferons (soluble factors) that acts on the neighbouring infected host cells. In the case of the adaptive immune response, the significant components are T and B lymphocytes which are primarily essential for cell-mediated and humoral immunity, respectively [5].

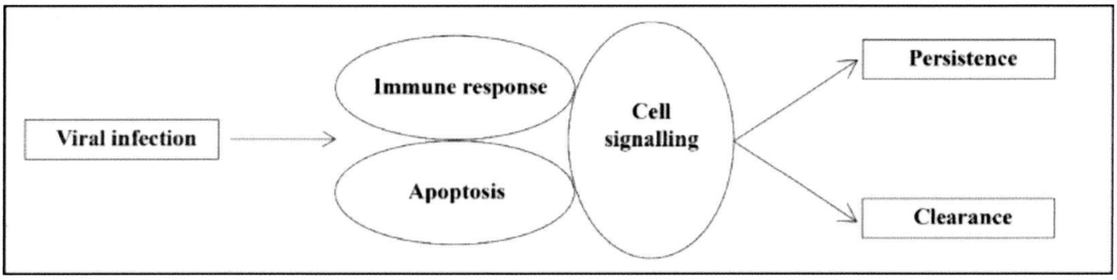

Fig. (2). Basic types of host-virus interactions.

DEMOGRAPHY AND EPIDEMIOLOGY OF PAST AND PRESENT VIRAL OUTBREAKS

The emergence of popular viral pandemic outbreaks in the ancient world occurred in 165-180 AD as the "Antonine Plague" which affected the population of Asia Minor, Egypt, Greece and Italy. The causative agent was either smallpox or measles virus. Its origin was believed to be Seleucia of the ancient world, and the death toll was counted as approximately 5 million [7]. The next eminent viral outbreak was the Japanese smallpox epidemic in the year 735-737 AD, which was caused by the Variola virus that mainly originated in Korea. The death toll was almost up to ~1 million [7]. This continued with consecutive waves of viral pandemic and epidemic outbreaks across the world in the 7th, 10th, 11th, 15th, 16th, 17th, 18th and 20th centuries [8]. In the year 1878, the yellow fever epidemic took up to 1.5 lakh lives in the United States of America (USA) [7]. This outbreak again hit the world in the year 2016 [9]. In the current scenario, the most common and well-known form is the Influenza pandemic which can be divided into several phases. In the first phase, the Russian flu (also known as Asiatic flu) outbreak occurred in 1889, which was caused by the H2N2 Influenza virus taking up to 1.1 million lives. A similar outbreak occurred after 59 years, named Asian flu, where the death toll was ~1 million. In the second phase, another outbreak took place in the year 1900, which was caused by the H3N2 influenza virus. The next phase occurred almost after 68 years, named as Hong Kong flu, where the death toll was again 1 million. In the third pandemic wave, the outbreak of the Spanish flu in the year 1918 was caused by the H1N1 influenza virus that took 40-50 million lives. Again, we found that there was a recurrence of the Russian flu in the year 1977 [1]. In the year 2004, Bird flu took place in which the causative agent was H5N1 [10] and then later in the year 2013, another avian influenza occurred in which the causative agent was H7N9 [11]. During 2009, Swine flu took place in which the causative agent was mainly the H1N1 virus, which further included viral strains like H1N2, H2N1, H3N2, H2N3 and H3N1 [12]. Interestingly, the advent of the Human Immunodeficiency Virus (HIV) causing Acquired Immunodeficiency

Syndrome in the year 2006 also became a pandemic condition [13]. The first case affected by HIV was reported in 1981 and caused 36 million deaths [7]. Since then the concurrent outbreaks due to Zika (year: 2013-15) [14], Ebola (year: 2014-16) [15], and Nipah (in the year 2018) viruses [16] have been widely distributed in the regions of Central Africa, South America and South-East Asia, respectively.

The present ongoing pandemic named COVID-19 had its origin in Wuhan, China. The current death toll has already risen to 696 thousand worldwide [7], and the causative agent is the well-known SARS-CoV-2 (Fig. **3**). Previously, another 2 outbreaks also came across different regions of the world; and these outbreaks can be divided into 3 phases:

• In the first phase, the SARS outbreak occurred in 2002-2003 in several Asian countries, and the death toll was 774.

• In the second phase, Middle East Respiratory Syndrome (MERS) outbreak occurred in 2012 in the Middle East region of the Asian continent, and the death toll was 866.

• And till now, the last phase included the novel Coronavirus outbreak that turned into the current pandemic COVID-19 [17].

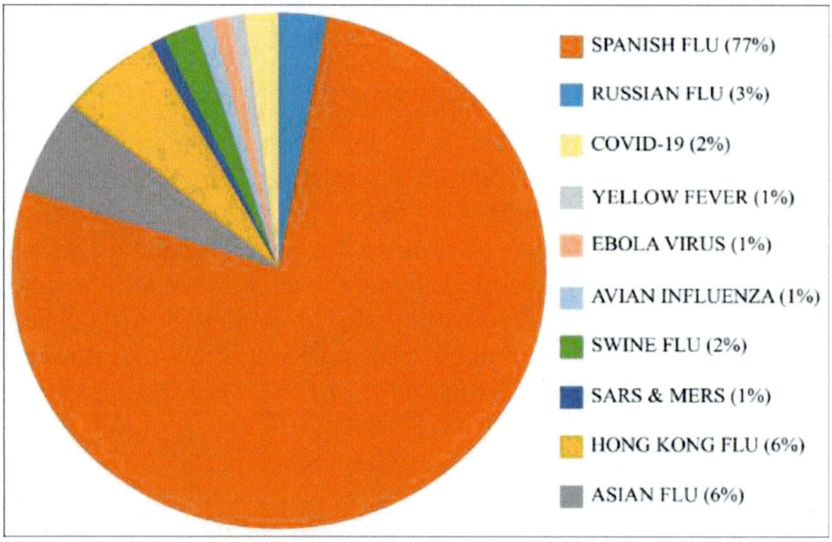

Fig. (3). Representative chart demonstrated the rate of major viral outbreaks.

ANTIVIRAL DRUG THERAPEUTICS - RESEARCH & DEVELOPMENT

Natural Products

The growing interest in natural products as sources of new drug moieties can be attributed to many factors. These include urgent therapeutic needs with adequate bioactive natural products that can be easily implemented as biochemical and molecular probes. As listed below in Table **2**, a wide range of different active constituents and secondary metabolites were screened as promising therapeutic agents against different classes and types of viruses.

Table 2. Common phytochemicals used in antiviral therapy.

Chemical Class	Active Ingredients	Virus	Mechanism of Actions	Natural Source	Refs.
Flavone	Pinocembrin	Zika virus (ZIKV)	Inhibits viral RNA production and envelope protein synthesis	*Turnera diffusa, Boesenbergia rotunda,* Honey, Propolis	[18]
Flavone	luteolin and hispidulin	Influenza A virus	Inhibition of glycoprotein neuraminidase in the H1N1 strain	*Salvia plebeian R. Br*	[19]
Isoflavone	Genistein	Herpes B Virus (HBV)	Blocks the late phase event in the life cycle of the Herpes B virus. Mainly inhibits the phosphorylation of glycoproteins	*Genista tinctoria*	[20]
Alkaloid	Dendrobine	Influenza A virus	Restraining the viral replication	*Dendrobium nobile*	[21]
Triterpenoid	3, 4-secodammarane and Dammarenolic acid	HIV	Inhibit the cell cycle proliferation	*Aglaia sp.*	[22]
Lignin	Honokiol	Dengue virus serotype-2 (DENV-2)	Inhibit the synthesis of dsRNA and nonstructural proteins	Genus Magnolia	[23]
Glycoprotein	Apolactoferrin (a synthetic form of lactoferrin) and lactoferrin	SARS-CoV-2	Interacts with heparan sulfate proteoglycans and angiotensin-converting Enzyme-2 (ACE-2) receptors of human	Bovine milk and other body fluids	[24]

There have been clinical trials involving natural products against SARS-CoV-2. Further, this segment of antiviral drug therapy needed wide exploration along with methodical isolation of the active ingredients. The specific mode of action, in particular for any isolated active ingredients, *i.e.*, used for different therapeutic categories, must be well approved by the drug regulatory authorities.

Chemical Entities and Different Types of Formulations

Chemically antiviral drugs are small entities that are used in widespread infectious diseases. A specific antiviral drug target can be designed based on any targeted receptor of the host cells or the enzymes needed for viral replication. Here we elucidated the classification of the antiviral drugs in detail:

Viral DNA Synthesis Inhibitors like Acyclovir, Valacyclovir, Maribavir and Letermovir

Acyclovir is an acyclic analogue of deoxiguanosine and Valacyclovir is an L-valyl ester prodrug [25]. Maribavir is a novel class of Benzimidazole riboside [26], whereas, Letermovir is an acetic acid derivative [27]. Acyclovir is very much effective against Herpes Simplex virus (HSV-1 and HSV-2) strains [25], Varicella-Zoster Virus (VZV) [28] and Epstein - Barr virus (EBV) [25]. Valacyclovir showed its effectiveness by eliminating high loads of herpes viruses in deep periodontal pockets and inflamed gingiva [29]. In contrast to Acyclovir, we find Maribavir [26] and Letermovir [27] are active against Cytomegalovirus (CMV). Exceptionally upon the other commonly used anti-CMV drugs like Ganciclovir, Foscarnet and Cidofovir, we find that Maribavir showed potent *in-vitro* activity against the resistant CMV strains within an acceptable therapeutic index of the drug [26]. These drugs show their action by the inhibition of the viral DNA polymerase. Notable exceptions were only seen in the cases of Maribavir and Letermovir, which inhibits the protein kinase UL97 and DNA terminase complexes respectively [26, 27].

Viral RNA-dependent RNA Polymerase (RdRp) Inhibitors such as Favipiravir, Remdesivir, Galidesivir and Ribavirin

Favipiravir is a pyrazinecarboxamide derivative [30] whereas Remdesivir [31], Galidesivir [32] and Ribavirin [33] are all nucleoside analogues. Favipiravir showed effective antiviral activity and reduced mortality rate against the positive- and negative-sense single-stranded RNA viruses, neuraminidase and M2 protein inhibitor-resistant strains, *i.e.*, specific for Influenza viruses [34] along with Lassa, Ebola [35] and Nipah [36] Viridae families. It can also be used in combination with Ribavirin against Asian and African strains of ZIKV [37]. Ribavirin is the only drug that was used against Lassa fever [38]. It was safe in geriatric patients on usage against Hepatitis E infection [39]. It majorly inhibits influenza A and B viruses [40], SARS and MERS Coronaviruses [41]. Hence it was extensively used in the initial phases of the COVID-19 breakout in China [41]. Remdesivir showed promising results and is active against RNA viruses like Corona, Ebola, and Nipah [42]. It was found even effective against the resistant strains of Coronaviruses [31]. Galidesivir is active (activity studied *in-vitro*) against several RNA viral pathogens that include Filo, Corona, Hepatitis C and Yellow Fever-causing viruses [32]. All of these drugs act by inhibiting the viral RdRp, which is responsible for the replication of viral RNA-based genomes (Fig. **4**) [30 - 33].

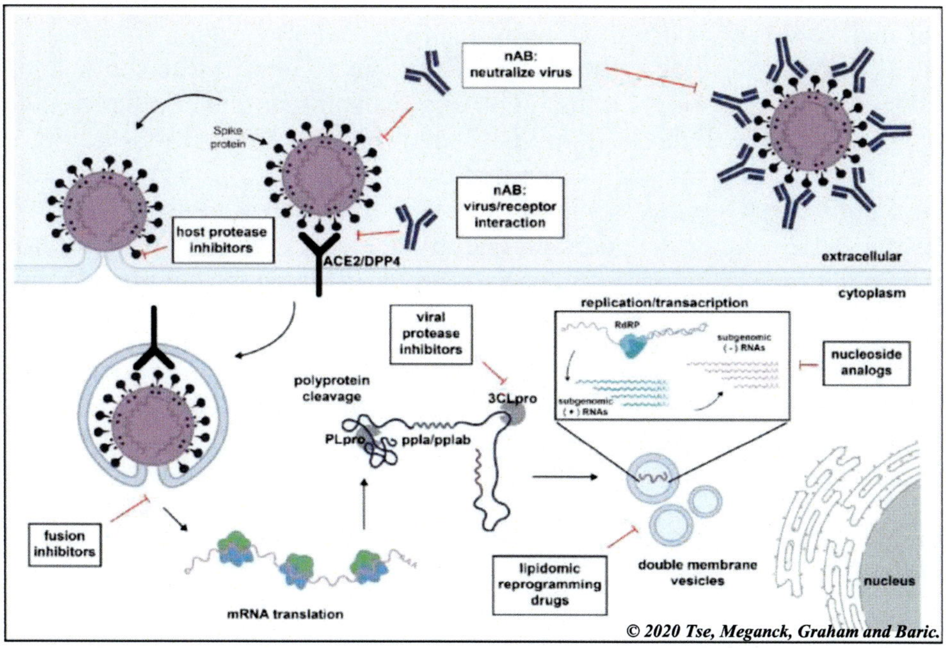

Fig. (4). Schematic Representation of Multiple Antiviral Drug Targets (mostly specified for COVID-19).

Viral RNA Synthesis inhibitor - Sofosbuvir

Sofosbuvir is a nucleotide analogue that is active against DENV, ZIKV and Hepatitis C Viruses (HCV). It blocks viral RNA synthesis by inhibiting RNA polymerase. It also induces antiviral action *via* increased mutation of alanine to glycine in the viral genome [43].

Fusion Inhibitor - Arbidol

Arbidol (ARB) is an indole-based broad-spectrum antiviral agent that mainly impairs several steps in the life cycle of the viruses, including virus attachment to cells, the fusion of viral and cellular membranes during viral entry. The drug is active against different Viridae families, including orthomyxo, paramyxo, picorna, bunya and rhabdo. ARB was also found effective against several ZIKV strains and protected the host cell from severe cytopathic effects (CPE) mostly caused due to the virulent attack of African and Asian strains of ZIKV [44].

Endonuclease Inhibitor - Baloxavir

Baloxavir marboxil is a prodrug, which is active as Baloxavir acid. Baloxavir is a substituted pyridine derivative that is highly active against influenza A and B, avian influenza subtypes H5N1 and H7N9, and resistant strains of neuraminidase inhibitor antiviral drugs like Oseltamivir, Zanamivir and Laninamivir *etc*. It acts by selectively inhibiting the cap-dependent endonuclease, which is essential for the initiation of mRNA synthesis, mainly in cases of influenza viruses. Hence, this drug shows a synergistic effect and significantly decreases the mortality rate when used in combination with other resistant antiviral drugs like Oseltamivir [45, 46].

Viral Protease Inhibitor –Lopinavir

Lopinavir is a valine derivative and a protease inhibitor that inhibits HIV-1 protease and 3-Chymotrypsin-like-Proteases (3CLpro) enzyme complex, *i.e.* , responsible for the viral RNA processing in the SARS-CoV-2 [47, 48].

Inhibitors of Viral Penetration into the Host Cell such as Amantadine and Rimantadine

These drugs are cyclic primary amines. Amantadine is active against influenza strains of H3N2 and H1N1. It can downregulate Cathepsin L1, which is a lysosomal cysteine protease, that can cleave the SARS-CoV-2 specific Coronaviral spike proteins and further inhibit the viral entry to the host cell [49, 50].

Neuraminidase Inhibitors for e.g., Oseltamivir, Zanamivir, Peramivir and Laninamivir

These drugs are sialic acid derivatives and are used basically against both Influenza A and B viruses. Here, Oseltamivir significantly decreases the load of mortality affected by Influenza, whereas, Zanamivir specifically prevents the viral release from the host cells. They primarily act by inhibiting the viral

Neuraminidase, a transmembrane protein that is responsible for the activation of the Influenza virus [51 - 53].

All the above-mentioned drugs with a different mode of action are clinically approved by the different drug regulatory bodies for their extensive use as antiviral drugs.

Biologicals – Vaccine & Sera Therapy

Vaccines are biological products that are capable of generating acquired immune responses against any infectious disease. Other emerging biological products include convalescent plasma, monoclonal antibodies and interferons, which are on trials that can modulate subjects' immune responses. In parallel, there is a rapid development of vaccines undergoing different clinical trial phases I, II and III to combat the desired SARS-CoV-2 pandemic outbreak.

Types of Biological Products

Vaccines

Vaccine is the best alternative to fight against any viral diseases. There are several types of vaccines that can be used against a wide range of viral diseases.

Inactivated Vaccines

Inactivated vaccines contain the inactivated viral form. There are techniques like UV rays or chemicals such as formalin, formaldehyde and β-propiolactone *etc.*, to inactivate the virus. In some cases, the occurrence of hypersensitive reactions was reported due to the usage of adjuvants like alum or formaldehyde in the vaccine preparation. For example – mice were immunized with double inactivation vaccines having alum as an adjuvant; there was an increased infiltration of eosinophils, neutrophils and other inflammatory cells in the lungs [54]. Thus, an increased rate of infiltration aggravated the hypersensitivity reactions *via* T helper immune cell responses. By replacing alum with toll-like receptor agonists, hypersensitivity reactions could be alleviated [55].

Live Attenuated Viral Vaccines

Live attenuated viral vaccines are created by reducing the virulence property of a virus. These vaccines can produce strong T and B cell immune responses [56]. One of the notable examples under this category is the development of genetically exhausted coronavirus vaccines by either excising or mutating different viral proteins. For example, "rSARS-CoV-ΔE" is a live attenuated SARS-CoV vaccine against complete pulmonary viral replication in which the proteinE is excised. This vaccine was formulated in the intranasal form using the *in-vivo* Golden Syrian hamster model [57].

Protein-based Subunit Viral Vaccines

Protein-based subunit vaccines can be defined as fragmented or isolated proteins of viruses to produce an acquired immune response. They have low immunogenicity. A few examples of protein-based subunit vaccines are different forms of Coronaviral spike proteins, like Receptor-binding Domain (RBD) and RBD-Fc (fusion of RBD and Fc region of human IgG). They are used as subunits for immunization in Non-Human Primate (NHP) models that can neutralize the produced antibodies to generate subsequent immune responses. This finding was further extended to clinical trials [58 - 61].

VLP and Nanoparticle Vaccines

Coronavirus-like particles (VLPs) are produced from the cells co-expressing the spike (S) protein, membrane (M) protein and envelope (E) protein. Different novel methods, such as the baculovirus infection cell system, were used to produce high levels of recombinant or chimeric VLP vaccine that contain the S protein along with avian influenza matrix 1 protein. These chimeric VLPs are of the same size and structure in comparison to the wild-type Coronaviruses and showed effective results against SARS-CoV-2 [62, 63].

Vector-based Vaccines

Vector-based vaccines contain a chemically weakened live virus, used as a vector to produce a specific immune response. They have some advantages over the use of conventional protein or DNA-based subunit vaccines, such as they are more coherent at delivering the genetic materials into cells [64]. There are a wide variety of vector systems established against multiple infectious diseases such as Measles Viruses, Modified Vaccinia Virus Ankara (MVA) [65], Chimpanzee Adenovirus (ChAdOx1) [66] and Para Influenza Virus type 3 [67].

Newly Emerged Approaches – mRNA, siRNA and SAM-based Vaccines

The latest intervention of using a full-length nucleic acid in the vaccine therapy might be a groundbreaking discovery in the arena of immunization. Here we highlight all the possible approaches that have been implemented to date, like mRNA, self-amplifying mRNA (SAM) and small-interfering RNA (siRNA) vaccines. In the current COVID-19 pandemic situation, the thorough quest for a safe and effective therapy to encounter the health crisis showed a success rate of the clinical outcome by the newly emerged mRNA vaccines. Mainly, there are two major types of mRNA vaccines that are developed against infectious pathogens: non-amplifying or non-replicating mRNA vaccines and self-amplifying or self-replicating mRNA vaccines. The non-replicating mRNA vaccines do not include additional protein sequences to induce self-amplification of the inserted mRNA sequence. After being delivered into the host cells, this mRNA vaccine is captured by Antigen-Presenting Cells (APCs) at the infection site [68], where the RNA sequence present in the vaccine is translated into proteins as antigens that further stimulate immune response [69]. Recently in the year 2020, Beissert et al. described a strategy that is based on two RNA vectors - one retaining the replicase-encoding gene and the other encoding the antigen called the trans-amplifying approach. Though it does not produce significant immune responses, it improved the mRNA vaccine technology through various *in-situ* modifications [70]. As the mRNA is fragile and there is a possibility of mRNA breakdown into pieces by the human natural enzymes, they are being wrapped into lipid nanoparticles which are delivered as injectables [71].

Self-amplifying mRNA (SAM) vaccine is derived from the backbone of an alphavirus genome. It contains 2 particular genes, where the genes encoding the viral RNA replication remained intact and another encoding the viral structural proteins gets replaced by the antigen encoding gene of the vaccine [69, 70, 72]. After being delivered into the host cells, the viral 'positive sense (+) RNA strand' encodes the gene for the formation of the RdRp complex, and produces a complementary strand of RNA [69, 72]. This newly formed RNA strand gets translated into proteins that further stimulate the immune response. With the aid of RdRp synthesis, these vaccines can generate multiple copies of RNA. Hence, they are named as SAM vaccine [72]. A self-amplifying RNA vaccine can be delivered in the form of plasmid DNA or virus-like RNA particles. These *in-vitro* transcribed RNA forms can elicit substantially stronger immune responses than non-replicating mRNA vaccines [70].

Small interfering RNAs (siRNAs) are double-stranded non-coding RNA molecules containing 20-25 nucleotide base pairs in length. They regulate the expression of genes by RNA interference, where the dsRNA is cleaved into fragments to form the siRNAs, which incorporate the antisense strand into RNA Induced Silencing Complex. The identified siRNAs can easily target the required sequences that code for the viral RNA-dependent RNA polymerase, helicase, proteolytic enzymes, and the nucleoprotein of SARS-CoV, leading to a decrease in the viral loads [73, 74].

Convalescent Plasma

Convalescent plasma therapy was initiated from the blood plasma of infected patients, who have recovered and eventually generated antibodies against the SARS-CoV-2 viral proteins in their plasma sample by the course of their normal immune responses. Despite its sudden use in the time of pandemics, the major limitation of this innovative therapy was that it highly lacked a well-defined and well-designed clinical study protocol which resulted in a low titre value of the available antibodies and thus resulted in the failure of the newly emerged Convalescent plasma therapy [75].

Monoclonal Antibodies

Monoclonal antibodies (mAbs) are prepared by the immune cells that are identical to their parent cells. There are several isolation methods and clinical trials reported for the usage of monoclonal antibodies against RSV, Ebola virus and Coronavirus [76, 77]. A phase II randomized controlled therapeutic trial on humans demonstrated the safety and efficacy of Nivolumab, a mAb used to improve the clinical outcome of hospitalized COVID-19-positive adult obese patients [78].

Interferons (IFNs)

Under the process of pathogenesis, likely the Interferons get released from the host cell showing their immunomodulatory action. It can activate immune cells like natural killer cells, and macrophages with an increase in the major histocompatibility complex (MHC) antigen expression of the host body. In cases of any pathogen or viral attack, the viral proteins are synthesized from the viral nucleic acids inside the host cells and are degraded into viral peptides. These peptides form complexes with MHC molecules known as peptide-MHC complex and travel towards the host cell surface, where it gets recognized by different immune cell responses. It is found that among the three types of IFNs, the IFN-λ is more potent and restricts Coronavirus transmission [79].

Translational Medicines

Translation medicine is the domain that deals with the metamorphosis of a molecular target from bench to bedside drug medicament. The lead generation through high throughput screening, isolation, purification, and Pharmacokinetic/Pharmacodynamics modelling (PK/PD) are major parts of the drug discovery process. After the successful preclinical stages, drug entities are validated as per their standards for safety, quality and efficacy laid down by the different regulatory authorities.

Repurposed Drugs

Drug repurposing can be termed as a "saviour" or an alternative to lifesaving therapeutics in a time of crisis for the newer drugs on the market. Here we like to discuss on few examples which are repurposed to combat the current pandemic SARS-CoV-2 for their antiviral activity and its known safety profile. One of the emerging examples is the 1930's β-D-N4-hydroxycytidine (NHC; EIDD-1931), a ribonucleoside analogue that reduces the viral load by inactivating mutation, thus causing viral error catastrophe against SARS and MERS Coronavirus diseases [80]. There are also two antimalarial drugs, Chloroquine and Hydroxychloroquine, repurposed as autophagy inhibitors in cases of COVID-1--infected patients [81].

Combined Drugs

Repurposing "combination of drugs" can be another alternative in the antiviral research domain due to the unavailability of newer drugs in the market. One of the notable examples in this area is the use of the anti-allergic OTC drug Chlorcyclizine Hydrochloride in Hepatitis C. It showed a synergistic effect when used with other anti-HCV drugs such as Ribavirin, Sofosbuvir, Cyclosporine A and Interferon-α. Such combinations of drugs have a safety profile when used against HCV. It specifically blocks the "late-stage viral entry" into the host but is unable to block the "viral replication, assembly or secretion" in the life cycle of HCV [82].

Prediction of Drug Targets

Viral diseases necessitate the prediction of newer drug targets. One of these predictable drug targets includes a common nucleophilic site that is present in Venezuelan Equine Encephalitis Virus (VEEV) and Coronaviruses. This nucleophilic site is important for undergoing the protein-substrate binding interactions which are required for both VEEV and Coronaviruses. Hence targeting and inhibiting this nucleophilic site by peptide-like inhibitors can be a valid strategy against coronaviral infection [83].

Assay Development

Several assays are being developed by the scientists of the National Centre for Advancing Translational Sciences (NCATS) for predicting numerous drug targets, as well as the development of enhanced and high-tech methodologies to fight against the pandemic COVID-19. The assay developments were based on - "Spike/ACE-2 protein-protein interaction, Spike/ACE-2 binding, ACE-2 binding, ACE-2 enzymatic activity, 3CLpro enzymatic activity and SARS-CoV-2 CPE". Scientists were able to draw a few significant conclusive points from the data obtained from these assays. (i) SARS-CoV-2 inclusive S-protein can't inhibit the biochemical activity of ACE-2. Rather SARS-CoV-2 Spike S1 protein, after interacting with the amino acid histidine, activates the biochemical activity of ACE-2. The impact of SARS-CoV-2 S-protein on ACE-2target and its pros and cons need to be further explored [84]. (ii) In parallel, there have been assays undergoing novel AlphaLISA technology for determining the ability of the antiviral agents (small molecules, antibodies, peptides, *etc.*) to disrupt the Spike protein-ACE-2 interaction; disrupting this interaction might cripple the virulent activity of the SARS-CoV-2. The best candidate drugs obtained from this assay are further tested and counter-screened by the preliminary TruHit assay methods (a known prior similar assay procedure) to verify the results obtained from the AlphaLISA assay [84]. (iii) An assay developed termed as "SARS-CoV-2 CPE reduction assay," which indirectly monitors the ability of the drug candidates like Mefloquine and Hycanthone [85] to inhibit viral replication and infection through various molecular mechanisms [86].

In-Vitro Assay and Laboratory Findings

Correct diagnosis and appropriate treatment of diseases from the laboratory findings is a key factor for any drug developmental process. An arbitrary laboratory finding showed that the activation of the innate immunity system by external factors can treat the mutated strains of any particular virus. This fact was supported by targeting and activation of the host restriction factors [87].

Others

The flaws of viral pandemics could be controlled by a rapid screening of small molecules and biologicals relevant to virology. Here we enlisted the model development studies that can reduce the time duration in the drug discovery pipeline. One of the emerging examples is the current development of Tissue Chips by NCATS in collaboration with the National Institute of Health and the U.S. Food and Drugs Association (FDA) for the screening of drug candidates. The tissue chips are bioengineered devices that can model the structure and function of human organs such as the lungs, liver and heart to predict the safety and toxicity profile of the drugs in humans more rapidly and effectively [88]. A popularly known adaptive-design Human Challenge Model is significantly used, as a drug discovery molecule gets investigated at the end of phase IIa of a clinical trial directly switching from healthy adults to infants through the Accelerated Pathway, *e.g.*, small molecule investigation as an antiviral agent against the Respiratory Syncytial Virus [89].

Therefore, here, in a nutshell the emerging targets and their future in the drug discovery pipeline for a wide range of antiviral therapeutics is an outcome of the translational research. In this scenario, continuous translational methods lead to a) screening of drug repurposing to identify active compounds that reduce disease phenotypes, b) assay development and optimization for high-throughput screening, c) identification of drug targets or disease phenotypes for assay development, and finally d) confirmation of the compound activity by using different *in-vitro* or *in-vivo* assays models. Translational medicine uses what can be gleaned from pre-clinical efforts to clinical studies to sharpen and improve the discovery of new medicines.

ADVANCED APPROACHES IN ANTIVIRAL THERAPY

Insights into Advanced Immunotherapies

A wide range of diversities in the genetic and antigenic properties of SARS-CoV-2 imposed a greater challenge to newer antiviral drugs under process. Recently, host-oriented therapeutic interventions were implied to treat viral attacks. Understanding the different types of host-virus interactions forms a progressive platform in the developmental process of various innovative immunotherapies [27]. Immunotherapy can be broadly classified into 2 categories, *i.e.* active immunotherapy, which is solely based on vaccines and passive immunotherapy, which involves direct infusion of the specified antibodies [27]. Here we like to give a brief overview of the current approaches to active and passive immunity, which are yet to be explored and approved for large-scale medical usage.

In the case of active immunity, the restoration and activation of T-cells at the infection site can treat several infections. But sometimes, virus-infected cells can find several ways to mislead them, either by disguising themselves as healthy cells or by sprouting so many antigens on their surface that the T cells get misled and cannot mount an effective attack. To treat these infected cells, Chimeric Antigenic Receptor (CARs) genes are inserted into the T cells which were previously collected from the patients. The gene of the CARs contains inactivated forms of viruses that can provide genetic information to the T cells. After the insertion into the patient body, the CAR-derived T cells can bind with the specific viral envelope proteins and kill the infected cells. These genetically engineered CARs can be used as a novel innovative therapy against HBV, HCV, influenza, HIV, CMV and EBV infections [90]. Another notable advancement includes the implementation of bispecific antibodies where two antibody fragments can simultaneously bind with viral and an immune cell antigen. For example, a bispecific antibody having single-chain variable fragment binding simultaneously with HIV envelope protein and CD4 cell, producing $CD4^+$ T cell immune responses [91].

In passive immunity, antibodies can easily bind with targets like (i) viral envelope proteins or (ii) viral receptors that can block the entry of a virus to the host cell by targeting specific lipid or protein domains which can also inhibit the development of viral resistance [92]. One such lipid domain is phosphatidylserine which gets targeted by a monoclonal antibody, *i.e.* , Bavituximab. It can inhibit the replication of HCV, as in HIV/HCV co-infected patients [93]. These antibodies are directly infused into the cytosol *via* several peptides commonly known as Cell-penetrating peptides, such as HIV-derived Trans-activator of Transcription peptide along with human-derived fusogenic peptides, which facilitates the cytosolic delivery system [94].

Ongoing Clinical Trials - New Antiviral Therapies

Ebola

There was an urgent need for vaccine development during the Ebola virus outbreak. Some of the Ebola virus vaccines are listed below. "Ad26.ZEBOV", an Ad26-based monovalent vaccine, can express glycoprotein of the Mayinga variant of the Ebola virus. As the Ebola virus has multiple strains, hence the development of a multivalent vaccine was essential. For example, the development of MVA was named "MVA-BN-Filo®". It contains Zaire strain of Ebola Virus, Sudan strain of Ebola virus, the Marburg Virus glycoprotein and Tai Forest strain of Ebola virus nucleoprotein. Currently, the combinations of these two vaccines are in a phase III clinical trial named "EBOVAC-Salone" [95].

Influenza

Currently, the development of the live, attenuated recombinant avian flu influenza vaccine against the H1N1 strain is under phase I clinical trials [96]. Introduction of the monovalent vaccines against the Influenza A H5N1 strain and quadrivalent vaccines popularly named as "FLULAVAL" and "FluMist QUADRIVALENT" that target two different Influenza A and B strains [97] are also marketed and delivered in the form of nasal spray between 2 to 49 years of age group [98].

SARS-CoV-2

The massive outbreak of COVID-19 in the year 2020 led to the rapid rollout of vaccines involving mRNA technology from the front-runners, including Pfizer Inc. Moderna Inc. and the University of Oxford in partnership with AstraZeneca Plc. The Pfizer and Germany-based company BioNTech collaboratively named their vaccine Comirnaty, showing a 95% of efficacy rate and was authorized for emergency use in the continents of the USA and Europe. These were followed with another mRNA vaccine named mRNA-1273, jointly developed by Moderna Inc. and the National Institute of Health with an efficacy rate of 94.5% that got its approval for use in Switzerland, United Kingdom, U.S.A and Europe. A third successful and emerging candidate in line was the Chimpanzee Adenovirus vector-based mRNA vaccine (AZD1222) with an effective rate of 82.4%, which was developed by the University of Oxford in partnership with AstraZeneca Plc. This got clinically approved in the UK, Europe and in countries like India (as marketed in the name of Covishield) for emergency purposes [71]. In parallel, the Imperial College of London worked with another alternative novel messenger RNA technology. This was popularly known as the SAM vaccine, which has a self-amplifying feature and is aimed at a much smaller dose to induce immunity. Unfortunately, this SAM vaccine technology is unlikely to go ahead with a late-stage trial in the U.K. to test its experimental Covid-19 shot. Currently, it was abandoned due to lackluster test results after the completion of phase I and II clinical trials [99]. There were also other limiting factors of the self-amplifying RNA technology over the fast track approved three mRNA vaccine candidates that had access to significant funding and already crossed their finish line. The SAM vaccine contained larger lengths of RNA and lacked the possibility of improvement in their mRNA technology by methods such as nucleoside modifications, stabilizing sequences, and codon optimization of the entire replicon gene. It was also not cost-effective in terms of technical and long-term storage facilities [70]. Apart from the vaccine therapy, effective drug therapy combinations cited with ACE receptor inhibitor losartan, DPP4 inhibitor Linagliptin, Arbidol and Ritonavir are presently in phase IV clinical trials [100, 101] to treat the COVID-19 pandemic, *i.e.*, caused by SARS-CoV-2.

Application of Artificial Intelligence

Artificial intelligence (AI) refers to the effective and prolonged simulation of human intelligence and mimicking their actions. With the use of AI, we can estimate the outbreak severity, detect the causative pathogen, analysis and data interpretation of vaccine/drug development process, and even can predict possible upcoming future outbreaks.

There are major applications of AI in the field of biomedicine:

Diagnostic Tool

Various AI systems can predict models for viral detection and diagnose different pathological states. An NVIDIA-built AI can detect visual signs of the diseases [102], whereas GOOGLE's AlphaFold can predict SARS-CoV-2-associated proteins [103].

Drug Development

AI can be used for the detailed characterization of different chemical moieties, their synthesis and computational modelling using the deep ML-based analysis platform that can form the molecular discovery loop. *E.g.*, Folding@home is a computer project based on protein dynamics simulation [104].

Immunotherapy

Using the reverse vaccinology tool, Vaxign-scientists can predict and design the vaccine targets [105]. A fully AI-dependent influenza vaccine got developed and is currently under clinical trial as per USFDA [106]. In these recent years, AI has become a major tool to develop new personalized vaccines or agents for prophylactic or therapeutic purposes [107, 108].

Clinical Trial

AI helps to generate data-driven protocols and virtual designing for the purpose and conduct of clinical trials, such as IBM Watson developed a clinical trial matching that can find patients' eligibility criteria for any particular trial by using the database of patient's medical record [109]. Finally, with the help of different machine learning methods, several AI-based databases are still functioning with the updated and modified algorithms in the field of immunology, such as, *e.g.*, Uniprot© database can provide data based on protein sequence along with their functional and structural information [107]. Even the virtual screening of targeted drugs for their wide range of purposes is performed through the AI platform using the DeepScreening database [110].

CONCLUSION

The cutting-edge challenge in the past and present viral pandemics to meet the demand for an appropriate antiviral therapy (it may be a drug or any biological product) has always been a revolutionary step in the drug discovery process. The safety and effectiveness of any antiviral therapeutic intervention were always estimated under rigorous Phase I, II, III and IV stages of the clinical trials, which finally led to its approval status by the different drug regulatory bodies. The wide application of chemical drug moieties having multiple antiviral drug targeted receptors on the host cell likewise RNA polymerase inhibitors such as Favipiravir, Remdesivir, viral protease inhibitor – Lopinavir or the most popularly known drugs like the DNA synthesis inhibitors - Acyclovir, Valacyclovir has shown its potency on the outcome of the disease and effectively limited the rate of morbidity. Long-term usage of these biosimilars leads to a higher rate of drug resistance with rapid viral mutations. Many times even to inhibit the stage of secondary infection, inappropriate empirical usage of broad-spectrum antibiotics leads to the worsening of the morbid conditions. The current SARS-CoV-2 pandemic outbreak deteriorated into a crisis where there is an immediate requirement for a validated and successful antiviral intervention to combat COVID-19. The need of the hour is made to explore newer alternative therapies *via* repurposing, different combinations of drugs and predicting susceptible drug targets. Even the primitive use of different phytochemicals from natural resources was used on a trial-and-error basis as potent immune boosters showing positive outcomes in fewer cases. Development of immune responses *via* vaccines as the source for active immunization or the direct intervention of targeted mAbs, the use of convalescent plasma therapy as part of passive immunity became the main approaches for host-oriented antiviral therapies. Scientists are yet closer to their groundbreaking achievements with the advanced development of mRNA and self-amplifying mRNA vaccines. Drastic endorsements with advanced *in-vitro* assay development tools and techniques like AlphaLISA technology, implementation of Artificial Intelligence on a wider scale for accelerated clinical studies, monitoring immune responses and bioinformatics tools paved the pathway for authentic diagnosis and appropriate treatment options. In the current outbreak of a severe viral pandemic, we see that under different ranges of clinical aspects and co-morbidities, with diverse antiviral therapeutic applications; overall, there are mixed responsive outcomes on the cure rate, which kindles "a ray of hope" on the ongoing drug discovery pipeline.

CONSENT FOR PUBLICATION

Not applicable.

CONFLICT OF INTEREST

The author declares no conflict of interest, financial or otherwise.

ACKNOWLEDGEMENT

The authors would like to thank Ms. Tiasha Das for her contribution in formatting and editing the figures as mentioned in the book chapter.

REFERENCES

[1] Cann AJ. Principles of molecular virology. London: Academic Press 1993; pp. 1-308.

[2] Norrby E. The morphology of virus particles. Classification of viruses. In: Textbook of Medical Virology Butterworth & Co (Publishers) Ltd 1983; 4-16.

[3] Baltimore D. Expression of animal virus genomes. Bacteriol Rev 1971; 35(3): 235-41. [http://dx.doi.org/10.1128/br.35.3.235-241.1971] [PMID: 4329869]

[4] Nina P, Schneegurt M, Tu A-HT, Lister P, Forster BM. The Viral Life Cycle - Microbiology | OpenStax. Microbiology. 2016 pp. 251–60. 1996.

[5] Klimpel G. Immune Defenses - Medical Microbiology - NCBI Bookshelf. University of Texas Medical Branch at Galveston.

[6] Bocharov G, Volpert V, Ludewig B, Meyerhans A. Principles of Virus–Host Interaction. 2018. [http://dx.doi.org/10.1007/978-3-319-72317-4_1]

[7] LePan. Visualizing the History of Pandemics. Vis Hist Pandemics 2020; 2: 1-16.

[8] The Spread and Eradication of Smallpox | Smallpox | CDC.

[9] World Health Organization. Managing epidemics. 2018; pp. 1–260..

[10] Events E, Events S. <H5N1_avian_influenza_update.pdf>(January) 2012; 1-5.

[11] WHO | Avian influenza A (H7N9) virus.

[12] Information on Swine/Variant Influenza | CDC.

[13] WHO | AIDS epidemic update: December 2006.

[14] https://www.who.int/news-room/fact-sheets/detail/zika-virus

[15] https://www.who.int/health-topics/ebola/#tab=tab_1

[16] https://www.who.int/southeastasia/outbreaks-and-emergencies/health-emergency-information-risk-assessment/surveillance-and-risk-assessment/nipah-virus-outbreak-in-kerala

[18] Lee JL, Loe MWC, Lee RCH, Chu JJH. Antiviral activity of pinocembrin against Zika virus replication. Antiviral Res 2019; 167: 13-24. [http://dx.doi.org/10.1016/j.antiviral.2019.04.003] [PMID: 30959074]

[19] Bang S, Quy Ha TK, Lee C, Li W, Oh WK, Shim SH. Antiviral activities of compounds from aerial parts of Salvia plebeia R. Br. J Ethnopharmacol 2016; 192: 398-405. [http://dx.doi.org/10.1016/j.jep.2016.09.030] [PMID: 27647011]

[20] LeCher JC, Diep N, Krug PW, Hilliard JK. Genistein has antiviral activity against herpes b virus and acts synergistically with antiviral treatments to reduce effective dose. Viruses 2019; 11(6): 499-511. [http://dx.doi.org/10.3390/v11060499] [PMID: 31159175]

[21] Li R, Liu T, Liu M, Chen F, Liu S, Yang J. Anti-influenza a virus activity of dendrobine and its mechanism of action. J Agric Food Chem 2017; 65(18): 3665-74. [http://dx.doi.org/10.1021/acs.jafc.7b00276] [PMID: 28417634]

[22] Esimone CO, Eck G, Nworu CS, Hoffmann D, Überla K, Proksch P. Dammarenolic acid, a secodammarane triterpenoid from Aglaia sp. shows potent anti-retroviral activity *in vitro.* Phytomedicine 2010; 17(7): 540-7. [http://dx.doi.org/10.1016/j.phymed.2009.10.015] [PMID: 19962871]

[23] Fang CY, Chen SJ, Wu HN, *et al.* Honokiol, a lignanbiphenol derived from the Magnolia tree, inhibits dengue virus type 2 infection. Viruses 2015; 7(9): 4894-910. [http://dx.doi.org/10.3390/v7092852] [PMID: 26378567]

[24] Clinical Assessment of Oral Lactoferrin as a Safe Antiviral and Immunoregulatory in Treating COVID-19 Disease - Full Text View - ClinicalTrials.gov 2020.

[25] Barnabas RV, Baeten JM, Lingappa JR, *et al.* Acyclovir prophylaxis reduces the incidence of herpes zoster among HIV-infected individuals: Results of a randomized clinical trial. J Infect Dis 2016; 213(4): 551-5.
[http://dx.doi.org/10.1093/infdis/jiv318] [PMID: 26142452]

[26] Maertens J, Cordonnier C, Jaksch P, *et al.* Maribavir for preemptive treatment of cytomegalovirus reactivation. N Engl J Med 2019; 381(12): 1136-47.
[http://dx.doi.org/10.1056/NEJMoa1714656] [PMID: 31532960]

[27] Heaton SM. Harnessing host–virus evolution in antiviral therapy and immunotherapy. Clin Transl Immunology 2019; 8(7): e1067.
[http://dx.doi.org/10.1002/cti2.1067] [PMID: 31312450]

[28] Wada-Shimosato Y, Tanoshima R, Hiratoko K, *et al.* Effectiveness of acyclovir prophylaxis against varicella zoster virus disease after allogeneic hematopoietic cell transplantation: A systematic review and meta□analysis. Transpl Infect Dis 2019; 21(3): e13061.
[http://dx.doi.org/10.1111/tid.13061] [PMID: 30756465]

[29] Slots J. Primer on etiology and treatment of progressive/severe periodontitis: A systemic health perspective. Periodontol 2000 2020; 83(1): 272-6.
[http://dx.doi.org/10.1111/prd.12325] [PMID: 32385884]

[30] Furuta Y, Gowen BB, Takahashi K, Shiraki K, Smee DF, Barnard DL. Favipiravir (T-705), a novel viral RNA polymerase inhibitor. Antiviral Res 2013; 100(2): 446-54.
[http://dx.doi.org/10.1016/j.antiviral.2013.09.015] [PMID: 24084488]

[31] Morse JS, Lalonde T, Xu S, Liu WR. Learning from the past: possible urgent prevention and treatment options for severe acute respiratory infections caused by 2019-nCoV. ChemBioChem 2020; 21(5): 730-8.
[http://dx.doi.org/10.1002/cbic.202000047] [PMID: 32022370]

[32] Taylor R, Kotian P, Warren T, *et al.* BCX4430 – A broad-spectrum antiviral adenosine nucleoside analog under development for the treatment of Ebola virus disease. J Infect Public Health 2016; 9(3): 220-6.
[http://dx.doi.org/10.1016/j.jiph.2016.04.002] [PMID: 27095300]

[33] Janowski AB, Dudley H, Wang D. Antiviral activity of ribavirin and favipiravir against human astroviruses. J Clin Virol 2020; 123(123): 104247-50.
[http://dx.doi.org/10.1016/j.jcv.2019.104247] [PMID: 31864069]

[34] Furuta Y, Komeno T, Nakamura T. Favipiravir (T-705), a broad spectrum inhibitor of viral RNA polymerase. Proc Jpn Acad, Ser B, Phys Biol Sci 2017; 93(7): 449-63.
[http://dx.doi.org/10.2183/pjab.93.027] [PMID: 28769016]

[35] Sissoko D, Laouenan C, Folkesson E, *et al.* Experimental Treatment with Favipiravir for Ebola Virus Disease (the JIKI Trial): A Historically Controlled, Single-Arm Proof-of-Concept Trial in Guinea. PLoS Med 2016; 13(3): e1001967.
[http://dx.doi.org/10.1371/journal.pmed.1001967] [PMID: 26930627]

[36] Dawes BE, Kalveram B, Ikegami T, *et al.* Favipiravir (T-705) protects against Nipah virus infection in the hamster model. Sci Rep 2018; 8(1): 7604.
[http://dx.doi.org/10.1038/s41598-018-25780-3] [PMID: 29765101]

[37] Kim JA, Seong RK, Kumar M, Shin O. Favipiravir and ribavirin inhibit replication of Asian and African strains of zika virus in different cell models. Viruses 2018; 10(2): 72-86.
[http://dx.doi.org/10.3390/v10020072] [PMID: 29425176]

[38] Raabe VN, Kann G, Ribner BS, *et al.* Favipiravir and ribavirin treatment of epidemiologically linked cases of lassa fever. Clin Infect Dis 2017; 65(5): 855-9.
[http://dx.doi.org/10.1093/cid/cix406] [PMID: 29017278]

[39] Péron JM, Abravanel F, Guillaume M, *et al.* Treatment of autochthonous acute hepatitis E with short-

term ribavirin: a multicenter retrospective study. Liver Int 2016; 36(3): 328-33. [http://dx.doi.org/10.1111/liv.12911] [PMID: 26179015]

[40] Liao S, Li Y, Lai Y, Liu N, Zhang F, Xu P. Ribavirin attenuates the respiratory immune responses to influenza viral infection in mice. Arch Virol 2017; 162(6): 1661-9. [http://dx.doi.org/10.1007/s00705-017-3291-7] [PMID: 28243801]

[41] Khalili JS, Zhu H, Mak NSA, Yan Y, Zhu Y. Novel coronavirus treatment with ribavirin: Groundwork for an evaluation concerning COVID-19. J Med Virol 2020; 92(7): 740-6. [http://dx.doi.org/10.1002/jmv.25798] [PMID: 32227493]

[42] Brown AJ, Won JJ, Graham RL, *et al.* Broad spectrum antiviral remdesivir inhibits human endemic and zoonotic deltacoronaviruses with a highly divergent RNA dependent RNA polymerase. Antiviral Res 2019; 169: 104541-50. [http://dx.doi.org/10.1016/j.antiviral.2019.104541] [PMID: 31233808]

[43] Sacramento CQ, De Melo GR, De Freitas CS, Rocha N, Hoelz LVB, Miranda M, *et al.* The clinically approved antiviral drug sofosbuvir inhibits Zika virus replication. Sci Rep 2016; 2017(7): 1-11.

[44] Fink SL, Vojtech L, Wagoner J, *et al.* The Antiviral Drug Arbidol Inhibits Zika Virus. Sci Rep 2018; 8(1): 8989. [http://dx.doi.org/10.1038/s41598-018-27224-4] [PMID: 29895962]

[45] Abraham GM, Morton JB, Saravolatz LD. Baloxavir: A Novel Antiviral Agent in the Treatment of Influenza. Clin Infect Dis 2020; 71(7): 1790-4. [http://dx.doi.org/10.1093/cid/ciaa107] [PMID: 32020174]

[46] Heo YA. Baloxavir: First Global Approval. Drugs 2018; 78(6): 693-7. [http://dx.doi.org/10.1007/s40265-018-0899-1] [PMID: 29623652]

[47] Anand K, Ziebuhr J, Wadhwani P, Mesters JR, Hilgenfeld R. Coronavirus main proteinase (3CLpro) structure: basis for design of anti-SARS drugs. Science 2003; 300(5626): 1763-7. [http://dx.doi.org/10.1126/science.1085658] [PMID: 12746549]

[48] Zhang L, Lin D, Sun X, *et al.* Crystal structure of SARS-CoV-2 main protease provides a basis for design of improved α-ketoamide inhibitors. Science 2020; 368(6489): 409-12. [http://dx.doi.org/10.1126/science.abb3405] [PMID: 32198291]

[49] Cheng PKC, Leung TWC, Ho ECM, *et al.* Oseltamivir- and amantadine-resistant influenza viruses A (H1N1). Emerg Infect Dis 2009; 15(6): 966-8. [http://dx.doi.org/10.3201/eid1506.081357] [PMID: 19523305]

[50] Jefferson T, Demicheli V, Di Pietrantonj C, Rivetti D. Amantadine and rimantadine for influenza A in adults (Unknown). Cochrane Database Syst Rev 2006; (3): 1-13.

[51] Li T, Chan M, Lee N. Clinical implications of antiviral resistance in influenza. Viruses 2015; 7(9): 4929-44. [http://dx.doi.org/10.3390/v7092850] [PMID: 26389935]

[52] Kumar S, Goicoechea S, Kumar S, *et al.* Oseltamivir analogs with potent anti-influenza virus activity. Drug Discov Today 2020; 25(8): 1389-402. [http://dx.doi.org/10.1016/j.drudis.2020.06.004] [PMID: 32554062]

[53] Russell RJ, Haire LF, Stevens DJ, *et al.* The structure of H5N1 avian influenza neuraminidase suggests new opportunities for drug design. Nature 2006; 443(7107): 45-9. [http://dx.doi.org/10.1038/nature05114] [PMID: 16915235]

[54] Bolles M, Deming D, Long K, *et al.* A double-inactivated severe acute respiratory syndrome coronavirus vaccine provides incomplete protection in mice and induces increased eosinophilic proinflammatory pulmonary response upon challenge. J Virol 2011; 85(23): 12201-15. [http://dx.doi.org/10.1128/JVI.06048-11] [PMID: 21937658]

[55] Agrawal AS, Tao X, Algaissi A, *et al.* Immunization with inactivated Middle East Respiratory

Syndrome coronavirus vaccine leads to lung immunopathology on challenge with live virus. Hum Vaccin Immunother 2016; 12(9): 2351-6.
[http://dx.doi.org/10.1080/21645515.2016.1177688] [PMID: 27269431]

[56] Zhao J, Alshukairi AN, Baharoon SA, *et al.* Recovery from the Middle East respiratory syndrome is associated with antibody and T cell responses. Sci Immunol 2017; 2(14): eaan5393.
[http://dx.doi.org/10.1126/sciimmunol.aan5393] [PMID: 28778905]

[57] Almazán F, DeDiego ML, Sola I, *et al.* Engineering a replication-competent, propagation-defective Middle East respiratory syndrome coronavirus as a vaccine candidate. MBio 2013; 4(5): e00650-13.
[http://dx.doi.org/10.1128/mBio.00650-13] [PMID: 24023385]

[58] Adney D, Wang L, van Doremalen N, *et al.* Efficacy of an adjuvanted Middle East respiratory syndrome coronavirus spike protein vaccine in dromedary camels and alpacas. Viruses 2019; 11(3): 212-24.
[http://dx.doi.org/10.3390/v11030212] [PMID: 30832356]

[59] Zhang N, Channappanavar R, Ma C, *et al.* Identification of an ideal adjuvant for receptor-binding domain-based subunit vaccines against Middle East respiratory syndrome coronavirus. Cell Mol Immunol 2016; 13(2): 180-90.
[http://dx.doi.org/10.1038/cmi.2015.03] [PMID: 25640653]

[60] Jiaming L, Yanfeng Y, Yao D, *et al.* The recombinant N-terminal domain of spike proteins is a potential vaccine against Middle East respiratory syndrome coronavirus (MERS-CoV) infection. Vaccine 2017; 35(1): 10-8.
[http://dx.doi.org/10.1016/j.vaccine.2016.11.064] [PMID: 27899228]

[61] Lan J, Deng Y, Chen H, *et al.* Tailoring subunit vaccine immunity with adjuvant combinations and delivery routes using the Middle East respiratory coronavirus (MERS-CoV) receptor-binding domain as an antigen. PLoS One 2014; 9(11): e112602.
[http://dx.doi.org/10.1371/journal.pone.0112602] [PMID: 25405618]

[62] Lokugamage KG, Yoshikawa-Iwata N, Ito N, *et al.* Chimeric coronavirus-like particles carrying severe acute respiratory syndrome coronavirus (SCoV) S protein protect mice against challenge with SCoV. Vaccine 2008; 26(6): 797-808.
[http://dx.doi.org/10.1016/j.vaccine.2007.11.092] [PMID: 18191004]

[63] Liu YV, Massare MJ, Barnard DL, *et al.* Chimeric severe acute respiratory syndrome coronavirus (SARS-CoV) S glycoprotein and influenza matrix 1 efficiently form virus-like particles (VLPs) that protect mice against challenge with SARS-CoV. Vaccine 2011; 29(38): 6606-13.
[http://dx.doi.org/10.1016/j.vaccine.2011.06.111] [PMID: 21762752]

[64] Ura T, Okuda K, Shimada M. Developments in viral vector-based vaccines. Vaccines (Basel) 2014; 2(3): 624-41.
[http://dx.doi.org/10.3390/vaccines2030624] [PMID: 26344749]

[65] Haagmans BL, van den Brand JMA, Raj VS, *et al.* An orthopoxvirus-based vaccine reduces virus excretion after MERS-CoV infection in dromedary camels. Science 2016; 351(6268): 77-81.
[http://dx.doi.org/10.1126/science.aad1283] [PMID: 26678878]

[66] Hashem A, Algaissi A, Agrawal A, Al-Amri S, Alhabbab R, Sohrab S, *et al.* A highly immunogenic, protective and safe adenovirus-based vaccine expressing MERSCoV S1-CD40L fusion protein in transgenic human DPP4 mouse model. J Infect Dis 2019; 220(10): 1558-67.
[http://dx.doi.org/10.1093/infdis/jiz137] [PMID: 30911758]

[67] Buchholz UJ, Bukreyev A, Yang L, *et al.* Contributions of the structural proteins of severe acute respiratory syndrome coronavirus to protective immunity. Proc Natl Acad Sci USA 2004; 101(26): 9804-9.
[http://dx.doi.org/10.1073/pnas.0403492101] [PMID: 15210961]

[68] Kowalski PS, Rudra A, Miao L, Anderson DG. Delivering the Messenger: Advances in Technologies for Therapeutic mRNA Delivery. Mol Ther 2019; 27(4): 710-28.

[http://dx.doi.org/10.1016/j.ymthe.2019.02.012] [PMID: 30846391]

[69] Zhang C, Maruggi G, Shan H, Li J. Advances in mRNA vaccines for infectious diseases. Front Immunol 2019; 10(MAR): 594.
[http://dx.doi.org/10.3389/fimmu.2019.00594] [PMID: 30972078]

[70] Fuller DH, Berglund P. Amplifying RNA Vaccine Development. N Engl J Med 2020; 382(25): 2469-71.
[http://dx.doi.org/10.1056/NEJMcibr2009737] [PMID: 32558474]

[71] Covid-19 Vaccine Tracker Updates: The Latest - The New York Times.

[72] Self-amplifying mRNA (SAM) Vaccine Platform - Creative Biolabs.

[73] Kundu AK, Iyer SV, Chandra S, Adhikari AS, Iwakuma T, Mandal TK. Novel siRNA formulation to effectively knockdown mutant p53 in osteosarcoma. PLoS One 2017; 12(6): e0179168.
[http://dx.doi.org/10.1371/journal.pone.0179168] [PMID: 28636657]

[74] Ghosh S, Firdous SM, Nath A. siRNA could be a potential therapy for COVID-19. EXCLI J 2020; 19: 528-31.
[PMID: 32398976]

[75] Arabi YM, Hajeer AH, Luke T, *et al.* Feasibility of using convalescent plasma immunotherapy for MERS-CoV infection, Saudi Arabia. Emerg Infect Dis 2016; 22(9): 1554-61.
[http://dx.doi.org/10.3201/eid2209.151164] [PMID: 27532807]

[76] Zumla A, Chan JFW, Azhar EI, Hui DSC, Yuen KY. Coronaviruses — drug discovery and therapeutic options. Nat Rev Drug Discov 2016; 15(5): 327-47.
[http://dx.doi.org/10.1038/nrd.2015.37] [PMID: 26868298]

[77] Corti D, Passini N, Lanzavecchia A, Zambon M. Rapid generation of a human monoclonal antibody to combat Middle East respiratory syndrome. J Infect Public Health 2016; 9(3): 231-5.
[http://dx.doi.org/10.1016/j.jiph.2016.04.003] [PMID: 27102927]

[78] Efficiency and Security of NIVOLUMAB Therapy in Obese Individuals With COVID-19 (Coronavirus Disease) Infection. 2020.

[79] Prokunina-Olsson L, Alphonse N, Dickenson RE, *et al.* COVID-19 and emerging viral infections: The case for interferon lambda. J Exp Med 2020; 217(5): e20200653.
[http://dx.doi.org/10.1084/jem.20200653] [PMID: 32289152]

[80] Sheahan TP, Sims AC, Zhou S, *et al.* An orally bioavailable broad-spectrum antiviral inhibits SARS-CoV-2 in human airway epithelial cell cultures and multiple coronaviruses in mice. Sci Transl Med 2020; 12(541): eabb5883.
[http://dx.doi.org/10.1126/scitranslmed.abb5883] [PMID: 32253226]

[81] Gorshkov K, Chen CZ, Bostwick R, Rasmussen L, Xu M, Pradhan M, *et al.* The SARS-CoV-2 cytopathic effect is blocked with autophagy modulators. bioRxivPreprServ Biol 2020.

[82] He S, Lin B, Chu V, *et al.* Repurposing of the antihistamine chlorcyclizine and related compounds for treatment of hepatitis C virus infection. Sci Transl Med 2015; 7(282): 282ra49.
[http://dx.doi.org/10.1126/scitranslmed.3010286] [PMID: 25855495]

[83] Compton JR, Mickey MJ, Hu X, Marugan JJ, Legler PM. Mutation of Asn-475 in the Venezuelan Equine Encephalitis Virus nsP2 Cysteine Protease Leads to a Self-Inhibited State. Biochemistry 2017; 56(47): 6221-30.
[http://dx.doi.org/10.1021/acs.biochem.7b00746] [PMID: 29064679]

[84] https://opendata.ncats.nih.gov/covid19/highlights

[85] Gorshkov K, Chen CZ, Bostwick R, *et al.* The SARS-CoV-2 cytopathic effect is blocked with autophagy modulators. Biorxiv 2020.
[http://dx.doi.org/10.1101/2020.05.16.091520]

[86] https://opendata.ncats.nih.gov/covid19/assay?aid=14

[87] Sivan G, Ormanoglu P, Buehler EC, Martin SE, Moss B. Identification of restriction factors by human genome-wide RNA interference screening of viral host range mutants exemplified by discovery of SAMD9 and WDR6 as inhibitors of the vaccinia virus k1L-C7L- mutant. MBio 2015; 6(4): e01122-15.
[http://dx.doi.org/10.1128/mBio.01122-15] [PMID: 26242627]

[88] About Tissue Chip | National Center for Advancing Translational Sciences.

[89] Consulting S. Consulting S (Certara), Biopharma A. Crafting a Translational Medicine Strategy for a Pediatric Antiviral Drug.

[90] Kruse RL, Shum T, Tashiro H, *et al.* HBsAg-redirected T cells exhibit antiviral activity in HBV-infected human liver chimeric mice. Cytotherapy 2018; 20(5): 697-705.
[http://dx.doi.org/10.1016/j.jcyt.2018.02.002] [PMID: 29631939]

[91] van der Burg SH, Arens R, Melief CJM. Immunotherapy for persistent viral infections and associated disease. Trends Immunol 2011; 32(3): 97-103.
[http://dx.doi.org/10.1016/j.it.2010.12.006] [PMID: 21227751]

[92] Soares MM, King SW, Thorpe PE. Targeting inside-out phosphatidylserine as a therapeutic strategy for viral diseases. Nat Med 2008; 14(12): 1357-62.
[http://dx.doi.org/10.1038/nm.1885] [PMID: 19029986]

[93] Slim J, Sulkowski MS, Shan JS. 1239 Escalating Repeat-Dose Study of Bavituximab in Patients Co-Infected With Chronic Hepatitis C Virus (HCV) and Human Immunodeficiency Virus. J Hepatol 2011; 54 (Suppl. 1): S489.
[http://dx.doi.org/10.1016/S0168-8278(11)61241-5]

[94] Zou L, Peng Q, Wang P, Zhou B. Progress in Research and Application of HIV-1 TAT-Derived Cell-Penetrating Peptide. J Membr Biol 2017; 250(2): 115-22.
[http://dx.doi.org/10.1007/s00232-016-9940-z] [PMID: 27933338]

[95] Ebola Vaccine Trial (Ad26.ZEBOV/MVA-BN-Filo) - Full Text View - ClinicalTrials.gov.

[96] Safety of and Immune Response to a Bird Flu Virus Vaccine (H5N1) in Healthy Adults - Full Text View - ClinicalTrials.gov

[97] FluLaval Quadrivalent | FDA [Internet].

[98] Live Attenuated Influenza Vaccine [LAIV] (The Nasal Spray Flu Vaccine) | CDC.

[99] Enesi Pharma and Imperial College London Collaborate to Develop Thermostable RNA Vaccines with Potential to Eliminate Cold Chain and Transform Global Vaccination Logistics for COVID-19.

[100] Do Angiotensin Receptor Blockers Mitigate Progression to Acute Respiratory Distress Syndrome with SARS-CoV-2 Infection - Full Text View - ClinicalTrials.gov.

[101] Guy R, DiPaola R, Romanelli F, Dutch R. Rapid repurposing of drugs for COVID-19. Science 2020; 368(6493): 829–30.

[102] AI Helps Doctors Diagnose the Coronavirus – NVIDIA Developer News Center.

[103] Computational predictions of protein structures associated with COVID-19 | DeepMind.

[104] Zimmerman M, Porter J, Ward M, Singh S, Vithani N, Meller A, *et al.* Zimmerman M, Porter J, Ward M, Singh S, Vithani N, Meller A, *et al.* Citizen Scientists Create an Exascale Computer to Combat COVID-19. bioRxivPreprServ Biol. 2020; 1-18.

[105] Ong E, Wong MU, Huffman A, He Y. COVID-19 coronavirus vaccine design using reverse vaccinology and machine learning. bioRxivPreprServBiol 2020; 1-24.

[106] U.S. clinical trials for more effective flu vaccine – News.

[107] Li Y, Niu M, Zou Q. ELM-MHC: An Improved MHC Identification Method with Extreme Learning Machine Algorithm. J Proteome Res 2019; 18(3): 1392-401. [http://dx.doi.org/10.1021/acs.jproteome.9b00012] [PMID: 30698979]

[108] Moghram BA, Nabil E, Badr A. Ab-initio conformational epitope structure prediction using genetic algorithm and SVM for vaccine design. Comput Methods Programs Biomed 2018; 153(153): 161-70. [http://dx.doi.org/10.1016/j.cmpb.2017.10.011] [PMID: 29157448]

[109] Mathewson P, Gordon B, Snowley K, Fennessy C, Denniston A, Sebire N. Study Data Element Mapping: Feasibility of Defining Common Data Elements Across COVID-19 Studies. medRxiv 2020. [http://dx.doi.org/10.1101/2020.05.19.20106641]

[110] Liu Z, Du J, Fang J, Yin Y, Xu G, Xie L. DeepScreening: a deep learning-based screening web server for accelerating drug discovery. Database (Oxford) 2019; 2019(1): baz104. [http://dx.doi.org/10.1093/database/baz104] [PMID: 31608949]

CHAPTER 7

Natural Products: Antibacterial, Anti-fungal, and Anti-viral Agents

Shipa Rani Dey[1], **Neha Gogia**[2,#] and **Prasann Kumar**[1, 3,*]

[1] *Department of Agronomy, School of Agriculture, Lovely Professional University, Punjab, India*

[2] *Department of Genetics and Neuroscience, Yale School of Medicine, New Haven, CT, United States*

[3] *Department of Trans-disciplinary Research; Division of Research and Development, Lovely Professional University, Punjab, India*

Abstract: The existence of substantial evidence about the development of resistance to a drug among microbes has gained a lot of attention from the scientific world. To address this problem, researchers have been conducting experiments and testing strategies, including screening various molecules and using plant-derived natural products to ascertain if these substances can serve as an untapped source of anti-bacterial, anti-viral, and anti-fungal agents. The non-toxic, non-synthetic, causing minimal side effects, and cost-effective nature of these substances make the development of new anti-microbials heavily dependent on the use of many of these existing products and increase the demand for finding new natural products that are yet to be discovered. These plant-based natural products offer great promises to provide the best protection against infections and pathogenesis in many diseases. Furthermore, the biodegradable nature of many of these products increases their chances of being chosen by farmers and plant biologists to use to combat microbial pathogenesis. This chapter covers the current insights on the conflicts and opportunities of popular plant-derived natural anti-microbial compounds containing a reservoir of secondary metabolites, *viz.*. flavonoids, alkaloids, terpenes, coumarins, phenols and polyphenols. The chapter lists natural vegetable products, which serve as potent anti-bacterial and anti-fungal agents, and describes various plant extracts, which exhibit bacterial quorum sensing, biofilm as well as efflux pump inhibitory activity. Previous studies have demonstrated the effectiveness of these plant-based natural products in the treatment of neurodegenerative diseases as well. This chapter also summarizes the neuroprotective activity of these products and their potential to serve as therapeutic agents to block or delay the progression of disorders.

* **Corresponding author Prasann Kumar**: Department of Agronomy and Trans-disciplinary Research, Lovely Professional University, Punjab, India; E-mail: prasann0659@gmail.com

Co-corresponding author Neha Gogia: Department of Genetics and Neuroscience, Yale School of Medicine, New Haven, CT, USA; E-mail: nehag0609@gmail.com

M. Amin-ul Mannan & Gaurav Kumar (Eds.)

Keywords: Anti-bacterial, Herbs, Medicinal plants, Phytochemicals, Plant-derived natural products, Secondary metabolites.

INTRODUCTION

Until the discovery of modern medicine, plants with medicinal values were extensively applied to cure anti-bacterial, anti-fungal, anti-viral and other diseases. People acknowledged the importance of using these medicinal herbs and herbal products (made from them) in medical practice, which has continued and is increasing worldwide [1 - 3]. The immense potential and great promises offered by natural products (in the form of herbal remedies) in the development of drugs effective in treating many diseases [4 - 7] greatly influenced pharmaceutical agencies worldwide in production as well as the development of these cost-effective plant-based green drugs [8 - 10]. A big challenge was to find plants containing novel medicinal agents. Scientists used a varied methodology for the same, but the four key methods or approaches profoundly used were (a) ethnopharmacological, (b) taxonomical, (c) phytochemical, and (d) random approach [11]. Ethnopharmacological methods are the practices where the primary information is more critical to collecting and evaluating the plants [12 - 14]. Taxonomical approaches involve selecting and evaluating plants based on either genus or family (present in different locations). Phytochemical approaches are the ones where the plants are collected and evaluated based on the biochemical properties of the secondary metabolite (such as alkaloids) present in them [15, 16]. These approaches offer great promises, however, the chemical complexities and difficulties in screening compounds for the identification of a new drug [17 - 21] make discovering a new compound from these scientific explorations difficult. Several compounds present in the crude extract (such as tannins and lipids) may cause a problem in the pharmacological assay [22, 23]. Therefore, the discovery of these compounds requires preliminary chemical analysis and separation of mixtures. Within ethnobotany, there is a popular branch of ethnopharmacology, which is dedicated to the medicinal use of plants, along with the screening of raw metabolites for future drug discovery. The branch also describes secondary metabolites as the key to the development of new drugs. Using best practices,109 anti-bacterial components were obtained and approved during the year1981 to 2006. Out of the 69% extracted from plant products, and at the same time, 21% of anti-fungal drugs were also found to be plant-based [24]. In studies conducted previously, plant families such as *Pinaceae, Cupressaceae, Apiaceae, Burseraceae, Anacardiaceae, Palmaceae, Euphorbiaceae, Dracenaceae and Fabaceae* have been known to release resins and natural gums, and these plant-derived resins may be used as a whole or in modified forms (such as specific extracts, fractions and essential oils) against many bacteria, fungi, viruses and for treatment of many diseases. Medicinal plants with rich aromatic compounds (menthe, fennel, thyme, and lavender) are the largest reservoir of essential oils. By nature, they are volatile and are chemically classified into terpenoids (C10 compounds, C15 compounds and their oxygenated derivatives), and phenylpropanoids. They are known as potential inhibitors of Gram-positive and Gram-negative bacteria and are known to be effective against many fungi and viruses. The essential oil of *Daucuscarota* seed is highly effective in the control of *Helicobacter pylori.* The growth inhibition was reported because of the disrupted intracellular cytoskeleton. Essential oil mixture, consisting primarily of aliphatic, aromatic and terpenoid, has been widely studied for inhibition of the growth and development of bacteria. These constituents have been found to inhibit the synthesis of exopolysaccharides, which results in less bacterial adhesion on the host.

ANTI-BACTERIAL, ANTI-FUNGAL AND ANTI-VIRAL ACTIVITY

Principle Antibacterial Phytochemicals

Secondary metabolites derived from plants and microbes are natural substances that can be used for the production of potential drugs. They have a wide range of activities, and their classification is based on the diversity of plant species, the topography, and the country's current climate. Another (second) set of classifications is based on a different class of active substance compounds [25]. Variation in the composition of compounds leads to a change in activity against microbes. The principal class of derived phytochemicals from medicinal plants has been described in the schematic diagram (Fig. **1**).

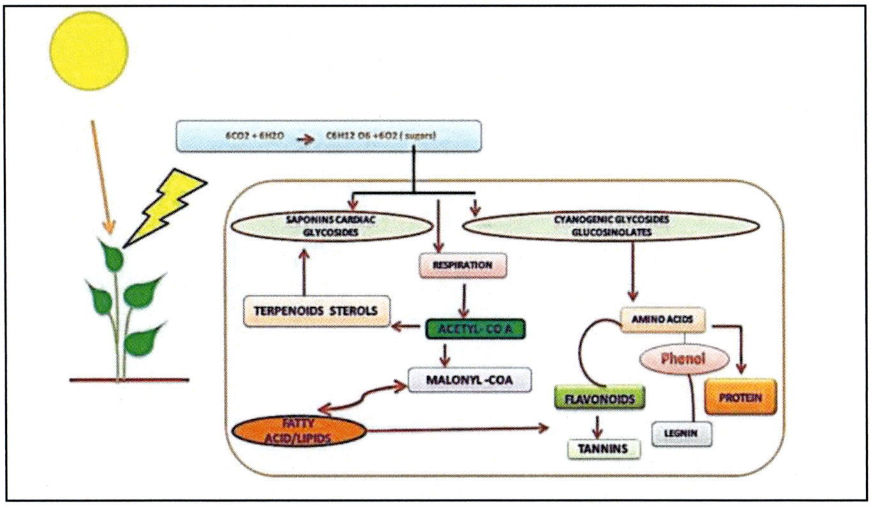

Fig. (1). Categories of classes for phytochemicals extracted from medicinal plants.

Terpenes

Terpenes are aromatic compounds unified by the fusion of five-carbon isoprene units and are known to possess anti-viral, anti-bacterial, anti-oxidative and anti-inflammatory characteristics. They are divided into several groups based on carbon numbers. A wide chemical complexity in terpenes comes from the number of simple units in the chain and how they are constructed. The creation of cyclic systems, the incorporation of oxygen-containing roles and the addition of sugars or other molecules contribute to their potential complexity. The term terpenoid comes from the idea that the first compounds of the groups were extracted from turpentine (Ger. terpentin), an essential oil (mainly pinene) obtained from coniferous trees. Scientists studied the anti-bacterial ability of monoterpenes (C10), diterpenoids (C15), triterpenoids, and their derivatives [26] and found that the modes of action of terpenes are based on the interruption of the membrane by the lipophilic compounds of the plasma membrane in bacteria. Diterpenoids and sesquiterpenes present the progress of gram-positive and *M. tuberculosis* bacteria.

Saponins are triterpenoids comprising one or more sugar units. Medigenic acid glucoside is triterpenoid alfalfa saponin (*Medicagosativa*). Diosgenin glycoside is a clover-isolated steroidal saponin (*Melilotus* spp.). Adding a hydrophilic sugar group to a normally hydrophobic terpenoid gives soap-like saponin surfactant properties. The main role of Saponins seems to be a preformed defence against fungal attacks. Studies have shown that saponins form complexes with sterols possessing an unsubstituted 3-β-hydroxyl ring. As saponins interfere with sterols in invading fungal hyphae membranes, the effect is a loss of membrane integrity. *E.g.,* the crop plant Oat (Avena) contains triterpenoid saponin, avenacin A-1, which is found in the root epidermal cells and effectively protects the cells against invasion by a fungal pathogen (*Gaeumannomycesgraminis* var. tritici) that infects both wheat (Triticum) and barley (Hordeum) roots. It is very interesting to note, at the same time, that other strains of *G. graminis* (var. *avenae*) produce an enzyme known as avenacinase. This enzyme detoxifies the effect of Avenacin A-1 and because of this, the pathogens can invade oats, wheat and rye again. The impact of saponins on eukaryotic membranes is extremely unspecific, and it is uncertain how plants shield their membranes from the deleterious effects of saponin. A possible explanation could be that saponins are processed as a biologically inert molecule called mosidsaponin, which has two sugar chains rather than one. At the time of the attack, the biological inert form is transformed into an active form. This activation is due to the hydrolytic removal of the second sugar chain. Finally, biologically active monodesmosidsaponins accumulate in the vacuoles and other organelles with a higher proportion of sterols.

Phenolic and Polyphenols

A broad class of secondary metabolites originating from aromatic amino acids are phenolic acids (also known as polyphenols and phenylpropanoids) [26]. Polyphenols are a complex chemical compound family, which are majorly divided into four groups: flavonoids, phenolic acids, polyphenolic amides and other polyphenols. Flavonoids are further divided into flavones, flavanones, flavonols, flavanols, isoflavones *etc.* Phenol (a hydroxylated aromatic ring) is the fundamental form of phenolics and is renowned for its essential function in many interactions between plant-herbivores [27]. Some act as important structural components (such as lignin), while others simply appear to be metabolic products without obvious function.

Flavones (comprises of the largest group of natural phenols) and their derivatives have a wide potential to inhibit the growth of bacteria by disrupting bacterial envelope [26]. Catechins (flavan-3-ols or flavanols) found in *Camellia sinensis* forms complexes, damaging bacterial cell walls and inactivating specific bacterial enzymes, which has wide applications on intestinal microorganisms [27].

Coumarins

Coumarins are known for their roles as potential feeding deterrents, anti-microbials agents along with germination inhibitors in their families. Coumarinsserves as an ingredient in bergamot oil as well, an important oil used to flavour tobacco pipes, tea and other products. While coumarins are only minimally toxic, many of their products can be highly toxic. Dicoumarol, a coumarin derivative, can usually be found in the clover moldy hay or silage (Melilotus spp.). It causes fatal haemorrhaging in cattle by inhibiting the synthesis of vitamin-K (an essential co-factor for the clotting of blood) [30]. Dicoumarol was discovered in the 1940s, and based on it, many synthetic coumarin derivatives, such as Warfarin™, (widely used as rodent poison), were developed [31]. Scopoletin, found in seed coats, is the most common coumarin in higher plants. This has known as being an inhibitor of germination, which must be released before germination can take place.

It is interesting to note that most toxic Coumarins products are not synthesised by either plants or fungi [32]. *Aspergillus flavus* fungus, typically related to the infection of food items such as cattle feed, peanuts and maize, forms a group of mycotoxins called aflatoxins, which are considered to be the most active and carcinogenic natural toxins, produced by the invading *Aspergillus*. Ingesting even 6-15 parts per billion of aflatoxin (or 6-15 $\mu g\ kg^{-1}$) of maize has reported death [31]. They are mutagenic, bind to DNA and resist RNA transcriptions, interact with hormones in the area of binding sites, weaken the immune system, and hampers the functioning of the liver and kidneys. Peanut products are a natural source of human aflatoxin toxicity. As per the Food and Drug Administration (FDA) regulations, raw peanuts must be tested for their aflatoxin content in the United States. Another specific concern is aflatoxins in cattle feed. Once cattle absorb the poison, it is more poisonous as it is secreted into milk. Drug resistance and toxic side effects are two serious drawbacks in the treatment of a person having a fungal infection. These drawbacks are more serious for people suffering from Acquired Immuno-Deficiency Syndrome (AIDS), cancer, and transplantation. Modern therapeutic methods need to be modified and applied based on plant molecules.

More than 600 plants (reported to date) possess anti-fungal properties [28]. The use of tea tree oil against fungus is popular nowadays. Candida species (from *in-vivo*, *in vitro* studies) have been found to be effective against mucosal *Candidiasis* [34]. One of the most active compounds, such as terpinene-4-ol, possesses anti-microbial activity. In *Menthasuaveolens* (Apple Mint), the principal microbial components include pulegone and piperitone, which exhibit anti-fungal properties [29]. The roots of *Bidens tripartite* (Bur Marigold Herb) are a reservoir of anti-fungal compounds such as pinene, bisabolene, cymene, hexanal, and linalool [30]. Extracts of *Coriandrumsativum*(commonly known as coriander) as found to be effective against the growth of germ tube formation [31]. Another compound, Caryophyllene, found in *Curcuma spp.,* possess anti-microbial activity against a fungal pathogen *Cryptococcus neoformans* [32]. The effective anti-dermatophytes compound against dermatomycosis was obtained from plant natural extract of *Fructuspsoralear*, *Folium eucalypti,* and the *Achilleamellefolium*. The growth of *Aspergillus spp.* has been found to be prevented by the application of anti-fungal compounds flavonoids isolated from *Mangiferaindica* leaves. Another compound isolated from the same leaves, schinol (a biphenyl compound), was isolated and found to be effective against *Paracoccidioidesbrasiliensis*. The list of plants and their respective parts used for therapeutic use are described in Figs. (**2a** - **f**).

2a.

Artocarpus heterophyllus Lam. Roots, leaves, fruits, seeds used for open wounds, skin diseases, having antidiarrheal properties.

2b.

Berberis aristate DC. Roots and barks are used for diarrhoea and jaundice treatment.

2c.

Chromolaena odorata L. Leaf juice used as antiseptic in cuts and open wounds.

(Fig. 2) contd.....

2d.

Embelia ribes Burm. Roots, bark and, fruits, are having diuretic, carminative and contraceptive properties.

2e.

Jasminum angustifolia L. Anti-poisionous, used for treating open wounds, ringworm infection.

2f.

Pluchea lanceolata DC. Stem and roots are used to treat rheumatic disorders, used for cough and athame as well.

Fig. (2a-f). Plant and their parts used for therapeutic use [33, 34].

EFFICACY OF PLANT-DERIVED NATURAL PRODUCTS IN MITIGATING NEURODEGENERATIVE DISEASES

The prevalence of neurodegenerative diseases such as Alzheimer's, Amyotrophic Lateral Sclerosis, Parkinson's, Huntington's, and other associated disorders continue to increase worldwide. Impairment of the structure and function of the neuronal cell population, which ultimately results in the death of these cells, is the hallmark of neurodegenerative diseases. It is known that aberrant signalling pathways, genetic mutations, increase in oxidative stress, mitochondrial dysfunction, genetic alterations, protein misfolding, aggregation and many other factors contribute to causing these neurodegenerative disorders, however, the exact molecular mechanisms of action, that trigger neurodegeneration remains unclear. The 1. Destructive nature of neurodegenerative diseases, 2. Several individuals are affected by them worldwide every year, 3. The cost of expenditure on treatments and, 4. The lack of availability of effective drugs puts an enormous burden on society and creates a desperate need for the development of reliable and cost-effective therapeutics that can efficiently cure these diseases. The use of plant-based natural products (NPs) (as a comparison to synthetic drugs), which are comparably inexpensive, safe, and non-synthetic, causing minimal or no side-effects, as therapeutics gained a lot of attention of scientists and researchers worldwide and increased their demand of development and use as therapeutics. This also emerged the need to search for more plant-derived NPs that have remained undiscovered. Plant-derived NPs have been known to possess anticholinesterase, anti-inflammatory, anti-oxidant, etc., properties and have widely been tested for their efficacy in the treatment of many neurodegenerative disorders in varied research models (*in-vitro* or *in-vivo*) such as *Drosophila melanogaster* (*a.k.a.* fruit fly), Zebrafish *(Danio rerio),* Mouse (*Mus musculus)*, Nematodes (*C. elegans)*, Cell culture and others. Among them, fruit flies serve as an excellent research model and have extensively been used to address research questions about the field of Biological Sciences. Fly eye comprises photoreceptor neuronal cells and has been used to answer questions related to growth regulation, pattern formation, conduct genetic screenings (to search for genetic modifiers of diseases), drug screening and elucidating cellular, molecular, and pathogenic mechanisms causing neurodegenerative diseases [35 - 40]. This chapter also summarizes the potential of NPs in blocking or ameliorating the progression of many neurodegenerative diseases.

Neurodegenerative diseases comprises many neurological disorders such as Alzheimer's Disease (AD), Amyotrophic Lateral Sclerosis (ALS), Parkinson's Disease (PD), Polyglutamine Diseases [such as Spinocerebellar Ataxia, Huntington's Disease (HD), Spinal and Bulbar Muscular Atrophy and many others]. The diseases share a commonality of progressive, irreversible death of the neuronal cell population. AD is caused by extracellular accumulation of amyloid beta-42 (Aβ42) plaques, hyper-phosphorylated tau neurofibrillary tangles, synaptic dysfunction, oxidative stress regulation, and other genetic or environmental factors that result in aberrant signalling, alterations in cellular homeostasis, mitochondrial dysfunction leading to memory impairment, loss of cognitive functions and ultimately leads to death of the patient.ALS involves the deterioration of upper, lower motor neurons in the brain and spinal cord. Prolonged muscle weakness in ALS results in paralysis, respiratory failure and ultimately, death of the patient. PD is characterized by the presence of intracytoplasmic inclusions (Lewy bodies), aggregation of misfolded α-synuclein and death of dopaminergic neurons in the substantial nigra (SN) region of the midbrain. Polyglutamine(or PolyQ) related diseases, also known as hereditary neurodegenerative diseases, include HD, dentatorubral-pallidoluysian atrophy (DRPLA), Spinal and Bulbar Muscular Atrophy (SBMA) and Spinocerebellar Ataxias (SCAs). These diseases share a commonality of repeated expansion of CAG (Cytosine-Adenine-Guanine) repeats mutation, which codes for amino acid glutamine (Q), within the coding region of their respective disease-causing genes. For example, HD is characterized by CAG repeats mutation within the Huntington (HTT) gene coding region, which leads to progressive degeneration of neuronal cells in the brain. Another neurodegenerative disease, Cerebral Ischemia (CI), is a disease condition caused by the destruction of the blood-brain barrier followed by serious consequences such as haemorrhage. Factors such as genetic mutations, oxidative stress, protein aggregation, impairment of mitochondrial dysfunction or axonal transport, and other factors contribute to the disease pathology of neurodegenerative diseases, however, the exact molecular mechanism of action remains unknown. A variety of synthetic drugs, such as

Riluzole, are widely used and are effective in the treatment of patients suffering from ALS [41], however, its effect on increasing the quality of life of patients from ALS remains unknown. Thus, there is a need to develop and use NPs to treat patients suffering from neurodegenerative diseases. Some of the plant NPs, when consumed alone or in combination with other NPs have demonstrated immense potential in ameliorating the onset or severity of many neurodegenerative diseases, thereby serving as excellent therapeutics. They broadly fall into the following categories: Lunasin, Polyphenols, Flavonoids, Alkaloids, Terpenes, Triterpenoids and Tannins.

Lunasin, a thermostable plant protein, extracted from soybean (*Glycine max*) and other seeds, is a part of the 2S albumin protein of soybean. Structural studies have shown that Lunasin is a 43 amino acid (a.a.) peptide without 5.5 kilo Daltons, comprised of residues from 1-22 (includes N-terminal domain having unknown function), residues 23-31 (codes for a predicted helical region), residues 32-34 (codes for Arginine-Glycine-Aspartate, RGD domain) and C-terminal region (containing 9 aspartic acid residues or poly-D tail) [42, 43] acts as H3-H4 histone acetylation inhibitor. It is a health-beneficial peptide, which possesses anti-antioxidant, anti-inflammatory, reducing stress and inflammation, anti-free radical, anti-cancer, anti-hypertension, hypocholesterolemic activity, anti-obesity and immunomodulation properties [42, 44]. This chemoprotective peptide, when taken orally, enters the bloodstream and reaches the region of interest in its stable and bioactive form [45]. Studies conducted using a *Drosophila* eye model of AD have reported Lunasina's potential to block or delay the progression of AD. The study has also shown that misexpression of UAS-*lunasin* in UAS-*Aβ42* (AD causing) background in the Glass Multiple Repeat (GMR) domain (comprising of developing retinal neurons of fly-eye), downregulates c-Jun N-terminal kinase (JNK) pathway and rescues Aβ42 mediated neurodegeneration [46, 47].

Many species of plants with rich polyphenol extract are available on the earth's heritage. They can play a role in the different metabolism of the biological system. Polyphenols have been reported to be known for their antitumour activities and to have the power of modulation in the biological response to revitalise and damage the immune system. Similarly, the defence mechanism presented by this compound now includes the elicitation of apoptosis, detoxification of xenobiotics, anti-oxidation, *etc.* It is very interesting to say that it can influence the nuclear factor (NF-kB or activator protein-1) and can control cancer pathogenesis. Polyphenols are known to possess neuroprotective properties. Extract (EGb761) collected from *Gingko biloba* (maidenhair tree), has been shown to play a neuroprotective role in AD and ALS [48, 49]. Curcumin (polyphenol) derived from rhizomes of turmeric (*Curcuma longa* L.), is effective in blocking or inhibiting hyperphosphorylation in JNK signalling, which further inhibits mitochondrial dysregulation and thereby prevents the death of dopaminergic neurons in the mice model of PD [50]. It also facilitates the inhibition of Aβ42 β-oligomers and plaques formation and prevents the formation of reactive oxygen species and neuroinflammation, thereby inhibiting or delaying the onset of AD [51, 52]. Dietary supplementary of another polyphenol-Resveratrol, exhibits neuroprotective properties by activating members of the Sirtuins (SRT) family and reducing the accumulation of Aβ plaques as seen in transgenic models of AD [53]. It is also been known to regulate the activity of microglia (macrophages that help to scavenge damaged neurons) in the brain and facilitate neuronal cell survival [54]. Another polyphenol, epigallocatechin gallate (EGCG), has also demonstrated significant potential to cross the human blood-brain (BBB) and exhibits neuroprotective ability against oxidative stress-induced death of neuronal cell population, thereby serving as a promising neuroprotective compound [55]. Flavonoids, low molecular weight, and plant phenolic bioactive compounds can target multiple sites in the brain and serve as novel therapies in treating many neurodegenerative diseases. Studies conducted in mutant human superoxide dismutase, SOD1 (G93A) mice model of ALS, showed anthocyanin (Flavonoid) extracted from leaves of the strawberry plant, to increase survival, enhance grip strength, preserve Neuro-Muscular Junctions (NMJs), and delay the progression of ALS [56]. Eupatilin (flavonoid) is extracted from herbs belonging to the *Artemisia* species. As seen in a mice model of PD, Eupatilin alleviates behavioural impairment and loss of dopaminergic neuronal cell population through inhibition of neuroinflammation as well as apoptosis and has been found to be an effective therapeutic in treating PD [57].

Alkaloids are nitrogen-containing organic compounds and have potential efficacy in ameliorating the pathophysiology of diseases, and serve as potential therapeutics to treat many neurodegenerative diseases. Huperzine A, Hup A (an alkaloid obtained from herb *Huperzia serrate)* used widely in Chinese medicine, acts as an acetylcholinesterase inhibitor, reduces glutamate neurotoxicity, can target several molecular sites, exerts neuroprotective effects and slows down the onset or progression of many neurodegenerative disorders. Another alkaloid, berberine obtained from *Berberis vulgaris* L., helps to prevent changes in oxidative stress, cholinesterase (ChE) activity, thereby preventing cognitive dysfunction and memory impairment as seen in rats treated with ethanol [58]. It is also known to downregulate the expression of the BACE-1 enzyme, thereby attenuating the production and deposition of Aβ plaques and ameliorating AD [59]. Berberine helps inhibition of apoptosis in the hippocampus area of the brain and prevents the degeneration of dopaminergic neurons. It improves motor balance and helps to alleviate memory loss and motor dysfunction in Parkinson'spatients [60]. Berberine was also found to upregulate the expression of gene: heme oxygenase-1 (HO-1, codes for an enzyme), and downregulates the production of reactive oxygen species, activation of caspase-3 and thereby, cell death in 6-hydroxydopamine (6-OHDA) induced neuronal death in an experimental model of PD [61]. It has also been found effective in preventing mutant huntingtin accumulation in cultured cells, alleviating motor dysfunction and increasing survival, as seen in the transgenic mice model of HD [62]. It possesses a varied pharmacological activity and is useful in treating many disorders related to the nervous system [63, 64].

Terpenes: EGb761 extract, which comprises flavonol glycosides, terpenes trilactones, etc., is obtained from the leaves of an ancient Chinese tree named *Ginkgo biloba.* The extract is considered to provide neuroprotection against many neurodegenerative diseases. Some of the major constituents of terpenes trilactones include Ginkgolides (especially GA and GB), and bilobalide which exhibits neuroprotective effects after cerebral ischemic strokes [65]. Another plant product, Ginseng (contains ginsenosides and saponins), extracted from dried roots of a perennial herb, *Panex ginseng* also carries favourable effects on the nervous system. It helps to improve learning ability, as seen in animal models and is used for AD, PD and HD treatment [66 - 70]. Similarly, studies conducted in the SOD1mice model of ALS have shown that extracts from *Withaniasomnifera*, helps improve longevity, motor performance as well as some motor neurons in the spinal cord [71, 72]. A triterpenoid, Platycodin D (extracted from *Platycodongrandiflorus*), stimulates neuritic outgrowth and inhibits Aβ42 mediated neuronal cell loss. It ameliorates cognitive ability impairments and increases synaptogenesis by phosphorylating and activating the ERK1/2 pathway mitogen-activated protein kinase pathway [73].

CONTINUING SEARCH FOR NATURAL PLANT RESOURCES IN ANTI-CANCER THERAPY

In cancer treatment, NPs found in therapeutic plants played an essential role. Diverse kinds of bioactive compounds from plant sources have been separated. Some of them have been tested or assisted in clinical studies or pre-clinical studies. The plant-derived anti-cancer compounds are four main classifications, *i.e.*, vinca alkaloids, lignans, subalkaloids, and vincaepipodophylllotoxin (Fig. **3a - h**).

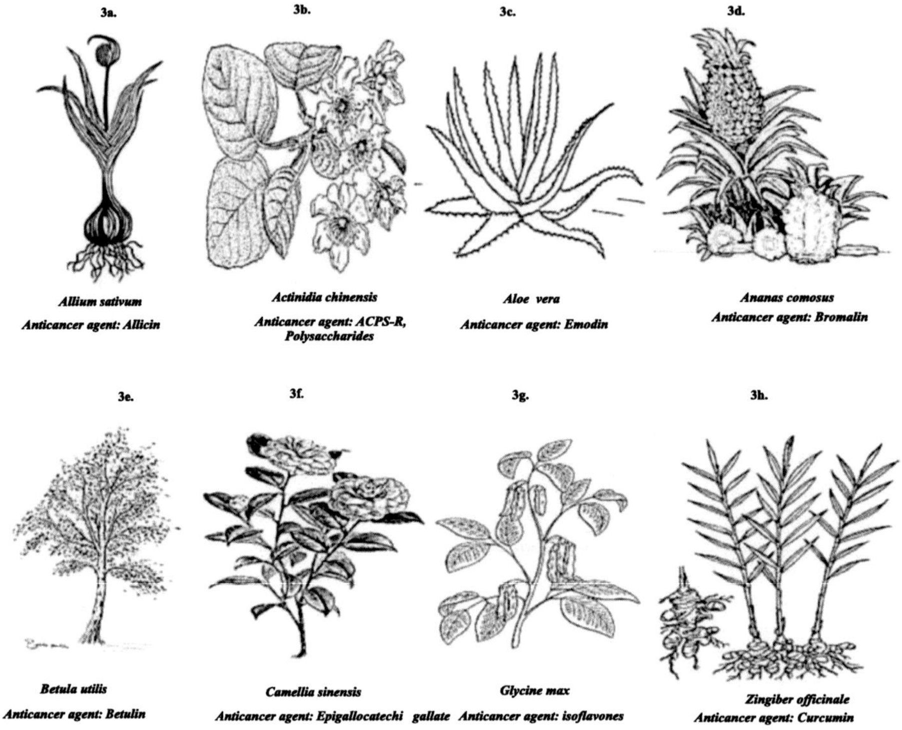

Fig. (3a-h). Plants with anti-cancerous activity and their active constituents (Source: Adapted from [74, 75]).

CONCLUSION

Plants have been blessed with naturally occurring secondary metabolites, which possess anti-bacterial, anti-fungal, anti-viral, and other pharmacological properties. As evidenced by previous studies, plant-derived SNPs and their bioactive compounds have also demonstrated their potential to inhibit or ameliorate the progression of many neurodegenerative diseases. Previous studies have also indicated that neurodegenerative diseases are multifactorial, and involve complex and multiple sets of reactions that results in the degeneration of neurons and, ultimately, death of the patients. Using therapeutic, which can only target a few sites or pathways, fails to provide significant results and limits their ability to be used as therapeutics. However, NPs, due to their multiple natural pharmacological properties, facilitation of consumption as cocktails formulation, and uncanny ability to act on multiple molecular targets/ sites, have demonstrated significant results and proven to serve as versatile naturally occurring neuroprotective therapeutics. Besides this, the ability of NPs to cross the human blood-brain barrier, cost-effective, with none or minimal side effects, increases their demand and triggers a shift of using them as therapeutics in place of currently available synthetic drugs. The pharmacological properties of the many plants derived NPs gained the attention of scientists who continued testing, screening and identification of natural compounds to treat many neurodegenerative diseases. Since then, many compounds have been discovered which can either block or delay the onset or severity of many neurodegenerative disorders, but more plant products still need to be identified. There is also an urgent need to design new methods or strategies (such as nanoparticles) to effectively deliver these plant products to the region of interest. Despite knowing so much about the neuroprotective role of plant-derived NPs, still, less is known about their role in increasing the quality of life of patients and their pharmacological actions. The transition from using them in clinical trials from an experimental lab set up to treat diseases is quite challenging. Thus, future studies should also focus on testing their efficacy in an animal as well as in human models of diseases.

CONSENT FOR PUBLICATION

Not applicable.

CONFLICT OF INTEREST

The author declares no conflict of interest, financial or otherwise.

ACKNOWLEDGEMENTS

The authors extend thanks to the members of Kumar Lab for providing comments on the manuscript.

REFERENCES

[1] Atilaw Y, Muiva-Mutisya L, Ndakala A, *et al.* Four prenylflavone derivatives with antiplasmodial activities from the stem of tephrosia purpurea subsp. leptostachya. Molecules 2017; 22(9): 1514. [http://dx.doi.org/10.3390/molecules22091514] [PMID: 28891957]

[2] Lemma MT, Ahmed AM, Elhady MT, *et al.* Medicinal plants for *in vitro* antiplasmodial activities: A systematic review of literature. Parasitol Int 2017; 66(6): 713-20. [http://dx.doi.org/10.1016/j.parint.2017.09.002] [PMID: 28890153]

[3] Awodiran MO, Adepiti AO, Akinwunmi KF. Assessment of the cytotoxicity and genotoxicity properties of *Uvaria chamae* P. Beauv (Annonaceae) and *Morinda lucida* Benth (Rubiaceae) in mice. Drug Chem Toxicol 2018; 41(2): 232-7. [http://dx.doi.org/10.1080/01480545.2017.1365884] [PMID: 28854821]

[4] Newman DJ, Cragg GM, Snader KM. Natural products as sources of new drugs over the period 1981-2002. J Nat Prod 2003; 66(7): 1022-37. [http://dx.doi.org/10.1021/np030096l] [PMID: 12880330]

[5] Kiraithe MN, Nguta JM, Mbaria JM, Kiama SG. Evaluation of the use of Ocimum suave Willd. (Lamiaceae), *Plectranthus barbatus* Andrews (Lamiaceae) and *Zanthoxylum chalybeum* Engl.

(Rutaceae) as antimalarial remedies in Kenyan folk medicine. J Ethnopharmacol 2016; 178: 266-71. [http://dx.doi.org/10.1016/j.jep.2015.12.013] [PMID: 26705665]

[6] Asnake S, Teklehaymanot T, Hymete A, Erko B, Giday M. Evaluation of the antiplasmodial properties of selected plants in southern Ethiopia. BMC Complement Altern Med 2015; 15(1): 448. [http://dx.doi.org/10.1186/s12906-015-0976-x] [PMID: 26698300]

[7] Lima RBS, Rocha e Silva LF, Melo MRS, *et al. In vitro* and *in vivo* anti-malarial activity of plants from the Brazilian Amazon. Malar J 2015; 14(1): 508. [http://dx.doi.org/10.1186/s12936-015-0999-2] [PMID: 26682750]

[8] Olanlokun JO, David OM, Afolayan AJ. *In vitro* antiplasmodial activity and prophylactic potentials of extract and fractions of Trema orientalis (Linn.) stem bark. BMC Complement Altern Med 2017; 17(1): 407. [http://dx.doi.org/10.1186/s12906-017-1914-x] [PMID: 28810849]

[9] Shen X, Wu M, Liao B, *et al.* Complete chloroplast genome sequence and phylogenetic analysis of the medicinal plant Artemisia annua. Molecules 2017; 22(8): 1330. [http://dx.doi.org/10.3390/molecules22081330] [PMID: 28800082]

[10] Alebie G, Urga B, Worku A. Systematic review on traditional medicinal plants used for the treatment of malaria in Ethiopia: trends and perspectives. Malar J 2017; 16(1): 307. [http://dx.doi.org/10.1186/s12936-017-1953-2] [PMID: 28764723]

[11] Katiyar C, Kanjilal S, Gupta A, Katiyar S. Drug discovery from plant sources: An integrated approach. Ayu 2012; 33(1): 10-9. [http://dx.doi.org/10.4103/0974-8520.100295] [PMID: 23049178]

[12] Lubbad MY, Al-Quraishy S, Dkhil MA. Antimalarial and antioxidant activities of Indigofera oblongifolia on Plasmodium chabaudi-induced spleen tissue injury in mice. Parasitol Res 2015; 114(9): 3431-8. [http://dx.doi.org/10.1007/s00436-015-4568-y] [PMID: 26109255]

[13] Girma S, Giday M, Erko B, Mamo H. Effect of crude leaf extract of Osyris quadripartita on Plasmodium berghei in Swiss albino mice. BMC Complement Altern Med 2015; 15(1): 184. [http://dx.doi.org/10.1186/s12906-015-0715-3] [PMID: 26077462]

[14] Tarkang PA, Appiah-Opong R, Ofori MF, Ayong LS, Nyarko AK. Application of multi-target phytotherapeutic concept in malaria drug discovery: a systems biology approach in biomarker identification. Biomark Res 2016; 4(1): 25. [http://dx.doi.org/10.1186/s40364-016-0077-0] [PMID: 27999673]

[15] Muthaura CN, Rukunga GM, Chhabra SC, Mungai GM, Njagi ENM. Traditional antimalarial phytotherapy remedies used by the Kwale community of the Kenyan Coast. J Ethnopharmacol 2007; 114(3): 377-86. [http://dx.doi.org/10.1016/j.jep.2007.08.033] [PMID: 17904318]

[16] Toma A, Deyno S, Fikru A, Eyado A, Beale A. In vivo antiplasmodial and toxicological effect of crude ethanol extract of Echinops kebericho traditionally used in treatment of malaria in Ethiopia. Malar J 2015; 14(1): 196. [http://dx.doi.org/10.1186/s12936-015-0716-1] [PMID: 25958112]

[17] Memvanga PB, Tona GL, Mesia GK, Lusakibanza MM, Cimanga RK. Antimalarial activity of medicinal plants from the Democratic Republic of Congo: A review. J Ethnopharmacol 2015; 169: 76-98. [http://dx.doi.org/10.1016/j.jep.2015.03.075] [PMID: 25862959]

[18] Mohanty S, Maurya AK, Jyotshna , *et al.* Flavonoids rich fraction of Citrus limetta fruit peels reduces proinflammatory cytokine production and attenuates malaria pathogenesis. Curr Pharm Biotechnol 2015; 16(6): 544-52. [http://dx.doi.org/10.2174/138920101606150407114023] [PMID: 25860065]

[19] Haslinda MS, Aiyub Z, Bakar NKA, *et al. In vitro* antiplasmodial activity, macronutrients and trace metals in the medicinal plants: Phyllanthus spp. and Alpinia conchigera Griff. Trop Biomed 2015; 32(1): 129-39.
[PMID: 25801263]

[20] Diarra N, Klooster C, Togola A, Diallo D, Willcox M, Jong J. Ethnobotanical study of plants used against malaria in Sélingué subdistrict, Mali. J Ethnopharmacol 2015; 166: 352-60.
[http://dx.doi.org/10.1016/j.jep.2015.02.054] [PMID: 25752587]

[21] Houël E, Fleury M, Odonne G, *et al.* Antiplasmodial and anti-inflammatory effects of an antimalarial remedy from the Wayana Amerindians, French Guiana: Takamalaimë (Psidium acutangulum Mart. ex DC., Myrtaceae). J Ethnopharmacol 2015; 166: 279-85.
[http://dx.doi.org/10.1016/j.jep.2015.03.015] [PMID: 25792015]

[22] Singh SV, Manhas A, Singh SP, *et al.* A phenolic glycoside from Flacourtia indica induces heme mediated oxidative stress in Plasmodium falciparum and attenuates malaria pathogenesis in mice. Phytomedicine 2017; 30: 1-9.
[http://dx.doi.org/10.1016/j.phymed.2017.04.010] [PMID: 28545664]

[23] Elbadawi M, Awadalla M, Hamid M, Mohamed M, Awad T. Valproic acid as a potential inhibitor of Plasmodium falciparum histone deacetylase 1 (PfHDAC1): an *in silico* approach. Int J Mol Sci 2015; 16(2): 3915-31.
[http://dx.doi.org/10.3390/ijms16023915] [PMID: 25679451]

[24] Newman DJ. Natural products as leads to potential drugs: an old process or the new hope for drug discovery? J Med Chem 2008; 51(9): 2589-99.
[http://dx.doi.org/10.1021/jm0704090] [PMID: 18393402]

[25] Arruda ALA, Vieira CJB, Sousa DG, Oliveira RF, Castilho RO. Jacaranda cuspidifolia Mart. (Bignoniaceae) as an antibacterial agent. J Med Food 2011; 14(12): 1604-8.
[http://dx.doi.org/10.1089/jmf.2010.0251] [PMID: 21663482]

[26] Cazarolli L, Zanatta L, Alberton E, *et al.* Flavonoids: prospective drug candidates. Mini Rev Med Chem 2008; 8(13): 1429-40.
[http://dx.doi.org/10.2174/138955708786369564] [PMID: 18991758]

[27] Friedman M, Henika PR, Levin C, Mandrell R, Kozukue N. Antimicrobial activities of tea catechins and theaflavins and tea extracts against Bacillus cereus. J Food Prot 2006; 69(2): 354-61.
[http://dx.doi.org/10.4315/0362-028X-69.2.354] [PMID: 16496576]

[28] Arif T, Bhosale JD, Kumar N, *et al.* Natural products – antifungal agents derived from plants. J Asian Nat Prod Res 2009; 11(7): 621-38.
[http://dx.doi.org/10.1080/10286020902942350] [PMID: 20183299]

[29] Agarwal V, Lal P, Pruthi V. Prevention of Candida albicans biofilm by plant oils. Mycopathologia 2008; 165(1): 13-9.
[http://dx.doi.org/10.1007/s11046-007-9077-9] [PMID: 17968673]

[30] Tomczykowa M, Leszczyńska K, Tomczyk M, Tryniszewska E, Kalemba D. Composition of the essential oil of Bidens tripartita L. roots and its antibacterial and antifungal activities. J Med Food 2011; 14(4): 428-33.
[http://dx.doi.org/10.1089/jmf.2010.0066] [PMID: 21370967]

[31] Silva F, Ferreira S, Duarte A, Mendonça DI, Domingues FC. Antifungal activity of Coriandrum sativum essential oil, its mode of action against Candida species and potential synergism with amphotericin B. Phytomedicine 2011; 19(1): 42-7.
[http://dx.doi.org/10.1016/j.phymed.2011.06.033] [PMID: 21788125]

[32] Kamazeri TSAT, Samah OA, Taher M, Susanti D, Qaralleh H. Antimicrobial activity and essential oils of Curcuma aeruginosa, Curcuma mangga, and Zingiber cassumunar from Malaysia. Asian Pac J Trop Med 2012; 5(3): 202-9.

[http://dx.doi.org/10.1016/S1995-7645(12)60025-X] [PMID: 22305785]

[33] Udayan PS, Balachandran I. Medicinal Plants of Arya Vaidya Sala Herb Garden 1st ed., 2009.

[34] Kirtikar KR, Basu BD. Kirtikar KR, Basu BD. Indian Medicinal Plants Inernational Book Ditributers, Dehra Dun, Edition 2 2005; 1: 102-3.

[35] Cutler T, Sarkar A, Moran M, *et al.* Drosophila Eye Model to Study Neuroprotective Role of CREB Binding Protein (CBP) in Alzheimer's Disease. PLoS One 2015; 10(9): e0137691. [http://dx.doi.org/10.1371/journal.pone.0137691] [PMID: 26367392]

[36] Sarkar A, Gogia N, Farley K, Payton L, Singh A. Characterization of a morphogenetic furrow specific Gal4 driver in the developing Drosophila eye. PLoS One 2018; 13(4): e0196365. [http://dx.doi.org/10.1371/journal.pone.0196365] [PMID: 29702674]

[37] Singh A, Gogia N, Chang CY, Sun YH. Proximal fate marker *homothorax* marks the lateral extension of stalk-eyed fly *Cyrtodopsis whitei.* Genesis 2019; 57(9): e23309. [http://dx.doi.org/10.1002/dvg.23309] [PMID: 31162816]

[38] Gogia N, Puli O, Raj A, Singh A. Gogia N, Puli O, Raj A, Singh A. Generation of Third Dimension: Axial Patterning in the Developing Drosophila Eye. Molecular Genetics of Axial Patterning, Growth and Disease in Drosophila Eye. 2020; II.

[39] Irwin M, Tare M, Singh A, *et al.* A Positive Feedback Loop of Hippo- and c-Jun-Amino-Terminal Kinase Signaling Pathways Regulates Amyloid-Beta-Mediated Neurodegeneration. Front Cell Dev Biol 2020; 8: 117.
[http://dx.doi.org/10.3389/fcell.2020.00117] [PMID: 32232042]

[40] Gogia N, Sarkar A, Mehta AS, *et al.* Inactivation of Hippo and cJun-N-terminal Kinase (JNK) signaling mitigate FUS mediated neurodegeneration *in vivo*. Neurobiol Dis 2020; 140: 104837. [http://dx.doi.org/10.1016/j.nbd.2020.104837] [PMID: 32199908]

[41] Lacomblez L, Bensimon G, Meininger V, Leigh PN, Guillet P. Dose-ranging study of riluzole in amyotrophic lateral sclerosis. Lancet 1996; 347(9013): 1425-31. [http://dx.doi.org/10.1016/S0140-6736(96)91680-3] [PMID: 8676624]

[42] Liu J, Jia SH, Kirberger M, Chen N. Lunasin as a promising health-beneficial peptide. Eur Rev Med Pharmacol Sci 2014; 18(14): 2070-5.
[PMID: 25027349]

[43] Galvez AF, Chen N, Macasieb J, de Lumen BO. Chemopreventive property of a soybean peptide (lunasin) that binds to deacetylated histones and inhibits acetylation. Cancer Res 2001; 61(20): 7473-8. [PMID: 11606382]

[44] Jones G, Srivastava A. Understanding Lunasin's biology and potential as a cancer therapeutic by utilizing *Drosophila* genetics. Exp Biol Med (Maywood) 2014; 239(5): 519-28. [http://dx.doi.org/10.1177/1535370214522180] [PMID: 24625440]

[45] Hsieh CC, Hernández-Ledesma B, Jeong HJ, Park JH, de Lumen BO. Complementary roles in cancer prevention: protease inhibitor makes the cancer preventive peptide lunasin bioavailable. PLoS One 2010; 5(1): e8890.
[http://dx.doi.org/10.1371/journal.pone.0008890] [PMID: 20126654]

[46] Singh A, Deshpande P, Gogia N. Exploring the efficacy of natural products in alleviating Alzheimer's disease. Neural Regen Res 2019; 14(8): 1321-9. [http://dx.doi.org/10.4103/1673-5374.253509] [PMID: 30964049]

[47] Sarkar A, Gogia N, Glenn N, *et al.* A soy protein Lunasin can ameliorate amyloid-beta 42 mediated neurodegeneration in Drosophila eye. Sci Rep 2018; 8(1): 13545. [http://dx.doi.org/10.1038/s41598-018-31787-7] [PMID: 30202077]

[48] Ferrante RJ, Klein AM, Dedeoglu A, Beal MF. Therapeutic efficacy of EGb761 (Gingko biloba extract) in a transgenic mouse model of amyotrophic lateral sclerosis. J Mol Neurosci 2001; 17(1): 89-

96.
[http://dx.doi.org/10.1385/JMN:17:1:89] [PMID: 11665866]

[49] Müller WE, Eckert A, Eckert GP, *et al.* Therapeutic efficacy of the Ginkgo special extract EGb761 ® within the framework of the mitochondrial cascade hypothesis of Alzheimer's disease. World J Biol Psychiatry 2019; 20(3): 173-89.
[http://dx.doi.org/10.1080/15622975.2017.1308552] [PMID: 28460580]

[50] Pan J, Li H, Ma JF, *et al.* Curcumin inhibition of JNKs prevents dopaminergic neuronal loss in a mouse model of Parkinson's disease through suppressing mitochondria dysfunction. Transl Neurodegener 2012; 1(1): 16.
[http://dx.doi.org/10.1186/2047-9158-1-16] [PMID: 23210631]

[51] Williams P, Sorribas A, Howes MJR. Natural products as a source of Alzheimer's drug leads. Nat Prod Rep 2011; 28(1): 48-77.
[http://dx.doi.org/10.1039/C0NP00027B] [PMID: 21072430]

[52] Lin CF, Yu KH, Jheng CP, Chung R, Lee CI. Curcumin reduces amyloid fibrillation of prion protein and decreases reactive oxidative stress. Pathogens 2013; 2(3): 506-19.
[http://dx.doi.org/10.3390/pathogens2030506] [PMID: 25437204]

[53] Karuppagounder SS, Pinto JT, Xu H, Chen HL, Beal MF, Gibson GE. Dietary supplementation with resveratrol reduces plaque pathology in a transgenic model of Alzheimer's disease. Neurochem Int 2009; 54(2): 111-8.
[http://dx.doi.org/10.1016/j.neuint.2008.10.008] [PMID: 19041676]

[54] Song J, Cheon S, Jung W, Lee W, Lee J. Resveratrol induces the expression of interleukin-10 and brain-derived neurotrophic factor in BV2 microglia under hypoxia. Int J Mol Sci 2014; 15(9): 15512-29.
[http://dx.doi.org/10.3390/ijms150915512] [PMID: 25184950]

[55] Pogačnik L, Pirc K, Palmela I, *et al.* Potential for brain accessibility and analysis of stability of selected flavonoids in relation to neuroprotection *in vitro*. Brain Res 2016; 1651: 17-26.
[http://dx.doi.org/10.1016/j.brainres.2016.09.020] [PMID: 27639810]

[56] Winter AN, Ross EK, Wilkins HM, *et al.* An anthocyanin-enriched extract from strawberries delays disease onset and extends survival in the hSOD1 ^{G93A} mouse model of amyotrophic lateral sclerosis. Nutr Neurosci 2018; 21(6): 414-26.
[http://dx.doi.org/10.1080/1028415X.2017.1297023] [PMID: 28276271]

[57] Zhang Y, Qin L, Xie J, Li J, Wang C. Eupatilin prevents behavioral deficits and dopaminergic neuron degeneration in a Parkinson's disease mouse model. Life Sci 2020; 253: 117745.
[http://dx.doi.org/10.1016/j.lfs.2020.117745] [PMID: 32376269]

[58] Patil S, Tawari S, Mundhada D, Nadeem S. Protective effect of berberine, an isoquinoline alkaloid ameliorates ethanol-induced oxidative stress and memory dysfunction in rats. Pharmacol Biochem Behav 2015; 136: 13-20.
[http://dx.doi.org/10.1016/j.pbb.2015.07.001] [PMID: 26159088]

[59] Yan R, Vassar R. Targeting the β secretase BACE1 for Alzheimer's disease therapy. Lancet Neurol 2014; 13(3): 319-29.
[http://dx.doi.org/10.1016/S1474-4422(13)70276-X] [PMID: 24556009]

[60] Kim , Cho KH, Shin MS, *et al.* Berberine prevents nigrostriatal dopaminergic neuronal loss and suppresses hippocampal apoptosis in mice with Parkinson's disease. Int J Mol Med 2014; 33(4): 870-8.
[http://dx.doi.org/10.3892/ijmm.2014.1656] [PMID: 24535622]

[61] Bae J, Lee D, Kim YK, Gil M, Lee JY, Lee KJ. Berberine protects 6-hydroxydopamine-induced human dopaminergic neuronal cell death through the induction of heme oxygenase-1. Mol Cells 2013; 35(2): 151-7.
[http://dx.doi.org/10.1007/s10059-013-2298-5] [PMID: 23329300]

[62] Jiang W, Wei W, Gaertig MA, Li S, Li XJ. Therapeutic effect of berberine on huntington's disease transgenic mouse model. PLoS One 2015; 10(7): e0134142. [http://dx.doi.org/10.1371/journal.pone.0134142] [PMID: 26225560]

[63] Ahmed T, Gilani AH, Abdollahi M, Daglia M, Nabavi SF, Nabavi SM. Berberine and neurodegeneration: A review of literature. Pharmacol Rep 2015; 67(5): 970-9. [http://dx.doi.org/10.1016/j.pharep.2015.03.002] [PMID: 26398393]

[64] Jiang W, Li S, Li X. Therapeutic potential of berberine against neurodegenerative diseases. Sci China Life Sci 2015; 58(6): 564-9. [http://dx.doi.org/10.1007/s11427-015-4829-0] [PMID: 25749423]

[65] Feng Z, Sun Q, Chen W, Bai Y, Hu D, Xie X. The neuroprotective mechanisms of ginkgolides and bilobalide in cerebral ischemic injury: a literature review. Mol Med 2019; 25(1): 57. [http://dx.doi.org/10.1186/s10020-019-0125-y] [PMID: 31864312]

[66] Chen X, Zhu YG, Zhu LA, *et al.* Ginsenoside Rg1 attenuates dopamine-induced apoptosis in PC12 cells by suppressing oxidative stress. Eur J Pharmacol 2003; 473(1): 1-7. [http://dx.doi.org/10.1016/S0014-2999(03)01945-9] [PMID: 12877931]

[67] Van Kampen J, Robertson H, Hagg T, Drobitch R. Neuroprotective actions of the ginseng extract G115 in two rodent models of Parkinson's disease. Exp Neurol 2003; 184(1): 521-9. [http://dx.doi.org/10.1016/j.expneurol.2003.08.002] [PMID: 14637121]

[68] Kim JH, Kim S, Yoon IS, *et al.* Protective effects of ginseng saponins on 3-nitropropionic acid-induced striatal degeneration in rats. Neuropharmacology 2005; 48(5): 743-56. [http://dx.doi.org/10.1016/j.neuropharm.2004.12.013] [PMID: 15814108]

[69] Wu J, Jeong HK, Bulin SE, Kwon SW, Park JH, Bezprozvanny I. Ginsenosides protect striatal neurons in a cellular model of Huntington's disease. J Neurosci Res 2009; 87(8): 1904-12. [http://dx.doi.org/10.1002/jnr.22017] [PMID: 19185022]

[70] Chen F, Eckman EA, Eckman CB, Chen F, Eckman EA, Eckman CB. Reductions in levels of the Alzheimer's amyloid β peptide after oral administration of ginsenosides. FASEB J 2006; 20(8): 1269-71. [http://dx.doi.org/10.1096/fj.05-5530fje] [PMID: 16636099]

[71] Dutta K, Patel P, Julien JP. Protective effects of Withania somnifera extract in $SOD1^{G93A}$ mouse model of amyotrophic lateral sclerosis. Exp Neurol 2018; 309: 193-204. [http://dx.doi.org/10.1016/j.expneurol.2018.08.008] [PMID: 30134145]

[72] Patel P, Julien JP, Kriz J. Early-stage treatment with *withaferin A* reduces levels of misfolded superoxide dismutase 1 and extends lifespan in a mouse model of amyotrophic lateral sclerosis. Neurotherapeutics 2015; 12(1): 217-33. [http://dx.doi.org/10.1007/s13311-014-0311-0] [PMID: 25404049]

[73] Kim J, Jeon S, Kim K, *et al.* Platycodon grandiflorus Root Extract Improves Learning and Memory by Enhancing Synaptogenesis in Mice Hippocampus. Nutrients 2017; 9(7): 794. [http://dx.doi.org/10.3390/nu9070794] [PMID: 28737698]

[74] Sakarkar DM, Deshmukh VN. Ethnopharmacological review of traditional medicinal plants for anticancer activity. Int J Pharm Tech Res 2011; 3(1): 298-308.

[75] Nirmala MJ, Samundeeswari A, Sankar PD. Nirmala MJ, Samundeeswari A, Sankar PD. Natural plant resources in anti-cancer therapy- A review. Res Plant Biol. 2011; 1(3): 01-14.

CHAPTER 8

Antimicrobial Drugs Obtained from Marine Algae

Rohan Samir Kumar Sachan[1], **Ritu Bala**[1], **Abdel Rahman M. Al-Tawaha**[2], **Samia Khanum**[3] and **Arun Karnwal**[1,*]

[1] *Department of Microbiology, School of Bio-engineering and Bio-sciences, Lovely Professional University, Phagwara-144411, Punjab, India*

[2] *Department of Biological Sciences, Al Hussein Bin Talal University, Ma'an, Jordan*

[3] *Department of Botany, University of the Punjab, Lahore, Pakistan*

Abstract: In recent years, the antimicrobial resistance to various synthetic or chemically formed antimicrobial agents in medicines and food products has been observed. The high preference of consumers for purchasing food products free from chemical preservatives has led to more exploration into using antimicrobial agents from natural sources like plants, fungi, algae, and bacteria. The marine ecosystem comprises microorganisms, plants, vertebrates, and invertebrates that are rich sources of diverse antimicrobial products and can be a significant potential for developing novel type therapeutic agents, as the major portion of the sea has still not yet been examined for the evaluation of natural molecules for their antimicrobial activity. Such marine ecological niches promise a great source of antibacterial agents against many drug-resistant strains of pathogenic microorganisms. Among the marine source, marine algae are a diverse group of organisms that includes brown, red, and green algae that have been targeted over the last few years for the secondary metabolites and a broad range of natural molecules for a broad spectrum of bioactivities beneficial to humans. Such bioactive compounds and secondary products possess a broad range of biological activities of antibacterial, antiviral, and antifungal properties. The class of compounds derived from marine algae, such as polysaccharides, fatty acids, phenolic compounds, pigments, lectins, alkaloids, terpenoids, and halogenated compounds, would be a new emerging area for unconventional drugs. Such classes of compounds will share a potent ability to control new diseases or tackling against multi-resistant strains of pathogens.

Keywords: Alkaloids, Antimicrobial activity, Antimicrobial resistance, Bioactive compounds, Fatty acids, Halogenated compounds, Marine algae, Phenolic compounds, Pigments, Polysaccharides, Terpenoids.

* **Corresponding author Arun Karnwal**: Department of Microbiology, School of Bio-engineering and Bio-sciences, Lovely Professional University, Phagwara-144411, Punjab, India; E-mail: arunkarnwal@gmail.com

M. Amin-ul Mannan & Gaurav Kumar (Eds.)

INTRODUCTION

The word algae are derived from the Latin word seaweed, and the study is known as phycology. This is a very diverse and large group of eukaryotic and photosynthetic organisms that are more likely plant-like because of the presence of chlorophyll for photosynthesis, but these are also differentiated from plants because of the absence of leaves, roots, flowers, vascular tissues and seeds. The photosynthetic pigments are also varied and different as compared to plants. They have unique features in their cells that are not present in animal and plant cells. Algae are usually varied from single-cell organisms to multicellular organisms. They have a variety of life cycles depending on the size and type of organism. The size of organisms also varies from microscopic Microsomonas to 60-mete--long marine algae known as Kelp [1].

If the algae are seen by the naked eye, multicellular and found in marine, they are known as seaweeds. The unicellular algae are not defined as seaweed. The seaweeds, also known as marine algae, are defined as macroscopic multicellular algae, benthic and mostly anchored at the bottom of the ocean and also attached to the solid structure at the bottom of the sea. Seaweeds are of many different types, having leaves and stems and look like plants, like moss, mushroom, leaf lettuce and a palm tree [1].

Seaweeds are also providing food and habitat for other marine organisms. They are also used as fertilizer and as a food source for humans, just like agar which is used for the culturing of microorganisms and as gelatin in soup, jellies, ice cream and many other products [1]. Their presence in the marine environment abundantly provides useful compounds such as dietary fibres, omega-3 fatty acids, carotenoids, vitamins and minerals. They also have therapeutic benefits, which are both internally and externally. They are also known as producers of oxygen and food in the marine ecosystem. They also produced economically important materials like crude oil, food, pharmaceutical, industrial, *etc.* [2].

The seaweed industry produces useful products worth 5.5 to 6 billion US dollars annually. About 5 billion US dollars are obtained from human food product consumption. The hydrocolloids produced by seaweeds are a major part of the left billion amount, and substances like miscellaneous and other smaller products like fertilizers, animal feed *etc.*, make up the rest. The seaweed industry uses about 7 to 8 million tons of wet seaweed annually both from natural and cultivated sources. Due to high demand, the farming of seaweed is expanded recently. Production of seaweed is carried out in 35 different countries at the commercial level ranging from the Northern and Southern hemispheres [3].

In the last three decades, the pharmaceutical industries produced a large number of antibiotics, but microorganisms developed resistance against these antibiotics, especially bacteria, because they have the genetic ability to develop resistance against drugs such as therapeutic agents [4]. Marine algae are very important in this aspect because these are useful sources for the production of bioactive agents for a very long time [5]. There are about 150,000 seaweed species found in marines around the globe, but only a limited number have been identified to date [6]. The bioactive compounds, primary or secondary, extracted from seaweed are of great importance to the pharmaceutical industries [7]. Substances produced from the seaweed are alginate carrageenan, and phycocolliods like agar have been used in medicine for decades [8]. There are many secondary metabolites produced by certain groups of microbes, sponges, seaweeds and other organisms in the marine ecosystem. Many types of research have reported the secondary metabolites like an antibiotic, antiviral, antimicrobial, anti-neoplastic, antifouling, anti-inflammatory, cytotoxic, and antimitotic activities of the chemical compounds produced by seaweeds.

DIVERSITY IN THE MARINE ENVIRONMENT

The world holds nearly 70% of the water; of that, 97% is marine water. The marine environment consists of a huge natural habitat for various microorganisms with different physiology. The main marine ecosystem is marshes, tidal zones, mangroves, seagrass, and coral reef [9, 10]. Marine life solely depends on the salinity of the sea. So, the marine habitat is divided into pelagic and demersal zones. More than 80% of plants and animal species, including microorganisms, are a part of the ocean [11]. The various factors, such as nutrient concentration, temperature and hydrostatic pressure, are most important for any microbial community to Inhabitat. The microorganisms present in the marine ecosystem emphasize the diverse collection of bacteria, fungi, algae, and viruses. The bacteria are present in huge numbers and perform wide activities, such as photosynthesis [12], nitrogen fixation [13], agarolysis [14], and production of enzymes and antibiotics such as chitinase, cellulase, protease, and phosphatase [15]. Different physiological groups of bacteria include Archaebacteriathat, sulfate-reducing bacteria *Desulphomonas* sp., *Desulphovibrio* sp., and *Desulphococcus* sp. These groups with chemoautotrophic, anaerobic, thermophilic, and mesophilic physiological properties play a crucial role in the marine environment, contributing to coastal marine sediments for over 50% of the carbon turnover. Another group of bacteria is methanogenic bacteria *Methanococcus* sp., *Methanogenium* sp., and *Methanobolus* sp., which are chemotrophic in nature that utilizes simple carbon compounds in limits such as hydrogen, formate, acetyl, and methanol for methanogenesis [16] and carbon and energy sources. They also utilize trimethylamine as a substrate to produce methane as an end product in the marine environment [17]. The last group of bacteria found in the marine environment is the halophiles like *Halococcus* sp., *Halobacterium* sp., *and Halofarex* sp., which requires sodium chloride of at least 12 to 15% for their survival. They can form red colonies in the marine environment due to the presence of a high amount of carotenoid content [18]. They have the capability to resist the denaturing effects of salt in the environment due to the regulation of the osmotic pressure and can dominate in high salt concentrations.

Viruses are more common in the marine environment, known as marine viruses. The most ubiquitous Piscine orthoreovirus (PRV), many DNA viruses, and virophage parasites are the viruses present in such an environment [19]. Many techniques have been used for the identification of viruses. Transmission electron microscopy was a standard method for the enumeration of viruses in seawater samples [20]. The DAPI (4, 6-di amidino-2-phenylindole) and SYBR Green I strains have been identified in recent years for the counting of virus particles from the marine environment by epifluorescence microscopy. It has been suggested that in deep-sea sediment, bacteria have a higher abundance than the virus, where viral growth depends upon both biotic and abiotic complex interaction factors, which include the water column for virus supply as well as a bacterial metabolic state [21].

Around 500 species of marine fungi that are important producers in the deep sea for carbonate detritus have been reported [22]. The marine fungi comprise species from basidiomycetes, ascomycetes, chytrids, and mitosporic. These fungi are capable of forming spores even in such saline water. However, The order *Halospaeriales* are a common marine species of fungi [23].

Marine fungi play an important role in ubiquitous organisms by the cycling of nutrients and in decomposition. Around 150 species are found in mangrove decaying wood, aerial roots, and seedlings, and these mangrove-inhabiting fungi are known as Manglicolous fungi [24]. There are two approaches for the ecology of fungal study: Terminal fragment length polymorphism analysis (T-RELF) and denaturing gradient gel electrophoresis (DGGE) [25]. With the help of these techniques, many samples can be processed in a short time which allows the fungal community profiling.

The marine environment is also comprised of many algal species that are equally important to the ecosystem. Marine algae are of two types microalgae, firstly, blue-green algae and dinoflagellates, diatoms, and secondly, macroalgae also known as seaweed, comprised of red, green, and brown algae [26]. These are the largest primary producers of food that circulate and feed the entire marine ecosystem. Marine algae are regarded as the main reservoir of secondary metabolites and have many industrial and pharmaceutical applications.

Much research regarding the exploitation of marine microorganisms, animals, and plants has been reported to be genetically and biologically important [27]. More than 15,000 natural compounds have been described as well as reported, from which the marine sponges have contributed 30% of products. However, the necessity to understand the phylogenetic perspective, the degradation mechanism of the marine organisms, or microorganisms is important for the development of novel strategies for the treatment of human diseases. Microbial diversity analysis helps in isolating and identifying new potential microorganisms which are having high specificity for recalcitrant compounds [28]. The utilization and conservation of biological marine diversity require knowledge about the species to keep the balance in the environment. Marine diversity plays a vital role in understanding the novel microorganism for bioactive compounds screening.

CLASSIFICATION OF SEAWEEDS

The macro algae are known as seaweeds. That's how seaweeds differ from micro algae which are mostly unicellular and microscopic in size. The most important example of microalgae is blue-green algae which are usually contaminated and bloom the rivers and streams. Based on pigments present in seaweeds, they are classified into three major groups, which are green, brown and red algae (Fig. **1**). The green algae are named because of the presence of chlorophyll a and b pigments, and they are smaller in size as compared to other groups. The brown algae are named because of the presence of carotenoids, xanthophyll, chlorophyll a and c. The brown algae mostly large in size ranges from large-sized kelp with 20 m in length to thick and leather-like brown algae with 2 to 4 m in length and smaller ones with 30 to 60 cm in length. The red algae are named because of the presence of red pigment r-phycoerythrin. The red algae are not always red, but their colour varies from purple to brownish red and red but classified as red algae because of other similar characteristics. They are not very large, and their length mostly varies from a few cm to one meter. These groups are named Chlorophyceae, Phaeophyceae and Rhodophyceae, respectively [29].

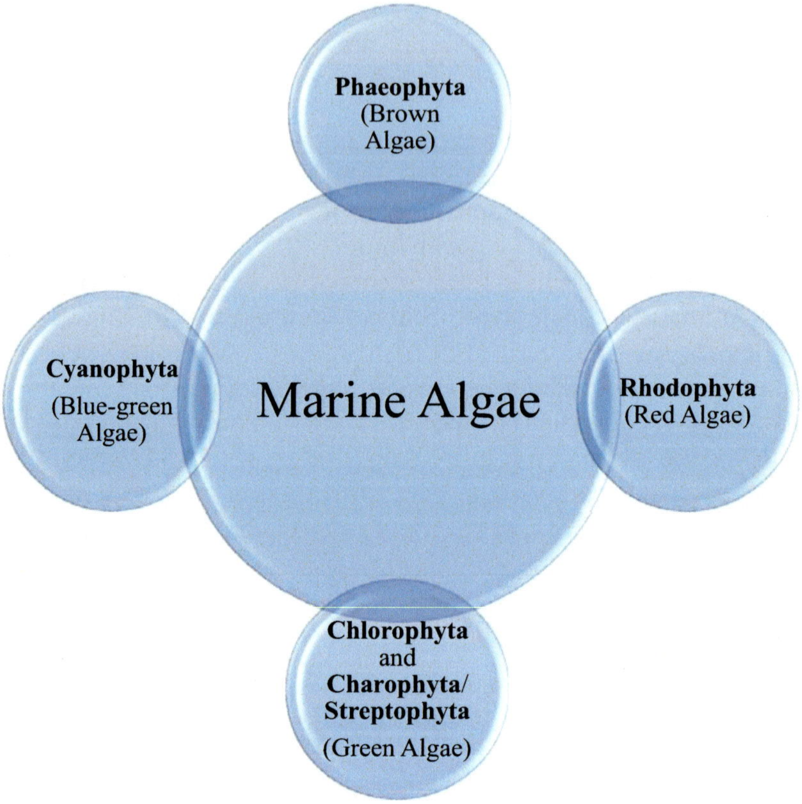

Fig. (1). Different groups of marine algae.

Green Algae

Green algae are one of the much-diversified groups of seaweeds. The habitat of the green algae ranges from marine to freshwater and is also found in terrestrial habitats. Based on DNA sequencing, the classification has been revised recently. According to recent classification, the green algae are not considered as forming homogeneous and coherent individuals but are considered as part of a large group named Viridiplantae. This group also includes land plants in it with green algae [30]. All the green algae present in the marine ecosystem are classified under the same group named Ulvophyceae. The group consists of 920 species which is diversely distributed in all seas of the world. This group shows variations in the body form but with simple morphology. The common growth form present in Cladophora and Chaetomorpha is thin filamentous, branched or unbranched. Two layers of cells form a sheet-like structure known as Ulva. They are known as sea lettuce because of their appearance. The high growth rate and nutrients absorption of green algae are striking features that make them famous in seawater. In eutrophic water, there is a high rate of growth of Ulva, which is a common phenomenon. When the growth at such a high rate is uncontrolled, it can lead to the formation of green tides. These green tides were then removed mechanically to clear the algal biomass.

Another unique growth form present in green algae is the siphonalean organization. These siphonalean algae are further classified into two major groups Bryopsidales and Dasycladales. This group is considered as ecologically successful seaweeds. Their body consists of one large single cell with so many nuclei. The genus Caulerpa is considered the best example of this group. They have creeping stolon which means they grow while attaching to the bottom of the rocky surface with many erected fronds with different shapes grows. The Caulerpa species are found both in tropical and warm temperate seas of the world. They are also very common in aquarium hobbyists and are also used in tropical aquariums widely because of their mind-blowing appearance [31]. The only drawback of this genus is its uncontrolled growth rate. One of the very remarkable examples of the invasion of marine organisms happened in the Mediterranean Sea when the *Caulerpa taxifolia* was released from Monaco Oceanographic Museum accidentally. After this event yet another population of the green alga Caulerpa racemosa var. cylindracea also spread in the same way from Australia that invaded the Mediterranean Sea [32, 33]. The other examples of the siphonalean are Codium, Acetabularia, Halimeda, Udotea and Valonia. The Halimeda is having a unique appearance. Its body is made up of connected segments which may be arranged in the form of coins, wedges or sausages in the form of branches. It also spread widely in the seas of the tropics. The calcium carbonate accumulated in their cell walls and form calcite. After the death of these algae, the calcite help in the formation of white sand, this is the typical feature of the atoll beaches.

Red Algae

The ancient eukaryotic algae are considered red algae. The fossils of Bangiomorpha pubescens which are 1.2 billion years old are considered the oldest red algae. The distinguishing feature of this group is the absence of flagella as they do not have centrioles but the other structures necessary for the flagellar apparatus are present at any stage of their life cycle [34]. There is a wide range of colours of the red algae which ranges from pinkish to bright red to purple and sometimes dark brown. This colour variation is because of the presence of phycobilins pigments. According to recent data about 6000 species of red algae have been identified. The life cycle of red algae is very complex and consist of an alternation of three generations: gametophyte, carposporophyte, and tetrasporophyte. The structure of sexual reproductive organs is very complex and sophisticated that's why used as the main criteria for studying taxonomy at the ordinal level for a long time. The advancement in molecular data improved the classification of the organisms belonging to the red algae. The single phylum is named Rhodophyta which is further divided into two subphyla named *Cyanidiophytina and Rhodophytina*. This group was further classified into seven classes: *Cyanidiophyceae, Bangiophyceae, Compsopogonophyceae, Florideophyceae, Porphyridiophyceae, Rhodellophyceae* and *Stylonematophyceae*, and other 33 orders [35, 36].

This group has a vast range of morphological variations. The single-celled simplest form like *Porphyridium* and thin filamentous red algae like *Bangia*. The expanded blades are a very common trait found in red algae especially present in one of the most spectacular ones *Delesseria, Polyneura* and *Halymenia*. Another example of economically important and widespread red algae that has a blade-like structure is *Porphyra*. The Japanese name nori *Porphyra yezoensis* and other species are recognized commonly used in eastern Asian countries as a food. These algae are also commonly used for wrapping sushi.

The body of other red algae is made up of crust which commonly attaches and grows on the rocky bottom. The classical example of this group is the species from the order *Corallinales*. They accumulate calcium carbonate in their cell wall in the form of aragonite and confer a hard rough coriaceous consistency to their body. *Lithophyllum, Lithothamnion* and *Phymatolithon* belong to this order are pink or red with calcified crusts. These algae have resistance against grazing and mechanical dislodgement that's why they are most commonly found on exposed rocky shores which are not suitable for the algae with soft tissues because they can be easily dislodged due to high waves.

Other species of red algae have branched plant-like forms and that's why they are looking like small bushes or even small trees. Very common genera like Chondrus, Gelidium, Gigartina, Gracilaria, Hypnea and Laurencia are showing such growth habits and body form. Species belonging to the Euchema and Kappahycus are a great source of carrageenans used in the food industry. This is why these algae are produced by farming on large scale, especially in the Philippines and Indonesia.

Brown Algae

There are 1780 species represented the brown algae which are recently classified in the phylum Ochrophyta under the class Fucophyceae also known as Phaeophyceae including 17 orders. They are not closely related to red and green algae instead they are macroscopic and live together on the rocky shores. Brown algae belong to another separate kingdom named Chromista. The close relative of these brown algae is microscopic algae found in the planktonic region of the sea and lakes. The common examples of these microscopic algae are diatoms, chryophytes and xanthophytes. Brown algae are found in all marine ecosystems in the world but most importantly they are abundant in cold sea water, especially their economically important and large-sized species from the orders *Laminariales* and *Desmaredtiales* are only found in the cold-temperate waters. Other than this general trend of living in cold water some other important members like Sargassum and Turbinaria are commonly found in tropical waters [37].

The variation in the shape or form found in the brown algae is diverse as that of green and red algae. They also have filamentous with thin branches like extensions and growing on the rock of also on the large seaweeds like *Ectocarpus* and *Pylaiella*. They usually grow in the intertidal zone of different regions in the World. The brown algae usually are larger and show branched ribbon-like arrangements or sometimes bushes or small trees.

The ecologically important members of the brown algae belong to the order *Fucales*. The reason behind that is they form dense belts in the intertidal zones of temperate seas, especially on rocky shores. Because of such a growth pattern they provide the habitat for supporting biological diversity and many other important species. The best examples of this are the *fucus* belt in the northern Atlantic region and the other one is the *Cystoseira* commonly found in Mediterranean regions. The only specie of this order which is free-floating never attached to any rocky surface and is the largest and found in the Sargasso Sea which is found in the North Atlantic Ocean. This area is restricted by high oceanic currents in which the Sargassum floats freely.

The largest seaweed is known as kelp also belongs to the group brown algae and is classified under the order Lminariales. *Macrocystis pyrifera* also recognized as Pacific kelp is the largest kelp in size with a maximum length of 60 meters. It formed the dense forest which is usually present on the Pacific shores of Canada and the USA. Other submerged forests are also formed by other brown algae in different regions of the world. These species include *Ecklonia, Eisenia, Laminaria and Lessonia.* Kelps are considered the most complex seaweed because of its anatomical and morphological characterizations. Their tissues have kinds of cells and tissue which are very comparable to the vascular plants.

ANTIMICROBIAL COMPOUNDS FROM MARINE ALGAE

The marine algae offer many useful products that are used worldwide in many sectors such as pharmaceuticals, medicines, agriculture, and food industries.

However, the detailed derived compounds from marine algae have been mentioned in the following points and Table **1**.

Table 1. Summarizes some of the marine-derived components along with their target microorganisms.

Marine Algae	Class of Compounds	Target Microorganisms	References
Brown Algae	Polysaccharides (Fucoidan)	Human Simplex Virus (HSV-1 and HSV-2) and Cytomegalovirus	[43]
Ulva rigia (Green algae)	Fatty acids	*Staphylococcus aureus* and *Enterococcus faecium*	[46]
Sargassum fusiforme, Sargassum vulgare (Brown algae)	Phlorotannins	*Staphylococcus aureus* and *Klebsiella pneumonia*	[52]
Padina boryana (Brown algae)	Phlorotannins	Methicillin-resistant *Staphylococcus aureus, Neisseria gonorrhoeae, Moraxella catarrhalia*	[51]
Undaria pinnatifida (Brown algae)	Flavonoids	*Staphylococcus aureus, Streptococcus faecium, Bacillus subtilis*	[57]
Dictyopteris membranacea (Brown algae)	Flavonoids	*Staphylococcus aureus, Enterococcus faecium*	[56]
Solieria filiformis (Red algae)	Lectin	*Serratia marcescens, Salmonella typhi, Klebsiella pneumoniae, Enterobacter aerogenes*	[64]
Eucheuma serra (Red algae)	Lectin	HIV and Influenza virus	[65]
Tetraselmis suecica (Green algae)	Antimicrobial peptides	*Escherichia coli, Salmonella typhimurium, Pseudomonas aeruginosa, Bacillus subtilis, Staphyloccocus aureus, Listeria monocytogenes, Micrococcus letus*	[69]
Laurencia obtusa (Red alage)	Terpenoids	*Bacillus subtilis, Staphylococcus aureus, Candida albicans*	[70]
Chrysophaeum taylorii (Brown algae)	Halogenated compound	Drug-resistant *Staphylococcus aureus* and Vancomycin-resistant *Enterococcus faecium*	[72]
Callophcus serratus (Red algae)	Halogenated compounds	*Enterococcus faecium*, Methicillin-resistant *Staphylococcus aureus*, Vancomycin-resistant *Enterococcus faecium*	[74]

Polysaccharides and Sulfated Polysaccharides

Marine algae components have always been a potential source of nutrients for human consumption and also the raw material for many potential pharmacological drugs with antimicrobial activity. The exploitation of marine sources in terms of their isolation has always been difficult to maintain the nutrient requirement on *ex-situ* sites. Many suggested component of the marine algae, polysaccharides offers more than just being a crucial component of the cell wall of algae. The cell wall of marine algae, as we know, consists of complex amorphous, crystallized or matrix polysaccharides, proteins, metal ions, lipids. Marine algae polysaccharides (MAPs) offer a much wide application for human health. The biological activities of such polysaccharides have been seen as immunomodulators, agents used as antitumor; antiviral; antimicrobial; antioxidant, and hypolipidemic and antioxidant properties. The marine algae also produce many derivatives of polysaccharides that are storage reservoirs inside the cells. The extracted polysaccharides would be useful for various purposes depending upon their three biological activities such as desulfation, depolymerization, and sulfation that occurs inside the marine algae. Such activities lead to the production of useful derivatives of polysaccharides and occurring of such derivatives depends on the group of algae and their particular requirement [38]. The main groups of marine algae known as Chlorophyta, Rhodophyta, and Phaeophyta are the potent producers of such polysaccharides in the marine ecosystem. These groups have different forms of carbohydrate reserved polysaccharides. The green algae have cellulose which is a homopolysaccharide that consists of repeating glucose monomers. The ulvan is another carbohydrate regarded as a complex sulfated-polysaccharides that consists of a disaccharide glucuronic acid and iduronic acid. The ulvans have an important property of forming hydro-reversible gels similar to that of agar. Such property has wide applications in the pharmaceutical, bio-medical, and agricultural sectors. The ulvans display a similar and essential role in eliciting an immune response because the polysaccharide resembles the surface antigen of many microorganisms. Such molecules can be useful in generating a wide range of antibodies against many pathogenic microorganisms. Yet another usefulness of ulvans is seen as an alternative to heparin due to risks associated with its uses like an allergic reaction and heparin-induced thrombocytopenia. The ulvan has high anticoagulant properties than the conventional heparin and had reported no cytotoxic activity in humans. The ulvans increases the stimulation of mucin secretion from epithelial cells in animal and human tissues that protect the colonic mucosa [39]. The marine algal polysaccharides are also useful in the formation of hydrogels due to their bio-compatible, adhesive, bio-degradable properties. Such hydrogels have wide applications in medical and pharmaceutical sectors like the controlled release of drugs, tissue engineering, the formation of contact lenses, and wound dressings. The hydrogels from *Falkenbergiaru folanosa* polysaccharides (FRP) which is sulfated polysaccharides have proven to absorb and retain biological fluids. Another biological function of FRP has been seen in the *in-vitro* release of insulin as an oral drug delivery carrier [40]. The sulfated galactans are the modified polysaccharides that are used as potent enhancers for increasing levels of class II MHC molecules in dendritic cells. This has been proven experimentally by Tamara and co-workers in the research that enhanced the maturation of dendritic cells in mice [38]. The polysaccharides 1, 2, 3 obtained from *Gracilaria lemaneiformis* have been shown to exhibit antitumor activity [41]. The sulfated polysaccharides like fucoidan, ulvan, carrageenan have applications in dental care. These polysaccharides prevent dental caries or dental plaque bacteria of the genus Streptococcus [42]. The red algae have agar, carrageenan polysaccharides, however, carrageenan has gained biomedical interest. The polysaccharides have properties like anticoagulant, antiviral. The carrageenan has been widely used in treating respiratory weakness like common cold or flu and maintaining a balance of cholesterol and triglycerides in the blood. The agar has a wide application in the cultivation of microorganisms owing to the property of forming gel at room temperature and no nutrient composition. The fucoidan is another yet important polysaccharide derived from brown algae that have anticoagulant properties. The polysaccharides also have antiviral properties against the human simplex virus (HSV-1 and HSV-2) and human cytomegalovirus [43]. A.H. and his co-workers isolated polysaccharides from red alga *Pterocladia capillacea* and brown alga *Dictyopteris membranancea* based on cold water and hot water extraction techniques. Both extracts of polysaccharides from both the algae upon characterization using HPLC showed the presence of different complex polysaccharides made from monomers of glucuronic acid, galactose, mannose, galacturonic acid, fructose, glucose, arabinose, ribose, *etc.* The polysaccharides from red and brown algae both confers high and potent antimicrobial activity against gram-negative and gram-positive bacteria than the activity against ampicillin and clotrimazole used as standard antibiotics. The polysaccharides also possess antitumor activity that was analyzed its efficacy against cervix carcinoma HELA cell lines [44]. The antimicrobial activities of polysaccharides like sulfated polysaccharides tend to depend on the distribution throughout the algae, the molecular weight of different compounds, charge density, sulfate content, and structural and conformation of the compounds.

Lipid, Glycerol, and Sterols

The lipid constituents are a crucial part of the cell membrane acting as a selectively permeable membrane in marine algae. Apart from being a constituent of the cell membrane, lipids are produced as a secondary metabolite in accordance to cope with the stress environment or to keep eaten away by the predators in the marine environment. The important function of lipids has been seen in the profiling of lipids of marine algae that are regarded as chemotaxonomic biomarkers. Such biomarkers help to position newly identified marine algae to their specific taxonomic and also helps in studying the organic geochemistry for analyzing their abundance and ecology in the marine system. This is also regarded as the signature profiling of marine algae. Although the lipids have broadened their use as antibacterial, antiviral, antitumor, anti-inflammatory, antiproliferative, antihelminth, and antioxidants activities. The lipids are the long chains of fatty acids and many reports have suggested that the fatty acids with a high range of polyunsaturation are beneficial to humans. The polyunsaturated fatty acids (PUFAs) such as omega-6 and omega-3 along with vitamins have served a great purpose in the prevention of atherosclerosis, and the frequency of heart and vessel diseases. The distribution of lipids in marine algae is somewhat similar to terrestrial micro-algae consisting of fractions of neutral lipids, glycerophospholipids, and phospholipids. Such distribution and the accumulation of lipids depends solely on climate and geographical area. The common fatty acids found in marine algae lipids are palmitic acid, stearic acids, eicosaterienoic acids, docosapentaenoic acids, arachidonic acids, oleic acids, *etc.* Such fatty acids have shown a broader range of antimicrobial activities against gram-negative as well as gram-positive bacteria. Karl-Gunnar and his coworker isolated several species of brown algae that were characterized to have lipid molecules consisting of myristic (14:0), palmitic (16:0), oleic (18:1), arachidonic (29:4) and ecisapentaenoic (20:5) acids. The lipid content showed a broad range of antibacterial activity against both gram-negative *Escherichia coli*, *Proteus vulgaris*, *Pseudomonas fluorescens*, *Serratia marcescens*, *Branhamella catarrhalis* bacteria and gram-positive *Bacillus cereus*, *Bacillus subtilis*, *Micrococcus luteus*, *Staphylococcus aureus* bacteria [45]. Ahmel and his associate isolated and characterized green marine algae *Ulva rigia* from Cap Zabib rocky shore and Ghar El Melh lagoon in Tunisia. The fatty acids fraction of oleic (18:1), linoleic (18:2), palmitic (16:0), stearic (14:0) acids were identified using NMR (nuclear magnetic resonance) and Gas chromatography techniques. These fatty acids fractions showed antimicrobial activity against *Staphylococcus aureus* ATCC 25923 and *Enterococcus faecalis* ATCC 29212 among the 19 indicator microorganisms taken under their experiment [46].

The sterols are important organic molecules and other types of lipids occurring naturally in plants, animals, algae, fungi, and certain species of bacteria. The biological function of sterols has been seen in maintaining the fluidity of the cell membrane, signal transduction, and cholesterol-lowering agents. There are different sterols present in the marine algae depending upon the different classes such as fucosterols are present in brown algae, cholesterol in red algae, and β-sitosterol in green algae. The sterols also play an important in limiting the pathogenic microbial load. The cholesterol 24-propylidene cholest-5-en-3 β-ol was identified and extracted from marine algae *Laurenica papillosa* that belongs to Rhodophtya class using GS-MC (Gas chromatography-Mass Spectrometer). The sterol showed promising a broad range of antimicrobial activities against clinical samples of *Escherichia coli*, *Klebsiella pneumoniae*, *Pseudomonas aeruginosa*, *Shigella flexineri* [47]. Kamariah and the associates isolated antimicrobial fractions of sterols like campesterol, stigmasterol, and β-sitosterol from marine algae *Dictyota dichotoma* and *Sargassum granuliferum*. The sterols showed antibacterial activity against marine gram-negative bacteria such as *Vibrio mimicus*, *Vibrio parahaemolyticus*, *Pseudomonas aeruginosa*, *Bacillus subtilis* [48]. A marine microalgae *Nannochloropsis oculata* derived oil named fatty acid methyl esters (FAMEs) consist majorly of three fatty acids palmitic acid (16:1), oleic acid (18:1), and arachidonic acid (18:2). The antimicrobial activity of FAMEs showed higher efficacy against gram-negative bacteria *Escherichia coli* and *Pseudomonas aeruginosa* than gram-positive bacteria *Bacillus subtilis* and *Staphylococcus aureus* where streptomycin was used as positive control and the antimicrobial activity against yeast *Candida albicans* was also achieved rather than the low activity of the positive control amphotericin B [49].

So the important aspect of lipid as a potent antimicrobial drug depends on the fatty acid composition along with the degree of unsaturation.

Phenolic Compounds and their Derivatives

The phenolic compounds are a group of phytochemicals produced by marine algae, especially brown algae, as secondary metabolites. Chemically, these are aromatic benzene ring-structured compounds having more than one hydroxyl group attached to the ring. The phenols are hydrophilic or may be embedded in the cell wall as its composition [50]. Manal and his fellow associates quantified total phenolic content from two brown algae isolates *Padina boryana* thivy and *Enteromorpha* sp. using ethanol extracts and only *Padina boryana* thivy showed promising antimicrobial activity against Methicilin resistance *Staphylococcus aureus*, *Moraxella catarrhalis*, and *Neisseria gonorrhoeae* [51]. The diethyl ether extract of *Sargassum vulgare* and ethanol extract of *Sargassum fusiforme,* upon characterization using GS-MS (Gas chromatography-Mass spectrometer), showed one compound as phenols among different compounds to possess antimicrobial activity against *Staphylococcus aureus* and *Klebsiella pneumoniae*, respectively [52]. The phlorotannins groups are derivatives of the polyphenols made of repeating units of phloroglucinol that have antimicrobial activities against a broad range of microorganisms [53]. The phlorotannins such as eckol, dieckol, dioxinodehydroeckol, fucofuroeckol-A, 7-phloroeckol, and phlorofucofuroeckol-A have a broad range of antimicrobial hosts like *Escherichia coli*, *Salmonella typhimurium*, Methicilin resistant *Staphylococcus aureus*, *Vibrio parahaemolyticus*, *Campylobacter jejuni*, *Staphylococcus aureus* [53 - 55]. The acetone fraction of *Dictyopterismembranacea* revealed a high amount of phenol and tannins. The acetone fraction confers high antimicrobial activity against *Staphylococcus aureus*, *Streptococcus agalactiae*, *Bacillus subtilis*, and *Enterococcus faecium* rather than the antimicrobial activity observed using ampicillin antibiotic. The acetone fraction also showed promising potent antitubercular activity against *Mycobacterium tuberculosis* [56].

Pigments

Marine algae possess pigments that are visible to human perception based on the wavelength of visible light that is reflected. The various pigments present in classes of algae are β-carotene, fucoxanthin, chlorophyll, phycoerythrin, and phycocyanin. Such pigments are of great economically and biologically important to humans as they possess antioxidants, antiobesity, antidiabetic, and anticancer properties. However, among the listed pigments, fucoxanthin fulfils the criteria of having antimicrobial activity. The fucoxanthin belongs to a class of xanthophyll commonly found only in the brown algae (Phaeophyceae) that gives a brown to olive-green colour to the algae [57, 58]. The fucoxanthin pigments tend to inhibit gram-positive bacteria or aerobic bacterial species and no has no antimicrobial activity against the gram-negative bacteria [57]. The fucoxanthin extracted from *Undaria pinnatifida* strongly inhibits gram-positive pathogenic bacteria such as *Enterococcus* sp., *Staphylococcus aureus*, *Streptococcus faecalis*, *Bacillus subtilis* as compared to gram-negative bacteria such as *Pseudomonas aeruginosa*. The effect of fucoxanthin was also documented in the reports of Liu and his associates on the growth of human microbiota bacteria like *Lactobacillus* sp. and *Escherichia coli* in *in-vitro* and *in-vivo* experiments. The high concentration of fucoxanthin adversely increased the growth of the probiotic bacteria *Lactobacillus* sp. and decreased or inhibited the growth of *Escherichia coli*. Conversely, the effect of a low concentration of fucoxanthin reported minimal growth of both *Lactobacillus* sp. and *Escherichia coli* [59]. Mohamed work on fucoxanthin from *Turbinaria triquetra* that was isolated from the Red Sea coast of Egypt, possesses a cytotoxicity effect against *Bacillus subtilis* infection. The pigment was extracted using chloroform: methanol (1:1) that was used to analyze the antimicrobial activity against *Bacillus subtilis* through dilution of the extract using ethanol [60]. The antitubercular effect or bacteriostatic effect of fucoxanthin leads to its application against the *Mycobacterium tuberculosis* that causes autoimmune diseases. The bacteriostatic effect was seen using cellular toxicity through the MIC (minimum inhibitory concentration) of 2.8 to 4.1 μM [61]. The ethanol fraction of *Dictyopteris membranacea*, brown algae showed the presence of a high concentration of flavonoid pigment was a potent bacteriostatic effect on *Staphylococcus aureus* and *Enterococcus faecium* than the activity showed by ampicillin antibiotic [56].

Lectin

The lectin is proteinaceous in nature capable of binding to sugar molecules. There is a wide range of lectin that binds to different sugar residues such as mannose specific-, galactose specific-, N-acetyl glucosamine-, fucose specific-, and sialic specific-lectin. Most of the sugar specific-lectins are antiviral in nature and other lectins possess antibacterial properties. The mannose specific-lectin are the potent inactivators of HIV proteins as these lectins tend to interact with the envelope glycoproteins gp 120, which are required for attachment to the host CD4 T lymphocytes cells through binding with gp 120 and block the virus to enter the host cell [62]. Hyum and his coworkers isolated and extracted lectin from red algae *Grateloupia chiangii* known as *Grateloupia chiangii* lectin (GCL) have binding specificity towards β-glc-sp, β-gal-sp, α-man-sp, maltohexose-β-sp1, maltoheptose-β-sp1. The GCL was recorded to have an antiviral property and showed microbicide towards the Influenza virus and Human simplex virus (HSV). The lectin binds to the specific glycoprotein and thus blocks the entry of the virus into the host cells [63]. The bacteriostatic effect of the lectin from red algae *Solieria filiformis* has been reported against gram-negative bacteria like *Serratia marcescens*, *Salmonella typhi*, *Klebsiella pneumoniae*, *Enterobacter aerogenes*. However, no activity was seen against the gram-positive bacteria as the glycol conjugate mannan, a hapten, is commonly found in gram-negative bacteria to which the lectin interacts [64]. A novel microbicide lectin, ESA-2 (*Eucheuma serra* agglutinin) was isolated and extracted from red algae *Eucheuma serra*. The lectin has a high mannose (HM) binding role as an anti-HIV lectin that plays an important role in the inhibition of the Influenza virus through suppression of glycoprotein that bears HM glycans. The ESA-2 was first observed to have a role in the inhibition of HIV through the suppression of specific glycoprotein-bearing glycan and thus prevent the virus from entry into the host cells [65].

Alkaloids

The alkaloids are secondary metabolites produced by many marine algae. These are chemically heterocyclic halogen and nitrogen-containing compounds. In the marine system, marine algae produce three groups of alkaloids: phenylethylamine alkaloid compounds, indolic compounds, halogenated-indole alkaloid compounds, and other alkaloid compounds. However, indole based alkaloids have played an indispensable role as an antimicrobial agent against a wide range of microorganisms [66, 67]. The indole-based alkaloids extracted from green, red, and brown marine algae from the Western Coast of Libya showed a broad range of antimicrobial activity against many pathogenic microorganisms such as *Escherichia coli*, *Salmonella typhi*, *Bacillus subtilis*, *Klebsiella pneumoniae*, *Pseudomonas aeruginosa*, *Staphylococcus aureus*, *Streptococcus epidermidis*, *Bacillus* sp. However, two species of brown algae *Cystoseirabarbata* and *Dictyopteris membranacea* extract of alkaloids showed high alkaloid content and high bacteriostatic effect on *Klebsiella pneumoniae* and *Salmonella typhi,* respectively, than the control compounds of ciprofloxacin and chloramphenicol [67].

Cyclic Peptides

Antimicrobial bioactive peptides are protein fragments formed upon enzymatic digestion using proteolytic enzymes. The efficiency of antimicrobial activity depends on both structural properties and amino acid compositional aspects [68]. Such aspects not only confer antimicrobial role but also have a wide range of activities such as antioxidant, anticancer, and antihypertensives. Antimicrobial peptides (AMP) are widely consumed as diet supplements by humans. According to the report published by Fanny and her associates, the extracted peptide from green algae *Tetraselmis suecica* had alanine and lysine residues that confer high and potent antimicrobial activity towards some gram-negative strains like *Escherichia coli* ML35 ATCC 25922, *Salmonella typhimurium* ATCC 14028, *Pseudomonas aeruginosa* ATCC 27853 and some gram-positive strains like *Bacillus cereus* ISP B7/13, *Staphylococcus aureus* ATCC 25933, *Listeria monocytogenes* ATCC 19115, *Micrococcus luteus* ATCC 9341 [69].

The antimicrobial peptides' activity lies in the nature and charge of the amino acids present in the peptides. The structural analysis of peptides to form alpha or beta structures also plays an important role to confer the potency as an antimicrobial agent.

Terpenes and its Derivatives

Terpenes are the group of secondary metabolites that are found mostly in red algae of class Phaeophyceae. These are the complex structured compounds made from a simple unit of isoprene unit from which different forms or types of terpenes are produced. All the terpenes follow the same general formula of $(C_5H_8)_n$. There are three groups of terpenes present in the marine ecosystem: Mono-terpenes; Di-terpenes; and Sesquiterpenes. The terpenes have been widely explored for their beneficial aspects as antimicrobial and antitumor properties for human health. The reports of Walied *et al.* discovered three novel Laurene-type terpenes that are sesquiterpenes by nature: 12-hydroxy isolaurene; 8,11-dehydr--12-hydroxy isolaurene; and isolauraldehyde from the red alga *Laurencia obtuse.* The three terpenes showed potent antimicrobial activities against *Bacillus subtilis, Staphylococcus aureus* (Gram-positive bacteria), and *Candida albicans* (Fungi) [70]. Another novel sesquiterpenespeyssonic acid A-B (1-2) which was a hydroquinone based terpenes extracted from red alga *Peyssonnelia* sp. showed promising antibacterial and antifungal properties against marine *Pseudoalteromans bacteriolytica* and *Lindra thalassiae*, respectively. The sesquiterpenes hydroquinones also have antineoplastic properties against ovarian cancer cell lines [71].

The content of terpenes depends on the habitat of the marine algae which includes light exposure and temperature. Mostly the biological role of terpenes has been seen as cytotoxicity property against different bacterial cells and viral cells that make the terpenes a highly useful compound for the development of drugs against many human diseases.

Halogenated Compounds

The halogenation of secondary metabolites such as peptides, polyketides, indoles, terpenes, acetogenins, phenols, and volatile hydrocarbons increases the biological activities of these compounds inside the marine algae. The halogen molecules like bromide and chloride ions are present in the seawater, however, bromide-based halogenation tends to be more common in such compounds. Such halogenated secondary compounds are of pharmacological importance. The chrysophaentineis 2-chlorobutene connects two biphenyl ether units derived from red alga *Chrysophaeum taylorii* [72]. There were eight chrysophaetine (A-H) derivatives isolated from the algae and are potent inhibitors of gram-positive drug-resistance bacteria such as drug-resistant *Staphylococcus aureus* and vancomycin-resistant *Enterococcus faecium*. These derivatives tend to inhibit the activity of the FtsZ bacterial cytoskeleton by interfering with GTPase activity and thus blocking cell division [73]. Maria and her co-workers isolated another halogenated compound known as 8-bromophycolides (J-Q) from red alga Fijian *Callophycus serratus* that possesses antimalarial properties against *Plasmodium falciparum.* The antimalarial tendency of the compound was found due to the presence of a macrolide motif in the chemical structure of the compound. The compound also showed a bactericidal effect against methicillin-resistant *Staphylococcus aureus* and vancomycin-resistant *Enterococcus faecium* [74]. Four (1-4) brominated phenolic derivatives were isolated from red alga *Symphyocladialatiuscula* from the coast of Qingdao, Shandong Province, P. R. China:(1)1-[2,5-dibromo-3,4-dihy droxy-6-(2,3,6- tribromo-4,5-dihydroxybenzyl)benzyl]pyrrolidin-2-one, (2)methyl 4-{(2,3,6-tribromo-4,5-dihydroxybenzyl)[(2,3,6-tribromo4,5-dihydroxy benzyl) carbamoyl]amino}butanoate, (3)methyl 4-{(2,5-dibromo-3,4-dihydroxy benzyl) [(2,3,6-tribromo-4,5-dihydroxybenzyl)carbamoyl]amino} butanoate, and (4)2,5-dibromo-3,4-dihydroxy-6-(2,3,6-tribromo-4,5-dihydroxybenzyl)benzyl me thyl ether. The two bromophenols (3, 4) showed antifungal activity against *Candida albicans* ATCC 10231 as compared to the positive control fluconazole antibiotic [75].

The antimicrobial activity of the halogenated secondary metabolites solely lies in the structural conformation of the compound as well as the position of the halogen atom in the structure.

ANTIMICROBIAL ASSESSMENT OF COMPOUNDS FROM MARINE ALGAE

The antimicrobial assessment is the performance of the compounds that possess the ability to inhibit or decrease pathogenic or non-pathogenic microorganisms. In other words, the ability of the compounds to show bacteriostatic or bactericidal effects against the microorganisms.

In-vitro Assessment

Disc Diffusion Method

One such method of analyzing the antimicrobial assessment of the compounds is the disc diffusion method. The disc of 6 mm diameter is cut from Whatman's filter paper no. 1, which is soaked in the extracts of the compounds. The discs are kept or placed in the media swabbed with bacterial, fungal, or yeast cultures. The different media used depending upon the cultures used, like Mueller-Hinton (MH) agar or nutrient agar for bacterial cultures and Sabouraud dextrose agar for fungal or yeast cultures. The plates are incubated at 37° C overnight for bacteria and 30° C for 48 hours for fungi and yeast. The zone of inhibitions is calculated as the diameter in mm of the zones and is compared with the positive controls of the antibiotics like chloramphenicol, ampicillin, fluconazole, *etc.* The experiments are often done thrice for concordant results.

Sreenivasan and his co-workers performed disc diffusion of the methanol extract of marine algae *Gracilaria changii* against bacterial species of *Pseudomonas aeruginosa,* and the zone of inhibitions was compared to the chloramphenicol antibiotic used as the positive control [76]. Xavier and his co-workers worked on the antimicrobial assessment of hexane, ethyl-acetate, acetone, methanol extract of three marine algal species (*Sargassum wightii*, *Chaetomorpha linum*, *Padina gymnospora*) against ten human pathogenic bacterial species. The maximum antimicrobial activity in terms of zone of inhibition was recorded from acetone extract of *Sargassum wightii* against methicillin-resistant *Staphylococcus aureus* as compared to other extract and streptomycin as the positive control [77]. The work published by Akremi worked on the antimicrobial activity assessment using crude, acetone, and ethanol extract fractions of *Dictyopteris membranacea* showed the highest zone of inhibition against *Streptococcus agalactiae*, *Bacillus subtilis*, and *Enterococcus faecium* [56].

Minimum Inhibitory Concentration (MIC)

The MIC is defined as the lowest and minimum possible concentration required to inhibit the visible growth of pathogenic microorganisms. There are two techniques for MIC: macro dilution and microdilution.

The macro dilution is done through serial dilution of the antimicrobial compound extracts from marine algae to obtain the desired concentration (mg) per mL. The bacterial or fungal suspension of 20 µL was added to each of the dilutions and incubated at 37°C for a fraction of time. From each of the incubated dilutions, 50µL was spread or swabbed on the sterile nutrient agar plates and incubated at 37°C overnight. The results are interpreted as the minimum concentration of the dilution that showed no microbial growth. Xavier's work on the MIC of antimicrobial compound extract from marine algae showed MIC of 2.5, 5, and 0.625 for ethyl-acetate, hexane, and acetone *Sargassum wightii* against *Pseudomonas aeruginosa*, *Enterococcus faecium*, and *Staphylococcus aureus* [77]. The MIC for methanol extract of *Gracilariachangii* was found to be 6.25 mg/mL against the bacteria *Pseudomonas aeruginosa* [76].

The microdilution technique involves the dissolution of antimicrobial compound extracts from marine algae in 10% DMSO and performing geometric dilution on 96-well micro-titer plates. 10 µL of the microbial culture having 5.105 CFU/mL maintained on MH broth for bacteria and SD broth for yeast or fungus are mixed with the extracts. The micro-titer plates were incubated at 37°C for 24 hours for bacteria and 30°C for 48 hours for yeast or fungi. The bacterial growth analysis was done using 3-(4,5-dimethylthiazol-2-yl)-2,5-diphenyltetrazolium bromide (MTT) assay in which the bacteria reduces tetrazolium dye (MTT) to purple colour insoluble formazan. The MIC was calculated using the microdilution method recorded MIC for crude, ethanol, acetone, and methanol/CH_2Cl_2 fraction of *Dictyopteris membranacea* as 62.5 µg/mL, 15.62 µg/mL, 15.62 µg/mL, and 250 µg/mL against the gram-positive bacteria *Staphylococcus aureus* [56].

In-vivo Assessment

Very few studies for *in-vivo* assessment have been conducted for seaweeds as antimicrobials. Feeding trials and Immersion methods have been widely studied. The feeding trial method provides the rationale of the dose standardized for seaweed for effective antimicrobial activity. Manilal et al. studied this method on commercial shrimp which were fed orally with seaweed of differing concentrations and simultaneously infected with bacterial strains. The method provides insight into the dose concentration of the seaweed for its therapeutic potential [97].

In the next immersion method, Thanigaivel et al. analyzed the antimicrobial activity of seaweeds using fishes that were infected with bacterial doses. The immersed solution of seaweeds of varying concentrations was added to analyze the effect of seaweeds [96].

PRODUCTION OF ANTIMICROBIAL COMPOUNDS FROM MARINE ALGAE

There are different kinds of antimicrobial compounds obtained from marine algae. The maximum production of such compounds depends on the different seasons, as algal cells have an alternation of generation. Such reproductive and vegetative phases of marine algae have differing variations in the production of antimicrobial compounds. So optimized conditions for maximal production, extraction, and recovery of antimicrobial compounds need to be established for each particular marine algae. The following sections describe different abiotic and biotic factors and processing effects on algal growth and their production of antimicrobial compounds.

Factors Affecting their Production

Marine microbial communities such as seaweed are highly diverse and have an exceptional secondary metabolites structure. Potential antimicrobials have different chemical natures, which can lead to their maximum production, extraction, and recovery process [78]. Antimicrobial activity and the chemical composition of marine algae vary with species and on some physiological parameters, such as the thallus region rich in compounds. The depolymerization effects on seaweed polysaccharides produce oligosaccharides which can initiate protection through the accumulation of proteins and compounds which are having antimicrobial activities against infections [79]. There are some abiotic factors of environmental aspects such as temperature, pH, climate, salinity and location, and growth conditions that affect their production [80]. The variation in antimicrobial action and the chemical composition have been confirmed by various studies, which are precisely according to the climatic seasons. The antimicrobial potential detected maximum in the spring season of some active compounds [81]. Some studies show some seasons reflect the lack of activity like antifungal, antibacterial activities, and a different pattern in algal classes. Climatic variations have been shown by Rhodophyceaeclass and the absence of activity in some seasons in the case of Phaeopyceaewhereas throughout the year Chlorophyceae were active for the production of bioactive compounds [82]. The association relationship also affects the growth and production of bioactive compounds in marine ecosystems. Over a long period, there are established associations among marine microorganisms that share their common environment. In the case of algal cells, many endophytic bacterial species have been found to be present inside or outside the cell, which also produces some other bioactive compounds to protect against pathogenic organisms or predators [83].

Extraction Techniques and Solvent Effect

The extraction of antimicrobials from seaweeds is a very crucial stage, particularly when dealing with effective techniques and the kind of solvent used. However, conditioning methods can also affect the bioactive yield. The loss of volatile antimicrobials in fresh algae, such as hydrogen peroxide, terpenoid, and volatile fatty compounds, can occur at high temperatures. The process is also called the drying stage, which is important for volatile antimicrobials [84]. It has been reported, without altering the antimicrobial activity, that increased Co-60 gamma rays radiation at 1-30 KGy with Ecklonia cave ethanol extract in which the hunter L value is also increased. This was seen in a dose-dependent manner due to a, b values of irradiation and antioxidant properties. The irradiated E. cava extract was tested on mice and a cytotoxicity test on cell line RAW264.7 of murine macrophage for its safe use by humans at moderate doses [85].

Some extraction techniques have been used for the extraction of bioactive compounds from marine microorganisms, such as seaweed, using organic solvents. The antimicrobials and extractable yield from seaweed species are solvent-dependent. Accurate and reproducible extracts are necessary and prepared by optimization and systematic evaluation [86]. Several studies have been followed to select the best solvent from the following such as water, ethanol, methanol, chloroform, and hexane. Different families of seaweed have been compared by different authors regarding their antibacterial activity [87]. The organic solvents for extract production are more efficient. The higher antibacterial activity was confirmed in the polar extracts by the workers. In the extraction process, the time and energy process can be reduced by intensification [88] and also proposed the extraction of the essential oil from *Enteriomorpha linza* by microwave-assisted hydrodistillation. The growth of pathogenic bacteria inhibition is obtained by some major active compounds such as n-hexanal, tetradecanoic acid, and azetidine. The acetone extracts of *Sargassum wightii*, *Chaetomorpha linum*, and *Padina gymnospora* showed the highest antimicrobial activity than methanol, hexane, and ethylacetate extracts of marine algae against *Pseudomonas aeruginosa*, *Enterococcus faecium*, and *Staphylococcus aureus* [77]. The ethanol and acetone extracts of marine algae *Dicytopterismambranacea* showed the best bactericidal, fungicidal and antitubercular activity than methanol/CH_2Cl_2 extracts. Both the extracts showed a good significant correlation between phytochemical compositions [56]. The dichloromethane/methanol extracts of *Pudina* sp. and *Ulva* sp. proved the highest antimicrobial activity against the common food borne-pathogens [89]. The mixture of methanol: toluene (3:1) was considered the best organic solvent for the extraction of bioactive compounds from fourteen seaweeds that were isolated from the Intertidal Zone of the Southwest Coast of India. The methanol: toluene extracts showed comparatively high antimicrobial properties against the multi-resistant pathogenic bacteria and fungi *Candida albicans* (ATCC 90027) [90]. The mixture of dichloromethane/methanol and diethyl acetate-based extraction of marine algae *Caulepa taxifolia* and *Chaetomorpha antennina*, respectively, that were isolated from the West African Coast showed the best inhibition of *Staphylococcus aureus* than ethanol, and chloroform/methanol-based extractions [91].

IMPACT ON HUMAN HEALTH

Seaweeds as Food

Seaweeds draw an extraordinary wealth from the sea minerals, which are about 36% of their total dry mass. The macronutrients in these minerals required for growth are sodium, calcium, magnesium, potassium, chlorine, sulfur, and phosphorous. Iodine, zinc, copper, selenium, molybdenum, fluoride, manganese, boron, nickel, and cobalt are the micronutrients found in these mineral nutrients growth [92]. Seaweed has a large amount of iodine in it, so primarily known as the major source of iodine. The maximum iodine content found in brown algae is 1500-1800 ppm in dry kelp and 500-1000 ppm in dry rockweed [93]. There is a low concentration of iodine found in green and red algae which normally ranges from 100-300 ppm with dry seaweeds. A very small quantity of seaweed can easily fulfill the daily requirement of an adult, which is 150 μg /day, for instance, one gram of brown algae in dried form gives 500- 8000 μg of iodine content; similarly, dried red and green algae gives 100-300 μg [94].

In Japan, brown seaweed is normally consumed as food or in supplement form is about 1 gram per day. The human body is modified to intake higher concentrations of iodine, but only the thyroid gland uses the iodine. Overall world population does not have sufficient consumption of iodine because the most commonly used dietary elements are from land plant and animal sources, which have a very low concentration of iodine. To fulfill this iodine requirement, it is added to table salt. But still, in many countries, people suffer from iodine deficiency, especially in China which is number one for not having sufficient iodine for their population, led by India in second.

Seaweed is also a very important source of the important element calcium. In seaweeds, the average range of the calcium content is 4-7% of dry mass. But the calcium content in dry seaweed is not sufficient to fulfill the daily requirement because one gram of 7% seaweed only gives 70mg of calcium, but the daily requirement is 1000mg per day. No doubt 70mg is still a sufficient amount as compared to non-milk food items.

The content of protein found in seaweed is varied. The protein content is low in brown alga, which is 5-11% of dry mass, but in red algae, it is 30-40% of dry matter, which is almost equal to the legumes. There is only 20% dry matter protein content present in green algae as well. The highest protein content, which is up to 70% of dry mass, is present in a microalga Spirulina.

Many different vitamins are also present in seaweeds. The brown and red algae have a high amount of carotenes ranging from 20 to 170 ppm in them, and these carotenes are used as a dietary supplement as a source of natural carotenes. There is about 500-3000 ppm vitamin C found in red and brown algae. Vitamin B12 is usually absent in many land plants but is present in seaweeds.

The essential fatty acids are high in content in lipids of seaweeds rather than in land plants. The range of fat present in seaweeds is 1-5%. The fatty acids present in green algae are similar to land plants, but the most common forms of fatty acids present in green algae are oleic acid and alpha-linoleic acid. The EPA, which is very common in animals and fish, are also present in a high concentration of red algae. The fibre content is also very high in seaweeds which are up to 32 to 50% of the dry mass. The soluble fibre content present in green and red algae is 51 to 56%, but a very high amount of about 67 to 87% is present in brown algae. The soluble fibres are known for their low content of cholesterol and hypoglycemic effect. The seaweeds are harvested for the production of a gelatinous compound defined as phycocolloids. The common examples of these substances are alginate, agar and carrageenan. The food additive hydrocolloids also have significant use on the commercial level [29].

Medicinal Use of Seaweeds

Many seaweeds are of medicinal importance. The following list provides typical seaweeds which are potential medicine.

Seaweed as Medicines

Kunbu (Laminaria and Ecklonia)

• These are salted and cold in colour and flavour.

• The channel through which these seaweeds enter is the liver, kidney and stomach.

• It softens the hardness and disperses the accumulation. It also helps resolve phlegm, a form of soft textured masses that includes goitre, which is caused by iodine deficiency and leads to the swelling of the thyroid [95] and heat cleanses.

• It can be applied to the Scrofula, goitre, tumour, and oedema, pain in the testicles and swelling.

Haizo (Sargassum)

• It is bitter, salted and cold in essence and flavour.

• The channels through which it can enter are the liver, stomach and kidney.

• It helps in the dispersion of accumulated phlegm, goitre and tumour. It also helps in delivering water and help in cleansing heat.

• It can be applied to the Scrofula, goitre, tumour, and oedema, pain in the testicular and swelling.

Ziaci (Porphyra)

• These are salted, sweet and cold in colour and flavour.

• The channel through which these seaweeds enter is the lung.

• It helps in resolving the phlegm and softens the hardness. It also promotes diuresis.

• It can be applied to the goitre, oedema, beriberi, urinary infection and sore throat.

The haizao and kunbu are quite similar in their description; both are salty and cold, can enter in liver and kidney, clear the heat, hardness softening, promote urination, transform the phlegm, dissolve the nodules and help in oedema reduction. Both of these are used to treat goitre and scrofula in clinical practices. Both of these are the most common seaweeds, but there are many differences Haizao is more efficient in treating nodules like goitre and scrofula and transforming the phlegm. On the other hand, kunbu is more efficient in the reduction of congealed blood, softening the hardness, liver and spleen enlargement treatment, cirrhosis of the liver and also for treating tumours.

The most common formulas are Haizao Yuhu Tang and Sargassum Decoction of Jade flask. These formulas are consists of 12 ingredients and the major ingredients are Sargassum, Ecklonia and Laminaria. These are used to treat severe goitre. These can also be used for treating other light swellings, such as ovarian cysts, breast lumps, lymphatic nodes, lipomas, and accumulated fat because of obesity. For the production of dental moulds and dressing wounds, alginates have been used. The agar is most excessively used for the culture media in microbiology research. The best iodine source for the treatment of goitre is seaweed. Seaweeds are also used for the cure of tuberculosis, arthritis, tumours, and infections caused by worms, cold and influenza [29].

Respiratory Treatment

Carrageenan, found in red algae, has been used for treating respiratory ailments for ages. The most commonly treated ailments are infection of intractable sinus and lingering cases of pneumonia.

Hormones in Seaweeds

Melatonin

The amount of melatonin found in seaweeds is up to 1000 times the concentration in land plants. The seaweed harvested in the daytime has less melatonin than the seaweed harvested at night. There are therapeutic opportunities in using melatonin extracted from seaweed.

Thyroid Hormones in Seaweeds

The only source of thyroid hormones other than animals is brown algae. In brown seaweeds, the iodine is organically bound as thyroid hormone that's why explains why one should consume brown algae.

Estrogen Hormones Regulation

Brown algae have a structural composition of indigestible fibre. The alginate, fucoxanthin, sulfated polysaccharides, such as fucoidan and laminarin, and polyphenol compounds make brown seaweed a potential anticarcinogen. Japanese diets often include seaweeds. Brown algae have a strong binding effect on estrogen hormones. Dietary brown algae have been shown to regulate hormonal serum levels through the excretion of urinary estrogen metabolites and phytoestrogens [98].

Di-iodothyronine

For the treatment of thyroid disorder, the fucus is most commonly used. In fucus it is found in the form of di-iodothyronine but active weakly in the mammalian body. By esterification of two molecules of DIT results in the formation of tetraiodotyrosine. This organically bound iodine in the Fucus increases the formation of T4 Thyroxin by providing initial T4 products.

The therapeutic use of thyroxin treatments is goitre shrinking, weight loss, return of vigour and vim, resolution of symptomatic non-autoimmune hypothyroidism, resolving eczemas and decreasing the psychiatric disruptions. The T4 and T3 are the organically bound iodine found in the many brown algae, especially in Laminaria and Sargassum species. In laminaria, 10% of iodine is present in the form of DIT, MIT, T3 and T4. The most commonly used seaweed is Kombu in Japan and USA used to treat coronary artery disease, healthy liver function, increase the metabolic rate, increase the transit time of food, decrease the LDL cholesterol and increase the HDL cholesterol in the blood [29].

Essential Fat and Vitamins in Seaweeds

Many vitamins are present in seaweeds. Most common vitamins are B, B12, 1-3% omega-3 fatty acids. The 3% of omega-3 fatty acids, vitamins A and C, are mostly present in Nori. It is recommended to eat Nori on a daily basis for boys who have inherited baldness in Japan [29]. Vitamin C source seaweeds often help regulation of blood pressure, β-carotene (Vitamin A) seaweed protect from various cardiovascular diseases, whereas, vitamin E, C, and carotenoids are useful in the reduction of cancer [99].

CONCLUSION

The seaweeds are the potent producers of high biomass in the marine environment. Such biomass is very useful for the extraction of important products such as feedstock for animals with high nutrient values and the products such as agar, carrageenan, alginates have been exploited by humans over many decades for food, medicines, *etc.* This biomass also provides other important applications such as antibacterial, antiviral, antiprotozoal, antifungal, antifouling that can be used as an alternative in the future. The antimicrobial compounds from the marine algae can be used instead of chemically derived synthetic drugs as the marine algae-derived compounds have high specificity as a well broad range of antimicrobial activity against the many pathogenic microorganisms. The specificity lies in their structure, chemical composition, and new targeting values. The bioactive compounds present in the seaweeds, like polysaccharides, lipid, glycerol, sterols, phenolic compounds, various pigments, halogenated compounds and many more, are regarded as important sources for human usage. Even seaweed enzymes have shown crucial advantages in terrestrial usage, from bioremediation to medicinal drug formulations.

So more research concerning the exploitation of new targets of such antimicrobial compounds against microorganisms needs to work out. The production and innovative design of the new antimicrobial compounds from marine algae need to be focused on in the future in terms of low cost-effectiveness for the extraction. The chemical analysis, the safety of the compounds as well as the synergistic approach with a combination of the various antimicrobial compounds for targeting various sites of pathogenic microbes need to be studied more thoroughly.

CONSENT FOR PUBLICATION

Not applicable.

CONFLICT OF INTEREST

The author declares no conflict of interest, financial or otherwise.

ACKNOWLEDGEMENT

Declared none.

REFERENCES

[1] Round FE, Ed. The Biology of the Algae. 1973.

[2] Veluchamy C, Palaniswamy R. A Review on marine algae and its applications. Asian J Pharm Clin Res 2020; 13(3): 21-7.
[http://dx.doi.org/10.22159/ajpcr.2020.v13i3.36130]

[3] Ertürk Ö, Taş B. Antibacterial and antifungal effects of some marine algae. Kafkas Univ Vet Fak Derg 2011; 17 (Suppl. A): S121-4.

[4] Cohen ML. Epidemiology of drug resistance: implications for a post-antimicrobial era. Science 1992; 257(5073): 1050-5.
[http://dx.doi.org/10.1126/science.257.5073.1050] [PMID: 1509255]

[5] Chesters CGC, Stott JA. The production of antibiotic substances by seaweeds. International Seaweed Symposium. 49-53.

[6] Harvey W. Cracking open marine algae's biological treasure chest. Nat Biotechnol 1988; 6(5): 487-92.
[http://dx.doi.org/10.1038/nbt0588-487]

[7] Attaway DH, Zaborsky OR, Eds. Marine biotechology: pharmaceutical and bioactive natural products. 1993.
[http://dx.doi.org/10.1007/978-1-4899-2391-2]

[8] Francisco MEY, Erickson KL, Erickson KL. Ma'iliohydrin, a cytotoxic chamigrene dibromohydrin from a Philippine Laurencia species. J Nat Prod 2001; 64(6): 790-1.
[http://dx.doi.org/10.1021/np0005053] [PMID: 11421745]

[9] Alavandi SV. Relationship between heterotrophic bacteria and suspended particulate matter in the Arabian Sea. Indian J Microbiol 1990; 30(1): 89-92.

[10] Donia M, Hamann MT. Marine natural products and their potential applications as anti-infective agents. Lancet Infect Dis 2003; 3(6): 338-48.
[http://dx.doi.org/10.1016/S1473-3099(03)00655-8] [PMID: 12781505]

[11] El Gamal AA. Biological importance of marine algae. Saudi Pharm J 2010; 18(1): 1-25.
[http://dx.doi.org/10.1016/j.jsps.2009.12.001] [PMID: 23960716]

[12] Hawksworth DL, Colwell RR. Microbial Diversity 21: biodiversity amongst microorganisms and its relevance. Biodivers Conserv 1992; 1(4): 221-6.
[http://dx.doi.org/10.1007/BF00693760]

[13] Holguin G, Vazquez P, Bashan Y. The role of sediment microorganisms in the productivity, conservation, and rehabilitation of mangrove ecosystems: an overview. Biol Fertil Soils 2001; 33(4): 265-78.
[http://dx.doi.org/10.1007/s003740000319]

[14] Hyde KD, Jones EBG, Leaño E, Pointing SB, Poonyth AD, Vrijmoed LLP. Role of fungi in marine ecosystems. Biodivers Conserv 1998; 7(9): 1147-61.
[http://dx.doi.org/10.1023/A:1008823515157]

[15] Kathiresan K, Bingham BL. Biology of mangroves and mangrove Ecosystems. Adv Mar Biol 2001; 40: 81-251.
[http://dx.doi.org/10.1016/S0065-2881(01)40003-4]

[16] Kumaran KP, Shindikar M, Limaye RB. Fossil record of marine manglicolous fungi from Malvan (Konkan) west coast of India; 2004.

[17] Margulia L, Schwartz KV. Five kingdoms: an illustrated guide to the phyla of life on earth. Freeman 1998.

[18] Middelboe M, Brussaard C. Marine viruses: key players in marine ecosystems. Viruses 2017; 9(10): 302.

[http://dx.doi.org/10.3390/v9100302] [PMID: 29057790]

[19] Mohanraju R, Natarajan R. mangrove sediments. In: Jaccarni V, Martens Els, Eds. The Ecology of Mangrove and Related Ecosystems. Dordrecht: Springer 1992; pp. 187-93. [http://dx.doi.org/10.1007/978-94-017-3288-8_20]

[20] Nikolcheva LG, Cockshutt AM, Bärlocher F. Determining diversity of freshwater fungi on decaying leaves: comparison of traditional and molecular approaches. Appl Environ Microbiol 2003; 69(5): 2548-54. [http://dx.doi.org/10.1128/AEM.69.5.2548-2554.2003] [PMID: 12732520]

[21] Oliver JD. Taxonomic scheme for the identification of marine bacteria. Deep-Sea Res A, Oceanogr Res Pap 1982; 29(6): 795-8. [http://dx.doi.org/10.1016/0198-0149(82)90007-3]

[22] Richman S, Loya Y, Sloboclkin LB. The rate of mucus production by corals and its assimilation by the coral reef copepod Acartia negligens1. Limnol Oceanogr 1975; 20(6): 918-23. [http://dx.doi.org/10.4319/lo.1975.20.6.0918]

[23] Sherr EB, Sherr BF. Temporal offset in oceanic production and respiration processes implied by seasonal changes in atmospheric oxygen: the role of heterotrophic microbes. Aquat Microb Ecol 1996; 11(1): 91-100. [http://dx.doi.org/10.3354/ame011091]

[24] Shome BR, Rajeswari S, Ahlawat SP, *et al.* Agar depolymerizing (agarolytic) bacteria isolated from mangrove soil samples of Andaman. Curr Sci 2000; 79(6): 696-7.

[25] Toledo G, Bashan Y, Soeldner A. Cyanobacteria and black mangroves in Northwestern Mexico: colonization, and diurnal and seasonal nitrogen fixation on aerial roots. Can J Microbiol 1995; 41(11): 999-1011. [http://dx.doi.org/10.1139/m95-139]

[26] Tudhope AW, Risk MJ. Rate of dissolution of carbonate sediments by microboring organisms, Davies Reef, Australia. J Sediment Res 1985; 55(3): 440-7.

[27] Vethanayagam RR. Purple photosynthetic bacteria from a tropical mangrove environment. Mar Biol 1991; 110(1): 161-3. [http://dx.doi.org/10.1007/BF01313103]

[28] Ward DM. A natural species concept for prokaryotes. Curr Opin Microbiol 1998; 1(3): 271-7. [http://dx.doi.org/10.1016/S1369-5274(98)80029-5] [PMID: 10066488]

[29] Kandale A, Meena AK, Rao MM, *et al.* Marine algae: An introduction, food value and medicinal uses. J Pharm Res 2011; 4(1): 219-21.

[30] Lewis LA, McCourt RM. Green algae and the origin of land plants. Am J Bot 2004; 91(10): 1535-56. [http://dx.doi.org/10.3732/ajb.91.10.1535] [PMID: 21652308]

[31] Stam WT, Olsen JL, Zaleski SF, Murray SN, Brown KR, Walters LJ. A forensic and phylogenetic survey of Caulerpa species (Caulerpales, Chlorophyta) from the Florida coast, local aquarium shops and e-commerce: establishing a proactive baseline for early detection. J Phycol 2006; 42(5): 1113-24. [http://dx.doi.org/10.1111/j.1529-8817.2006.00271.x]

[32] Verlaque M, Durand C, Huisman JM, Boudouresque CF, Le Parco Y. On the identity and origin of the Mediterranean invasive *Caulerpa racemosa* (Caulerpales, Chlorophyta). Eur J Phycol 2003; 38(4): 325-39. [http://dx.doi.org/10.1080/09670260310001612592]

[33] Piazzi L, Meinesz A, Verlaque M, *et al.* Invasion of Caulerpa racemosa var. cylindracea (Caulerpales, Chlorophyta) in the Mediterranean Sea: an assessment of the spread. Cryptogam, Algol 2005; 26: 189-02.

[34] Maggs MA, Verbruggen H. De Clerck. Molecular systematics of red algae: building future structures

of firm foundations. In: Brodie J, Lewis J, Eds. Unravelling the algae: the past, present and future of algal systematics London & New York 2007; 103-21. [http://dx.doi.org/10.1201/9780849379901.ch6]

[35] Saunders GW, Hommersand MH. Assessing red algal supraordinal diversity and taxonomy in the context of contemporary systematic data. Am J Bot 2004; 91(10): 1494-507. [http://dx.doi.org/10.3732/ajb.91.10.1494] [PMID: 21652305]

[36] Yoon HS, Müller KM, Sheath RG, Ott FD, Bhattacharya D. Defining the major lineages of red algae. J Phycol 2006; 42(2): 482-92. [http://dx.doi.org/10.1111/j.1529-8817.2006.00210.x]

[37] De Reviers B, Rousseau F, Draisma SGA. Classification of the Phaeophyceae from past to present and current challenges. In: In Brodie J, Lewis J, Eds. Unravelling the algae: the past, present and future of algal systematics, London & New York 2007; 267-84.

[38] Barahona T, Encinas MV, Imarai M, *et al.* Bioactive polysaccharides from marine algae. Bioactive Carbohydrates and Dietary Fibre 2014; 4(2): 125-38. [http://dx.doi.org/10.1016/j.bcdf.2014.09.002]

[39] Cesário MT, da Fonseca MMR, Marques MM, de Almeida MCMD. Marine algal carbohydrates as carbon sources for the production of biochemicals and biomaterials. Biotechnol Adv 2018; 36(3): 798-817. [http://dx.doi.org/10.1016/j.biotechadv.2018.02.006] [PMID: 29428561]

[40] Feki A, Hamdi M, Jaballi I, Zghal S, Nasri M, Ben Amara I. Conception and characterization of a multi-sensitive composite chitosan-red marine alga-polysaccharide hydrogels for insulin controlled-release. Carbohydr Polym 2020; 236: 116046. [http://dx.doi.org/10.1016/j.carbpol.2020.116046] [PMID: 32172860]

[41] Shi F, Yan X, Cheong KL, Liu Y. Extraction, purification, and characterization of polysaccharides from marine algae Gracilaria lemaneiformis with anti-tumor activity. Process Biochem 2018; 73: 197-203. [http://dx.doi.org/10.1016/j.procbio.2018.08.011]

[42] Achmad H, Huldani YF. Antimicrobial Activity and Sulfated Polysaccharides Antibiofilms in Marine Algae Against Dental Plaque Bacteria: A Literature Review. Sys Rev Pharm 2020; 11(6): 459-65.

[43] Ahmadi A, Zorofchian Moghadamtousi S, Abubakar S, Zandi K. Antiviral potential of algae polysaccharides isolated from marine sources: a review. BioMed Res Int 2015; 2015: 1-10. [http://dx.doi.org/10.1155/2015/825203] [PMID: 26484353]

[44] Abou Zeid AH, Aboutabl EA, Sleem AA, El-Rafie HM. Water soluble polysaccharides extracted from Pterocladia capillacea and Dictyopteris membranacea and their biological activities. Carbohydr Polym 2014; 113: 62-6. [http://dx.doi.org/10.1016/j.carbpol.2014.06.004] [PMID: 25256459]

[45] Rosell KG, Srivastava LM. Fatty acids as antimicrobial substances in brown algae. In: Ragan MA, Bird CJ, Eds. Twelfth International Seaweed Symposium. 471-5. [http://dx.doi.org/10.1007/978-94-009-4057-4_69]

[46] Ismail A, Ktari L, Ben Redjem Romdhane Y, *et al.* Antimicrobial fatty acids from green alga Ulva rigida (Chlorophyta). BioMed Res Int 2018; 2018: 1-12. [http://dx.doi.org/10.1155/2018/3069595] [PMID: 30539008]

[47] Kavita K, Singh VK, Jha B. 24-Branched Δ5 sterols from *Laurencia papillosa* red seaweed with antibacterial activity against human pathogenic bacteria. Microbiol Res 2014; 169(4): 301-6. [http://dx.doi.org/10.1016/j.micres.2013.07.002] [PMID: 23910454]

[48] Bakar K, Mohamad H, Tan HS, Latip J. Sterols compositions, antibacterial, and antifouling properties from two Malaysian seaweeds: Dictyota dichotoma and Sargassum granuliferum. J App Pharm Sci 2019; 9(10): 047-53.

[49] Surendhiran D, Vijay M, Razack A, Subramaniyan T, Shellomith AS, Tamilselvam K. A green synthesis of antimicrobial compounds from marine microalgae Nannochloropsis oculata. J Coast Life Med 2014; 2(11): 862-59.

[50] Generalić Mekinić I, Skroza D, Šimat V, Hamed I, Čagalj M, Popović Perković Z. Phenolic content of brown algae (Pheophyceae) species: Extraction, identification, and quantification. Biomolecules 2019; 9(6): 244.
[http://dx.doi.org/10.3390/biom9060244] [PMID: 31234538]

[51] Sameeh M, Mohamed A, Elazzazy A. Polyphenolic contents and antimicrobial activity of different extracts of Padina boryana Thivy and Enteromorpha sp marine algae. J Appl Pharm Sci 2016; 6(9): 087-92.
[http://dx.doi.org/10.7324/JAPS.2016.60913]

[52] El Shafay SM, Ali SS, El-Sheekh MM. Antimicrobial activity of some seaweeds species from Red sea, against multidrug resistant bacteria. Egypt J Aquat Res 2016; 42(1): 65-74.
[http://dx.doi.org/10.1016/j.ejar.2015.11.006]

[53] Eom SH, Kim YM, Kim SK. Antimicrobial effect of phlorotannins from marine brown algae. Food Chem Toxicol 2012; 50(9): 3251-5.
[http://dx.doi.org/10.1016/j.fct.2012.06.028] [PMID: 22735502]

[54] Choi JG, Kang OH, Brice OO, *et al.* Antibacterial activity of Ecklonia cava against methicillin-resistant Staphylococcus aureus and Salmonella spp. Foodborne Pathog Dis 2010; 7(4): 435-41.
[http://dx.doi.org/10.1089/fpd.2009.0434] [PMID: 20001325]

[55] Nagayama K, Iwamura Y, Shibata T, Hirayama I, Nakamura T. Bactericidal activity of phlorotannins from the brown alga Ecklonia kurome. J Antimicrob Chemother 2002; 50(6): 889-93.
[http://dx.doi.org/10.1093/jac/dkf222] [PMID: 12461009]

[56] Akremi N, Cappoen D, Anthonissen R, Verschaeve L, Bouraoui A. Phytochemical and in vitro antimicrobial and genotoxic activity in the brown algae Dictyopteris membranacea. S Afr J Bot 2017; 108: 308-14.
[http://dx.doi.org/10.1016/j.sajb.2016.08.009]

[57] Karpiński TM, Adamczak A. Fucoxanthin—an antibacterial carotenoid. Antioxidants 2019; 8(8): 239.
[http://dx.doi.org/10.3390/antiox8080239] [PMID: 31344844]

[58] Alam T, Najam L, Al-Harrasi A. Extraction of natural pigments from marine algae. J Agri Marine Sci 2019; 23(1): 81-91.
[http://dx.doi.org/10.24200/jams.vol23iss1pp81-91]

[59] Liu Z, Sun X, Sun X, Wang S, Xu Y. Fucoxanthin isolated from Undaria pinnatifida can interact with *Escherichia coli* and *lactobacilli* in the intestine and inhibit the growth of pathogenic bacteria. J Ocean Univ China 2019; 18(4): 926-32.
[http://dx.doi.org/10.1007/s11802-019-4019-y]

[60] Abou MA. Antibacterial activity of some marine algal extracts against most nosocomial bacterial infections. Egypt J Exp Bio (Bot) 2013; 9(2): 281-6.

[61] Šudomová M, Shariati M, Echeverría J, Berindan-Neagoe I, Nabavi S, Hassan S. A Microbiological, Toxicological, and Biochemical Study of the Effects of Fucoxanthin, a Marine Carotenoid, on *Mycobacterium tuberculosis* and the Enzymes Implicated in Its Cell Wall: A Link Between Mycobacterial Infection and Autoimmune Diseases. Mar Drugs 2019; 17(11): 641.
[http://dx.doi.org/10.3390/md17110641] [PMID: 31739453]

[62] Barre A, Simplicien M, Benoist H, Van Damme EJM, Rougé P. Mannose-specific lectins from marine algae: Diverse structural scaffolds associated to common virucidal and anti-cancer properties. Mar Drugs 2019; 17(8): 440.
[http://dx.doi.org/10.3390/md17080440] [PMID: 31357490]

[63] Hwang HJ, Han JW, Jeon H, *et al.* Characterization of a Novel Mannose-Binding Lectin with

Antiviral Activities from Red Alga, *Grateloupia chiangii.* Biomolecules 2020; 10(2): 333. [http://dx.doi.org/10.3390/biom10020333] [PMID: 32092955]

[64] Holanda ML, Melo VMM, Silva LMCM, Amorim RCN, Pereira MG, Benevides NMB. Differential activity of a lectin from Solieria filiformis against human pathogenic bacteria. Braz J Med Biol Res 2005; 38(12): 1769-73. [http://dx.doi.org/10.1590/S0100-879X2005001200005] [PMID: 16302091]

[65] Sato Y, Morimoto K, Kubo T, *et al.* Entry inhibition of influenza viruses with high mannose binding lectin ESA-2 from the red alga Eucheuma serra through the recognition of viral hemagglutinin. Mar Drugs 2015; 13(6): 3454-65. [http://dx.doi.org/10.3390/md13063454] [PMID: 26035023]

[66] Güven KC, Percot A, Sezik E. Alkaloids in marine algae. Mar Drugs 2010; 8(2): 269-84. [http://dx.doi.org/10.3390/md8020269] [PMID: 20390105]

[67] Alghazeer R, Whida F, Abduelrhman E, Gammoudi F, Naili M. In vitro antibacterial activity of alkaloid extracts from green, red and brown macroalgae from western coast of Libya. Afr J Biotechnol 2013; 12(51): 7086-91.

[68] Kang H, Seo C, Park Y. Marine peptides and their anti-infective activities. Mar Drugs 2015; 13(1): 618-54. [http://dx.doi.org/10.3390/md13010618] [PMID: 25603351]

[69] Guzmán F, Wong G, Román T, *et al.* Identification of antimicrobial peptides from the microalgae Tetraselmis suecica (Kylin) Butcher and bactericidal activity improvement. Mar Drugs 2019; 17(8): 453. [http://dx.doi.org/10.3390/md17080453] [PMID: 31374937]

[70] Alarif WM, Al-Lihaibi SS, Ayyad SEN, Abdel-Rhman MH, Badria FA. Laurene-type sesquiterpenes from the Red Sea red alga Laurencia obtusa as potential antitumor–antimicrobial agents. Eur J Med Chem 2012; 55: 462-6. [http://dx.doi.org/10.1016/j.ejmech.2012.06.060] [PMID: 22819506]

[71] Lane AL, Mular L, Drenkard EJ, *et al.* Ecological leads for natural product discovery: novel sesquiterpene hydroquinones from the red macroalga Peyssonnelia sp. Tetrahedron 2010; 66(2): 455-61. [http://dx.doi.org/10.1016/j.tet.2009.11.042] [PMID: 20661312]

[72] Davison JR, Bewley CA. Antimicrobial chrysophaentin analogs identified from laboratory cultures of the marine microalga Chrysophaeum taylorii. J Nat Prod 2019; 82(1): 148-53. [http://dx.doi.org/10.1021/acs.jnatprod.8b00858] [PMID: 30623657]

[73] Plaza A, Keffer JL, Bifulco G, Lloyd JR, Bewley CA, Chrysophaentins A. Chrysophaentins A-H, antibacterial bisdiarylbutene macrocycles that inhibit the bacterial cell division protein FtsZ. J Am Chem Soc 2010; 132(26): 9069-77. [http://dx.doi.org/10.1021/ja102100h] [PMID: 20536175]

[74] Cabrita MT, Vale C, Rauter AP. Halogenated compounds from marine algae. Mar Drugs 2010; 8(8): 2301-17. [http://dx.doi.org/10.3390/md8082301] [PMID: 20948909]

[75] Xu X, Yin L, Gao J, Gao L, Song F. Antifungal bromophenols from marine red alga Symphyocladia latiuscula. Chem Biodivers 2014; 11(5): 807-11. [http://dx.doi.org/10.1002/cbdv.201300239] [PMID: 24827691]

[76] Sasidharan S, Darah I, Noordin MKMJ. In vitro antimicrobial activity against Pseudomonas aeruginosa and acute oral toxicity of marine algae Gracilaria changii. N Biotechnol 2010; 27(4): 390-6. [http://dx.doi.org/10.1016/j.nbt.2010.02.002] [PMID: 20170762]

[77] Rosaline XD, Sakthivelkumar S, Rajendran K, Janarthanan S. Screening of selected marine algae from

the coastal Tamil Nadu, South India for antibacterial activity. Asian Pac J Trop Biomed 2012; 2(1): S140-6.
[http://dx.doi.org/10.1016/S2221-1691(12)60145-2]

[78] Dussault D, Vu KD, Vansach T, Horgen FD, Lacroix M. Antimicrobial effects of marine algal extracts and cyanobacterial pure compounds against five foodborne pathogens. Food Chem 2016; 199: 114-8.
[http://dx.doi.org/10.1016/j.foodchem.2015.11.119] [PMID: 26775951]

[79] Shanmughapriya S, Manilal A, Sujith S, Selvin J, Kiran GS, Natarajaseenivasan K. Antimicrobial activity of seaweeds extracts against multiresistant pathogens. Ann Microbiol 2008; 58(3): 535-41.
[http://dx.doi.org/10.1007/BF03175554]

[80] Agabji-Daniels F, Adeleye A, Nwankwo D, Adeniyi B, Seku F, Beukes D. Antibacterial Activities of Selected green Seaweeds from West African Coast. EC Pharmacol Toxicol 2020; 8(4): 84-92.

[81] Hellio C, De La Broise D, Dufossé L, Le Gal Y, Bourgougnon N. Inhibition of marine bacteria by extracts of macroalgae: potential use for environmentally friendly antifouling paints. Mar Environ Res 2001; 52(3): 231-47.
[http://dx.doi.org/10.1016/S0141-1136(01)00092-7] [PMID: 11570804]

[82] Osman ME, Abushady AM, Elshobary ME. In vitro screening of antimicrobial activity of extracts of some macroalgae collected from Abu-Qir bay Alexandria, Egypt. Afr J Biotechnol 2010; 9(42): 7203-8.

[83] Patra JK, Kim SH, Baek KH. Antioxidant and Free Radical-Scavenging Potential of Essential Oil from *E nteromorpha linza* L. Prepared by Microwave-Assisted Hydrodistillation. J Food Biochem 2015; 39(1): 80-90.
[http://dx.doi.org/10.1111/jfbc.12110]

[84] Choi JS, Ha YM, Lee BB, Moon HE, Cho KK, Choi IS. Seasonal variation of antibacterial activities in the green alga Ulva pertusa Kjellman. J Environ Biol 2014; 35(2): 341-4.
[PMID: 24665759]

[85] Mendes M, Pereira R, Pinto IS, *et al.* Antimicrobial activity and lipid profile of seaweed extracts from the North Portuguese Coast. Int Food Res J 2013; 20(6): 3337-45.

[86] Freile-Pelegrín Y, Morales JL. Antibacterial activity in marine algae from the coast of Yucatan, Mexico. Bot Mar 2004; 47(2): 140-6.
[http://dx.doi.org/10.1515/BOT.2004.014]

[87] García-Bueno N, Decottignies P, Turpin V, *et al.* Seasonal antibacterial activity of two red seaweeds, *Palmaria palmata* and *Grateloupia turuturu*, on European abalone pathogen *Vibrio harveyi.* Aquat Living Resour 2014; 27(2): 83-9.
[http://dx.doi.org/10.1051/alr/2014009]

[88] Padmakumar K, Ayyakkannu K. Seasonal variation of antibacterial and antifungal activities of the extracts of marine algae from southern coasts of India. Bot Mar 1997; 40(1-6): 507-15.
[http://dx.doi.org/10.1515/botm.1997.40.1-6.507]

[89] Stabili L, Acquaviva MI, Biandolino F, *et al.* Biotechnological potential of the seaweed *Cladophora rupestris* (Chlorophyta, Cladophorales) lipidic extract. N Biotechnol 2014; 31(5): 436-44.
[http://dx.doi.org/10.1016/j.nbt.2014.05.002] [PMID: 24852224]

[90] El Amraoui B, El Amraoui M, Cohen N, Fassouane A. Anti- Candida and anti- Cryptococcus antifungal produced by marine microorganisms. J Mycol Med 2014; 24(4): e149-53.
[http://dx.doi.org/10.1016/j.mycmed.2014.04.004] [PMID: 25442916]

[91] Vera J, Castro J, Gonzalez A, Moenne A. Seaweed polysaccharides and derived oligosaccharides stimulate defense responses and protection against pathogens in plants. Mar Drugs 2011; 9(12): 2514-25.
[http://dx.doi.org/10.3390/md9122514] [PMID: 22363237]

[92] Seaweeds TD, Ed. Life Series 2002.

[93] Cole KM, Sheath RG, Eds. Biology of the red algae 1990.

[94] Hoek C, Mann D, Jahns HM, Jahns M, Eds. Algae: an introduction to phycology 1995.

[95] Hsu HY, Chen YP, Hsu CS, Shen SJ, Chen CC, Chang HC, Eds. Oriental materia medica: a concise guide New Canaan 1986.

[96] Thanigaivel S, Vijayakumar S, Gopinath S, Mukherjee A, Chandrasekaran N, Thomas J. In vivo and in vitro antimicrobial activity of *Azadirachta indica* (Lin) against Citrobacter freundii isolated from naturally infected Tilapia (Oreochromis mossambicus). Aquaculture 2015; 437: 252-5. [http://dx.doi.org/10.1016/j.aquaculture.2014.12.008]

[97] Manilal A, Selvin J, George S. In vivo therapeutic potentiality of red seaweed, Asparagopsis (Bonnemaisoniales, Rhodophyta) in the treatment of Vibriosis in Penaeus monodon Fabricius. Saudi J Biol Sci 2012; 19(2): 165-75. [http://dx.doi.org/10.1016/j.sjbs.2011.12.003] [PMID: 23961176]

[98] Teas J, Hurley TG, Hebert JR, Franke AA, Sepkovic DW, Kurzer MS. Dietary seaweed modifies estrogen and phytoestrogen metabolism in healthy postmenopausal women. J Nutr 2009; 139(5): 939-44. [http://dx.doi.org/10.3945/jn.108.100834] [PMID: 19321575]

[99] Škrovánková S. Seaweed vitamins as nutraceuticals. Adv Food Nutr Res 2011; 64: 357-69. [http://dx.doi.org/10.1016/B978-0-12-387669-0.00028-4] [PMID: 22054961]

CHAPTER 9

Mushroom and Related Fungi: A Natural Source of Anti-Microbial Compounds

Prabhjot Kaur[1], **Abhijit Dey**[2], **Vijay Kumar**[1], **Padmanabh Dwivedi**[3], **Tabarak Malik**[4], **R.M. Banik**[5], **Brijendra Pratap Mishra**[6] and **Devendra Kumar Pandey**[1,*]

[1] *School of Basic and Applied Sciences, Raffles University, Alwar, Rajasthan, India*

[2] *Department of Life Sciences, Presidency University, Kolkata, India*

[3] *Laboratory of Plant Tissue Culture and Stress Physiology, Department of Plant Physiology, Institute of Agricultural Sciences, Banaras Hindu University, Varanasi, India*

[4] *Department of Biochemistry, College of Medicine and Health Sciences, University of Gondar, Gondar, Ethiopia*

[5] *School of Biochemical Engineering, Indian Institute of Technology (BHU), Varanasi, India*

[6] *Department of Biochemistry, Autonomous State Medical College Bahraich, KGM University, Lucknow, UP, India*

Abstract: There is an urgent need to search for effective novel antibiotics due to the evolution of pathogen resistance towards the existing anti-microbial drugs. To fulfill the demand of pharmaceutical industries for novel drugs against pathogenic microbes, the potential source is nature, which is the largest repertoire for discovering biologically active drugs. Among the natural products, mushrooms are primary sources of diverse low and high-molecular-weight compounds that demonstrate anti-bacterial, anti-fungal, anti-parasitic and anti-viral activities. Mushrooms belonging to basidiomycetes or ascomycetes were classified into edible and non-edible and had high nutritive and medicinal properties due to the presence of bioactive compounds. The most common edible mushrooms comprise *Agaricus bisporus*, *Lentinus, Auricularia. Hericium, Grifola, Flammulina, Pleurotus, and Tremella* are potent sources of vitamins (thiamine, riboflavin, niacin, biotin and ascorbic acid, Vitamin A and D), lipids (mono, di, and triglycerides, sterols, phospholipids) and polysaccharides whereas non-edible mushrooms *Ganoderma lucidum* (Reishi), *Lentinus edodes* (Shiitake), *Inonotus obliquus* (Chaga), *Ganoderma, Trametes, Cordyceps* spp., *etc.*, are potent sources of alkaloids, terpenoids, steroids, anthraquinones, benzoic acid derivatives, and quinolines. The literature review suggests that mushrooms showed high anti-microbial activities against Gram-positive bacteria (*Bacillus spp., Listeria monocytogenes, Micrococcus spp., Staphylococcus spp. etc.*) and Gram-negative bacterial species

* **Corresponding author to Devendra Kumar Pandey:** School of Basic and Applied Sciences, Raffles University, Alwar, Rajasthan, India; Tel: +91 9779683414; E-mail: dkpandey1974@gmail.com

M. Amin-ul Mannan & Gaurav Kumar (Eds.)

(*Escherichia coli,Klebsiella spp. or Salmonella sp*) as well as anti-fungal (*Candida spp., Aspergillus spp., Penicillium spp. etc.*) and anti-viral (HIV-I, influenza) activities. The present chapter highlighted the mushrooms showing anti-microbial activity, techniques for appraisal of anti-microbial activity, anti-microbial bioactive compounds and last but not least, the downstream process of some selected compounds originally isolated from mushrooms.

Keywords: Antimicrobial, Antifungal, Antiviral, Ascomycetes, Basidiomycetes, Bioactive compounds, Bioactive potential, Edible and non-edible mushroom, Fatty acids, Green extraction methods, Microwave-assisted extraction (MAE), Multidrug resistance (MDR), Minimum inhibitory concentration (MIC), Phenols, Peptides, Sesquiterpenes, Supercritical fluid extraction (SFE), Steroids, Terpenoids, Ultrasound-assisted extraction (UAE).

INTRODUCTION

"Let your food be your medicine, and your medicine be your food" was adopted by Hippocrates, 2400 years ago. Since their discovery, anti-microbial drugs have been known to cure many infectious diseases and have played a vital role in human existence. It has been seen that medical practitioners misused anti-microbial agents, leading to drug resistance and multi-drug resistance (MDR) among pathogens [1]. There is an urgent need to search for effective novel antibiotics due to the evolution of pathogen resistance towards the existing anti-microbial drugs. To fulfill the demand of pharmaceutical industries for novel drugs against pathogenic microbes, the potential source is nature, which is the largest repertoire for discovering biologically active drugs. Worldwide, traditional and complementary systems of medicine are highly acknowledged and considered to be very effective in curing infectious diseases (WHO, 2019). Among the natural products, mushrooms are primary sources of diverse low and high-molecular-weight compounds that demonstrate anti-bacterial, anti-fungal, anti-parasitic and anti-viral activities. Mushrooms are popular low-calorie valuable foods for human consumption that have been cultivated and consumed in Asian countries such as China, Japan and India [2, 3]. Mushrooms belonging to basidiomycetes or ascomycetes were classified into edible and non-edible and had high nutritive and medicinal properties due to the presence of bioactive compounds [4]. The most common edible mushrooms (Fig. **1**) comprise *Agaricus bisporus, Lentinus, Auricularia, Hericium, Grifola, Flammulina, Pleurotus, and Tremella* are potent sources of vitamins (thiamine, riboflavin, niacin, biotin and ascorbic acid, Vitamin A and D), lipids (mono, di, and triglycerides, sterols, phospholipids) and polysaccharides whereas non-edible mushrooms (Fig. **2**) *Ganoderma lucidum* (Reishi), *Lentinus edodes* (Shiitake), *Inonotus obliquus* (Chaga), *Ganoderma, Trametes, Cordyceps* spp., *etc.*, are potent sources of alkaloids, terpenoids, steroids, anthraquinones, benzoic acid derivatives, and quinolones [5].

Fig. (1). Different sources of edible mushrooms with anti-microbial properties (Pictures are taken from Wikimedia common under the GNU Free Documentation License, Version 1.2 or any later version published by the Free Software Foundation).

The literature review suggests that mushrooms showed high anti-microbial activities against Gram-positive bacteria (*Bacillus sp., Listeria monocytogenes, Micrococcus* sp., *Staphylococcus* sp., *etc.*) and Gram-negative bacterial species (*Escherichia coli, Klebsiella* sp., *or Salmonella* sp.) as well as anti-fungal (*Candida* sp., *Aspergillus* sp., *Penicillium* sp., *etc.*) and anti-viral (HIV-I, influenza) activities [6]. The present chapter highlighted the mushrooms showing anti-microbial activity, techniques for appraisal of anti-microbial activity, anti-microbial bioactive compounds and last but not least, the downstream process of some selected compounds originally isolated from mushrooms.

Fig. (2). Different sources of Medicinal Mushrooms with anti-microbial properties (Pictures are taken from Wikimedia common under the GNU Free Documentation License, Version 1.2 or any later version published by the Free Software Foundation).

ANTI-MICROBIAL COMPOUNDS FROM MUSHROOMS

Mostly all types of pathogenic organisms cause infectious diseases that can spread in many ways (directly or indirectly). Pathogenic organisms such as bacteria, viruses, fungi and parasites that cause microbial infections are diverse and can survive under extreme environmental conditions. These tiny infectious agents are responsible for several deadly diseases (Pneumonia, Cholera, Malaria, Dengue, HIV/AIDS, Zika, SARS, Ebola, and Covid-19) that threaten and kill millions of people (WHO). The variety and number of infectious disease outbreaks are increasing worldwide; thus, effective anti-microbial drugs are required to prevent endemic diseases. On the other hand, anti-microbial drugs are losing their effectiveness against many human diseases (most commonly hospital-acquired and respiratory diseases) as many bacteria, fungi and viruses change their nature and become resistant. Therefore, the development of antibiotics from natural sources and the screening of bioactive compounds against multidrug-resistant strains has become a priority.

Although around 150000 mushroom varieties exist worldwide, only 68 genera and 158 species distributed in Bangladesh, Brazil, China, India, Japan, Nigeria, North

Serbia, Portugal, South Korea, Spain, and Turkey are known for their anti-microbial potential [6]. Anti-microbial (representing anti-bacterial, anti-fungal and anti-viral activities) bio-active compounds isolated from different sources of mushrooms have been tabulated in Tables **1** and **2**. The molecular structure of 68 bioactive compounds extracted from mushrooms that represented anti-microbial activities is presented in Fig. (**3**).

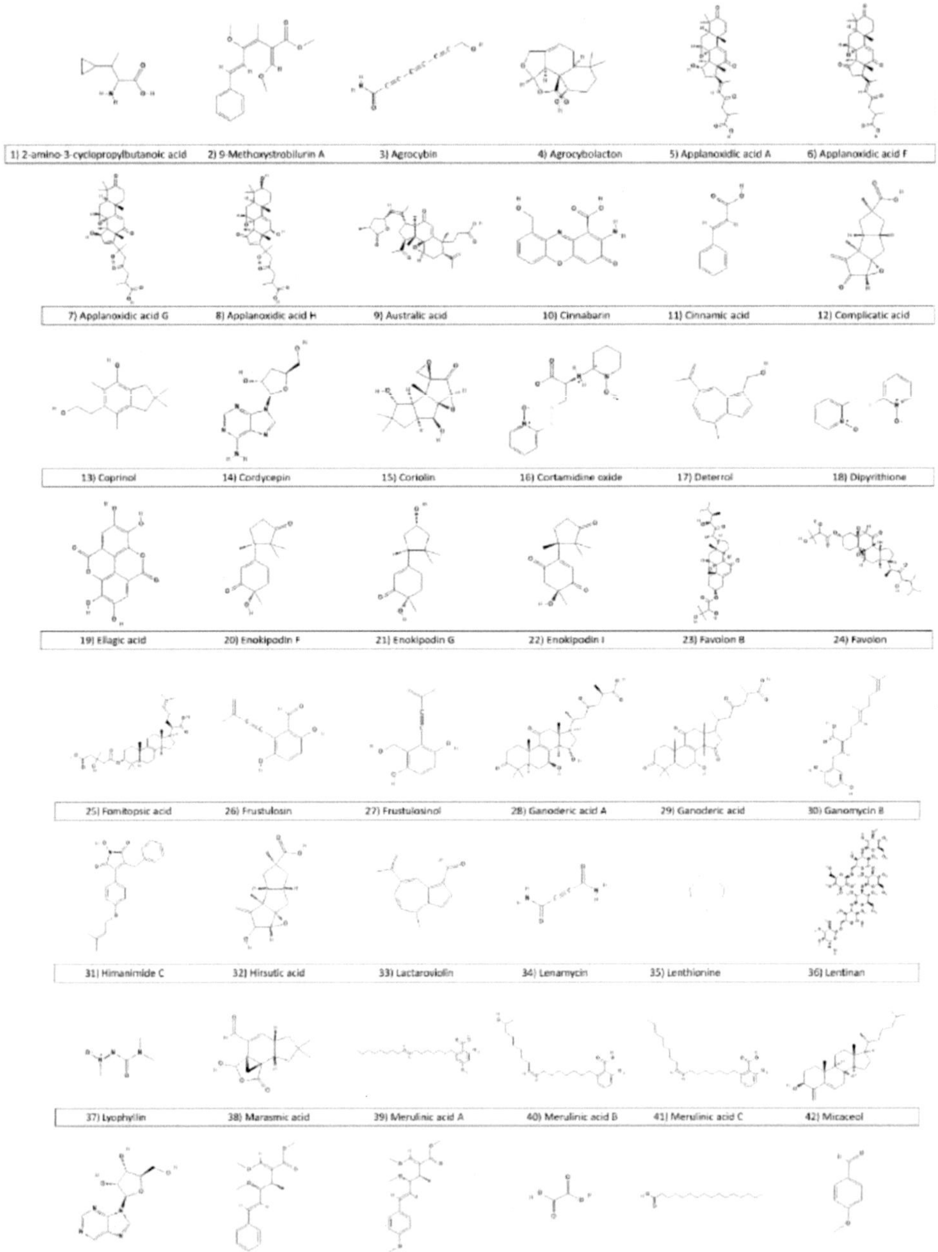

(Fig. 3) contd.....

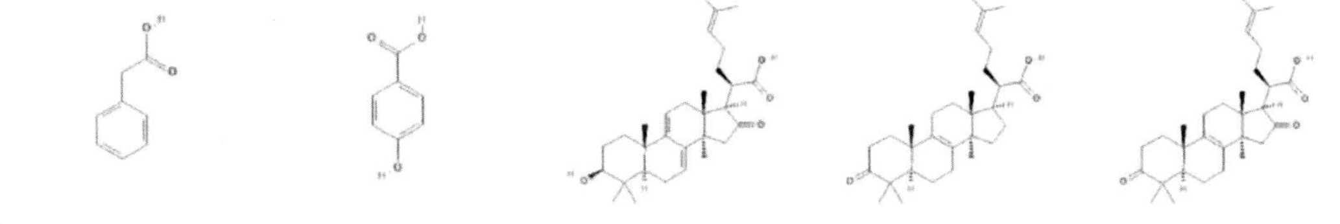

49) Phenylacetic acid | 50) p-hydroxybenzoic acid | 51) Pinicolic acid D | 52) Pinicolic acid A | 53) Pinicolic acid B

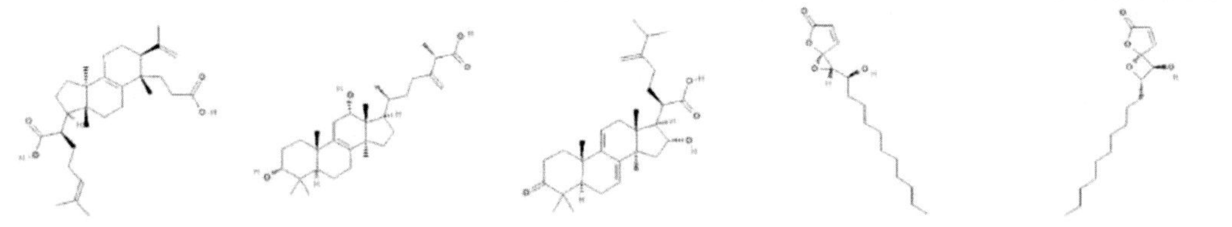

54) Pinicolic acid C | 55) Polyporenic acid A | 56) Polyporenic acid C | 57) Ramariolide A | 58) Ramariolide B

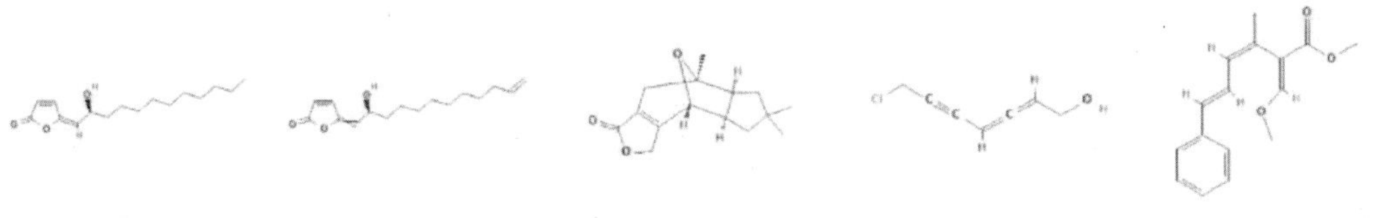

59) Ramariolide C | 60) Ramariolide D | 61) Rufuslactone | 62) Scorodonin | 63) Strobilurin A

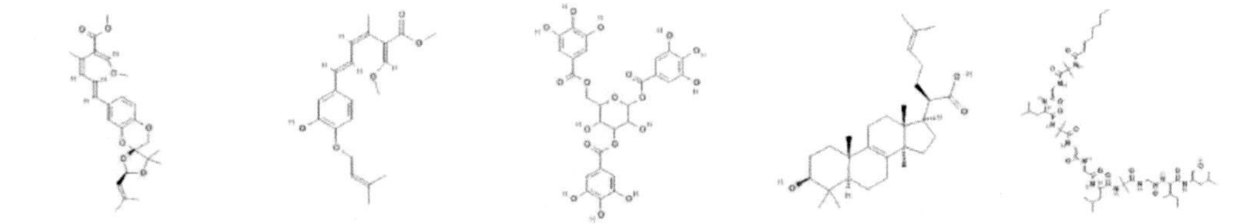

64) Strobilurin E | 65) Strobilurin F | 66) Tannic acid | 67) Trametenolic acid | 68) Trichogin A IV

Fig. (3). Bio-active compounds representing anti-microbial properties (Structures retrieved from PubChem).

Bio-active Compounds: Anti-Bacterial Properties

Several groups of bioactive compounds, phenols (Ellagic acid, *p*-Anisaldehyde, protocatechuic acid, vanillic acid, Tannic acid), terpenoids (Cuprinol,

Ganorbiformins A–G, Ganoderic acid), Sesquiterpenes (Coriolin, Enokipodins A-D, Marasmic acid, Scorodonin, Stereumamides A-D), steroids (Polyporenic acid C, Trametenolic acid B, and Fomitopsic acid), fatty acids (Palmitic acid Linoleic acid), and peptides (Agrocybin, Plectasin), extracted from numerous mushrooms showed strong anti-bacterial activities (Table **1**). Thus, the potential of these significant bioactive compounds can be beneficial in clinics.

Table 1. Bio-active compounds (representing Anti-bacterial properties) isolated from different sources of mushrooms.

S. No.	Mushroom	Bio-active Compound (s)	Group of Compounds	References
1.	*Agaricus* sp.	1-octen-3-ol and 10-oxo-trans-8-decenoic acid	Fatty acids	[15]
		Palmitic acid and Linoleic acid		[16]
		Protocatechuic acid	Phenols	[9, 10]
2.	*Agrocybe* sp.	Agrocybolacton	-	[17]
		Agrocybin	Peptide	[18]
3.	*Albatrellus* sp.	Confluentin, Grifolin, and Neogrifolin	Lanostanetype Triterpene	[19]
4.	*Amanita* sp.	2-amino-3-cyclopropylbutanoic acid	Nonprotein amino oxides	[20]
5.	*Auricularia* sp.	Vanillic acid	Phenols	[9, 10]
6.	*Coprinus* sp.	Coprinol	Terpenoid	[21]
		Micaceol	Sterol	[22]
		(Z,Z)-4-oxo-2,5-heptadienedioic acid	-	
7.	*Cordyceps* sp.	Cordycepin	Derivative of the nucleoside adenosine	[23]
8.	*Coriolus* sp.	Coriolin	Sesquiterpene	[24]
9.	*Cortinarius* sp.	Cortamidine oxide, 2,2'-Dithiobis(pyridine *N*-oxide)	Nonprotein amino oxides	[25]
10.	*Flammulina* sp.	Enokipodins A-D	Sesquiterpenoid	[26]
		Pinicolic acid (1-3)	Fatty acids	[27]
11.	*Fomitopsis* sp.	Polyporenic acid C, 3R-acetyloxylanosta-8, 24- dien-21-oic acid, Trametenolic acid B, and Fomitopsic acid	Steroids	[28]
		3α-(3′-butylcarboxyacetoxy)oxepanoquercinic acid and 3α-hydroxy-24-methylene-23-oxolanost-8-en-26-carboxylic acid	Triterpenes	[29]

(Table 1) cont.....

S. No.	Mushroom	Bio-active Compound (s)	Group of Compounds	References
12.	*Ganoderma* sp.	Methyl australate, Australic acid	Steroids	[30]
		Ganorbiformins A–G, Ganoderic acid	Triterpenoids	[31]
		Ganomycins A and B		[4, 30, 32]
		Ganosinensins A-C		[33]
		Palmitic acid	Fatty acids	[34]
13.	*Jahnoporus* sp.	3,11-dioxolanosta-8,24(Z)-diene-26-oic acid	Lanostanetype triterpene	[19]
14.	*Lactarius* sp.	Guaiol stearic acid Ester, Lactaroviolin, Deterrol	Sesquiterpenoids	[35]
15.	*Lentinula* sp.	Lentinan	Polysaccharide	[36]
		Oxalic acid		
		Ellagic acid, Tannic acid	Phenols	[37]
		-	Sesquiterpenoids	[7, 9, 10]
		Lenamycin	Fatty alcohol	[38]
		Lenthionine	Sulfurheterocyclics	[39]
16.	*Leucopaxillus* sp.	2-aminoquinoline	Alkaloid	[9, 10]
		Flavonoids	Phenols	[40]
17.	*Merulius* sp.	Merulinic acids A, B, and C	Polyketides	[41]
18.	*Marasmius* sp.	Scorodonin, Marasmicacid	Sesquiterpenoid	[42]
19.	*Pleurotus* sp.	Terpene	Terpenoids	[7, 9]
		p-Anisaldehyde	Phenols	[43]
20.	*Pseudoplectania sp.*	Plectasin	Peptide	[10]
21.	*Pycnoporus* sp.	Cinnabarin	Oxazine	[44]
22.	*Ramaria* sp.	Ramariolides A–D,	Butenolides	[45]
23.	*Stereum* sp.	Stereumamides A-D	Sesquiterpenes	[46]
		Hirsutic acid and Complicatic acid	Hirsutane derivatives	
		Frustulosin and Frustulosinol	Aromatic acetylene derivatives	

Several studies suggest that bioactive compounds extracted from various sources of mushrooms represent effective anti-microbial properties against multi-drug resistant strains *viz. Streptococcus* sp., *Staphylococcus* sp., *E. coli* [5, 7]. Several mushrooms *viz. Agaricus* sp., *Agrocybe aegerita, Boletus* sp., *Clitocybe alexandri, Coprinopsis atramentaria, Cordyceps militaris, Ganoderma* sp., *Laetiporus* sp.,

Lentinula sp., *Lycoperdon* sp., *Morchella* sp., *Phellinus* sp., *Pleurotus* sp., *Tirmania sp.* reported the anti-bacterial activities against Methicillin-sensitive bacterial strains such as *Staphylococcus aureus, S. mutans, S. saprophyticus, S. typhimurium, S. lutea, S. epidermidis and S. enteritidis* [2, 4, 8]. Several studies reported that methanolic, ethanolic and ethyl acetate extracts of these mushrooms also show anti-bacterial properties against other bacterial strains *viz. Actinomyces* sp., *Bacillus cereus, B. subtilis, Enterobacter aerogenes, Escherichia coli, Klebsiella pneumoniae, Lactobacillus* sp., *Micrococcus* sp., *Proteus vulgaris, Pseudomonas* sp., *Salmonella* sp., [8 - 10]. Another resistant gram-positive coccus- *Streptococcus* sp. was inhibited by the extracts of *Agaricus* sp., [11], *Flammulina velutipes* [12], *Lentinus edodes* [13] and *Pleurotus* mushrooms [14].

Bio-active Compounds: Anti-Fungal Properties

Fungal infections (Mycoses) are mostly not so dangerous, but some invasive fungal species can cause serious infections and also be deadly (bloodstream fungal infections). Some fungal infections affecting humans are caused by *Aspergillus, Candida, Pneumocystis, Sporothrix, Taloromyces.* Several scientific studies highlighted that mushroom extracts effectively against fungal infections caused by *Candida* yeast (the most common cause of fungal infections worldwide). Bio-active compounds isolated from Basidiomycetes mushrooms showing potential as anti-fungal properties are listed in Table **2**.

Table 2. Bio-active compounds (representing anti-fungal and anti-viral properties) isolated from different sources of mushrooms.

S. No.	Mushroom	Bio-Active Compound (s)	Group of Compounds	References
		Anti-fungal compounds		
1.	*Agaricus* sp.	1-octen-3-ol and 10-oxo-trans-8-decenoic acid	Fatty acids	[55]
		Strobilurin A, Strobilurin E, Strobilurin F	Polyketides	[56]
2.	*Agrocybe* sp.	Agrocybin	Peptide	[57]
3.	*Amanita* sp.	2-amino-3-cyclopropylbutanoic acid	Nonprotein amino oxides	[20]
4.	*Clitocybe sp.*	Phenylacetic acid	Phenol	[58]
5.	*Cortinarius* sp.	Cortamidine oxide, Dipyrithione	Nonprotein amino oxides	[25]
		Nebularine		[58]
6.	*Cordyceps* sp.	Cordymin	Peptide	[7]
7.	*Cyphellopsis* sp.	Strobilurin F, Strobilurin A	Polyketides	[56]

(Table 2) cont.....

S. No.	Mushroom	Bio-Active Compound (s)	Group of Compounds	References
8.	*Favolaschia* sp.	Favolon	Triterpenoid	[59]
		Strobilurin F, Strobilurin A, Strobilurin E, 9-Methoxystrobilurin A, Oudemansin A	Polyketides	[56]
9.	*Filoboletus* sp.	Strobilurin F, Strobilurin E, Strobilurin A		
10.	*Flammulina* sp.	Enokipodin F, G and I	Sesquiterpenes	[60]
11.	*Ganoderma* sp.	*p*-hydroxybenzoic and Cinnamic acids	Phenols	[61]
		Australic acid and methyl australate	Steroids	[30]
		5α-ergost-7-en-3β-ol, 5α-ergost-7,22-dien-3β-ol, and 5,8-epidioxy-5α,8α-ergosta-6,22-dien-3β-ol	Steroids	[62]
		Applanoxidic acid A, C, F, G, and H	Terpenes	
		Gonodermin	Protein	[63]
12.	*Hydropus* sp.	Strobilurin A	Polyketides	[56]
13.	*Hygrophorus* sp.	Chrysotriones A and B	Acylcyclopentenediones	[64]
14.	*Hypsizigus* sp.	Hypsin	-	[65]
15.	*Lactarius rufus*	Rufuslactone	Sesquiterpene	[66]
16.	*Lentinula* sp.	Lenthionine	Sulfurheterocyclics	[39]
		Ellagic acid, Tannins	Phenols	[37]
		Lentin	Protein	[65]
		Lentinamycin	Fatty alcohol	[38]
17.	*Leucopaxillus* sp.	2-aminoquinoline	Alkaloid	[67]
18.	*Lyophyllum* sp.	Lyophyllin	Protein	[65]
19.	*Mycena* sp.	Favolon B	Triterpenoid	[68]
		Chlorinated strobilurins B, Strobilurin A	Polyketides	[56]
20.	*Oudemansiella*sp.	Oudemansin A, Oudemansin X	Acrylates	[3]
21.	*Pleurotus* sp.	Ribonuclease	Protein	[69]
		Eryngin	Peptide	
		Pleurostrin	Peptide	[70]
		p-Anisaldehyde	Phenols	[43]
22.	*Polyporus* sp.	Alveolarin	Peptide	[71]

(Table 2) cont.....

S. No.	Mushroom	Bio-Active Compound (s)	Group of Compounds	References
23.	*Serpula* sp.	Himanimide C	Alkaloid	[68]
24.	*Strobilurus* sp.	Chlorinated strobilurin B, Strobilurin A	Polyketides	[56]
25.	*Tricholoma* sp.	Trichogin	Protein	[7]
26.	*Xerula* sp.	Chlorinated strobilurins B, Strobilurin A	Polyketides	[56]
27.	*Xylaria* sp.	Cloratin A	Derivative of benzoic acid	[72]
28.	*Agaricus* sp.	β-L-glutaminyl-3,4-benzoquinone	Quinone	[47, 50, 51,73]
		-	Sulphated Polysaccharides	
29.	*Agrocybe* sp.	AAL	Lectin	[1]
30.	*Auricularia auricula*	AAP	Sulphated Polysaccharides	[74]
31.	*Boletus* sp.	-	Polysaccharides	[48]
32.	*Flammulina* sp.	Velutin and Flammulin	Proteins	[52]
33.	*Ganoderma* sp.	Ganodermadiol, Lucidodiol, Ganoderic acid B and Aplanoxin acid G, Lutidinic acids, Methyl ganoderate F	Triterpenoids	[53, 54, 75]
34.	*Inonotus* sp.	Hispidin and Hispolon	Isoprenoid	[76]
35.	*Lentinula* sp.	-	Polysaccharides	[48]
36.	*Pleurotus* sp.			
37.	*Tremella* sp.	TPS	Sulphated Polysaccharides	[77]
		Glucuronoxylomannan (GXM)	Polysaccharide	[49]

Bio-active compounds isolated from different macromycetes genera *viz. Agaricus bisporus, Amanita cokeri, Clitocybe* sp., *Cortinarius nebularis, Dictyophora duplicate, Flammulina velutipes, Ganoderma austral, Ganoderma lucidum, Hypsizigus marmoreus, Hygrophorus chrysodon, Lentinus edodes, Lyophyllum shimeji, Pleurotus ostreatus, Tricholoma giganteum,* showed anti-fungal potential against several pathogenic fungi such as *Fusarium* sp., *Alternaria* sp., *Aspergillus* sp., *Candida* sp., *Microsporum* sp., *Mycosphaerella* sp., *Physalospora* sp., *Penicillium* sp., *Rhizoctonia* sp., *Trichoderma* sp., *Trichophyton* sp. [6, 7].

Different groups of bio-compounds: fatty acids, phenols, peptides, proteins, terpenoids, sesquiterpenes, and steroids (Table **2**) isolated from different sources of mushrooms are effective against several resistant pathogenic genera of fungi. Further *in vivo* and *in vitro* anti-fungal studies will discover novel anti-fungal drugs from mushrooms.

Bio-active Compounds: Anti-Viral Properties

Viruses are very small infecting agents that hijack, infect and multiply inside the normal organism's body, and these tiny viruses have been responsible for several pandemic diseases (Spanish flu, smallpox, HIV, COVID-19) throughout human history. Anti-viral drugs do not show broad-spectrum activities; thus, developing a wide range of anti-viral drugs is required. In the literature review, bioactive compounds isolated from several Basidiomycetes and one genus of Ascomycetes (*Ophiocordyceps sinensis*) reported anti-viral activities against the Influenza virus, Human Immunodeficiency Virus (HIV), Herpes Virus, Hepatitis Virus, West Nile Virus, Orth poxviruses, Poliovirus and Epstein-Barr virus [47].

Around 60 species of mushrooms are known to combat viruses. Bio-active compounds isolated from the extracts of different genera of mushrooms *viz.* *Agaricus* sp., *Agrocybe* sp., *Auricularia* sp., *Boletus* sp., *Flammulina* sp., *Ganoderma* sp., *Inonotus* sp., *Lentinula* sp., *Pleurotus* sp., *Tremella* sp. have been reported as potential anti-viral drugs (Table **2**). Different groups of fungal compounds such as Polysaccharides [48, 49], Sulphated Polysaccharides [50, 51], Lectins [1], Proteins [52], Triterpenoids [53, 54] and glycoproteins [47] are known to combat different types of viruses.

EXTRACTION OF BIOACTIVE COMPOUNDS FROM MUSHROOMS

The extraction yield of desired compound and bio-activity is largely affected by the selection of extraction solvents, and techniques, and consequently, extraction of bioactive compounds from raw material is the most vital step [78 - 83]. Conventional (Soxhlet or Heat reflux extraction) extraction methods in which water and organic solvents (methanol or ethanol) were most commonly used to extract the bio-active compounds in mushrooms [84, 85]. In the last few decades, several non-conventional and green extraction techniques have been used aiming to obtain the maximum yield of significant compounds from mushrooms (Table **3**).

In literature, the most commonly used extraction solvents in conventional extraction methods are hexane, aqueous ethanol, chloroform and methanol [16, 66, 86 - 88] for the extraction of phenols, flavonoids, tannins, polysaccharides, sesquiterpenoids, total oil content, fatty acids from different sources of mushrooms. On the other hand, a lot of studies (Table **3**) reported the green and cost-effective extraction methods such as Ultrasound-assisted extraction (UAE), Mechanically agitated extraction, Microwave-assisted extraction (MAE), Supercritical fluid extraction (SFE), Ultrahigh pressure extraction (UHPE), Enzymatic assisted extraction and Pulsed electric field extraction (PEF) *etc.*

Table 3. Extraction of bioactive compounds from different sources of Mushrooms.

Sl. No.	Source of Mushroom (s)	Target Compound (s)	Extraction Solvent/Conditions	Extraction Method	References
1.	*Agaricus sp.* mycelia	Sterols and phenolic compounds	96% ethanol	Mechanically agitated extraction and UAE	[81]
		Polysaccharides	Power:453 W; Time: 62 min; Temp: 70°C	Ultrasound-assisted extraction (UAE)	[89, 90]
		Sterols	Power:400 W; Temp: 132.8 °C; Time:19.4 min	Microwave-assisted extraction (MAE)	[91]
		Fatty acids, Sterols and total Phenolic content	Pure CO_2 and with 10% ethanol (co-solvent); Temp:50 °C; Pressure:30 MPa	Supercritical fluid extraction (SFE)	[16]
			Ethanol	Maceration	
		Proteins, Polyphenols and Polysaccharides	Electric field: 800 V/cm	Pulsed electric field extraction (PEF)	[92]
			Water and ethanol extraction		
		Total Phenolic content	Solvents: Cyclohexane, Dichloromethane, Methanol, Water; Time: 5 hr	Soxhlet extraction	[8]
2.	*Auricularia auricula*	Polysaccharides	24 kV/cm	High-intensity Pulsed electric field extraction	[93]
			Power: 500 W; Time: 1 min	Microwave-assisted extraction (MAE)	
			Power:125 W; Time: 5 min	Ultrasound-assisted extraction (UAE)	
		Alkaloids, saponins, tannins, flavonoids and phenols	Solvent: Methanol; Time: 2 hr	Heat reflux extraction (HRE)	[94]

(Table 3) cont.....

Sl. No.	Source of Mushroom (s)	Target Compound (s)	Extraction Solvent/Conditions	Extraction Method	References
3.	*Boletus edulis*	Polyphenolic compounds	Solvent: 80% ethanol; Time: 5 min; Temp: 80 °C	Microwave-assisted extraction (MAE)	[95]
		Fatty acids	Temp: 27 °C; Time: 3.6 h	Supercritical and subcritical carbon dioxide extraction	[88]
		Total oil content	Solvent: Hexane; Time: 10 hr	Soxhlet extraction	
4.	*Cordyceps militaris* mycelia and cultures	Polysaccharides	Pressure: 424 MPa, Time: 18 min, Temp: 43.5°C	Ultrahigh pressure extraction (UHPE)	[96]
			Temp: Below 50°C	Ultrasound-assisted extraction (UAE)	[97]
5.	*Coriolus versicolor*	Polysaccharides	Solvent: 95% ethanol; Temp: 70°C; Time: 3 h	Heat reflux extraction (HRE)	[86]
6.	Fruiting bodies of *Grifola frondosa*	Polysaccharides	Power: 7.2 to 40.3 W; Temp: 40 to 80°C; Time: 10 to 30 min	Ultrasound-assisted extraction (UAE)	[97]
7.	Fruiting bodies, mycelia and spores of *Ganoderma lucidum*	Tetranor-sesquiterpenoids	Solvent: 80% ethanol	Heat reflux extraction (HRE)	[87]
		Total phenolic content	Solvent: 40% ethanol; Power: 125 W; Time: 3.8 min	Microwave-assisted extraction (MAE)	[82]
				Soxhlet extraction	
			Temp: Below 40°C	Ultrasound-assisted extraction (UAE)	
			Distilled water Temp: 180°C; Time: 30 min	Microwave assisted extraction (MAE) and Accelerated Solvent Extractor (ASE)	[98]
		Triterpenoids	SCF-CO_2; Pressure 5-60 MPa; Temp: 32 °C-85 °C	Supercritical fluid extraction (SFE)	[99]

(Table 3) cont.....

Sl. No.	Source of Mushroom (s)	Target Compound (s)	Extraction Solvent/Conditions	Extraction Method	References
8.	*Inonotus* sp.	Polysaccharides	Intensity: 30 kV/cm; Pulse number:6; pH: 10	Pulsed electric field extraction (PEF)	[100]
		Total Phenolic content	Solvents: Cyclohexane, Dichloromethane, Methanol, Water; Time: 5 hr	Soxhlet extraction	[8]
9.	*Lactarius sp.*	Polyphenolic compounds	Solvent: 80% ethanol; Time: 5 min; Temp: 80 °C	Microwave assisted extraction (MAE)	[95]
		Sesquiterpene	Solvent: 80% ethanol; chloroform; chloroform: methanol (1:1)	-	[66]
10.	Fruiting bodies and mycelia of *Lentinula edodes*	Polysaccharides	Powder treated with disodium hydrogen phosphate-citric acid buffer and cellulose. Ultrasonics Power: 340 W; Time: 14 min	Ultrasonic extraction based on enzymatic treatment	[101]
			-	Ultrasound-assisted extraction (UAE)	[97]
		Fatty acids, Sterols and total Phenolic content	SCF-CO_2 (15% ethanol as co-solvent); Pressure 20 MPa; Temp: 40 °C	Supercritical fluid extraction (SFE)	[102]
		Sterols and vitamin-D derivatives	UV Irradiation: 254 nm for 1 hr	SFE followed by UV C-irradiation	[83]
11.	*Lycoperdon pyriform*	Steroids	Solvent: hexane; Time: 72 hours	Shaking	[103]

(Table 3) cont.....

Sl. No.	Source of Mushroom (s)	Target Compound (s)	Extraction Solvent/Conditions	Extraction Method	References
12.	*Pleurotusos treatus*	Polysaccharides	Distilled water Temp: 180°C; Time: 30 min	Microwave-assisted extraction (MAE)	[98]
				Accelerated Solvent Extractor (ASE)	
		Alkaloids, saponins, tannins, flavonoids and phenols	Solvent: Methanol; Time: 2 hr	Heat reflux extraction (HRE)	[94]
		Total Phenolic content	Solvents: Cyclohexane, Dichloromethane, Methanol, Water; Time: 5 hr	Soxhlet extraction	[8]
		Alkaloids, saponins, tannins, flavonoids and phenols	Solvent: Methanol Time: 30 min	Maceration	[104]
13.	*Phellinus linteus* mycelia	Polysaccharides	Ultrasonic intensity: 453 W/cm^2 and Time: 180 min	Ultrasound-assisted extraction (UAE)	[105]
14.	Fruiting bodies and mycelia of *Tricholoma* sp.	Polysaccharides	Power: 365 W; Time: 160 sec	Ultrasound-assisted extraction (UAE)	[106]
		Total Phenolic content	Solvents: Cyclohexane, Dichloromethane, Methanol, Water; Time: 5 hr	Soxhlet extraction	[8]
15.	*Terfezia boudieri*	Polyphenolic compounds	Solvent: 80% ethanol; Time: 5 min; Temp: 80 °C	Microwave-assisted extraction (MAE)	[95]

CONCLUSION

In the present chapter, the authors highlighted the significant bioactive compounds possessing strong anti-bacterial, anti-fungal and anti-viral activities. Several groups of bioactive compounds, phenols (Ellagic acid, *p*-Anisaldehyde, protocatechuic acid, vanillic acid, Tannic acid), terpenoids (Coprinol, Ganorbiformins, Ganoderic acid), Sesquiterpenes (Coriolin, Enokipodins, Marasmic acid, Scorodonin, Stereumamides), steroids (Polyporenic acid, Trametenolic acid, and Fomitopsic acid), fatty acids (Palmitic acid Linoleic acid), and peptides (Agrocybin, Plectasin), extracted from numerous mushrooms showed strong anti-bacterial activities against multi-drug resistant strains *viz.*

Streptococcus sp., *Staphylococcus* sp., *E. coli.* Moreover, different groups of bio-compounds: fatty acids, phenols, peptides, proteins, terpenoids, sesquiterpenes, and steroids isolated from different sources of mushrooms are effective against several resistant pathogenic genera of fungi, such as *Fusarium* sp., *Alternaria* sp., *Aspergillus* sp., *Candida* sp., *Microsporum* sp., *Mycosphaerella* sp., *Physalospora* sp., *Penicillium* sp., *Rhizoctonia* sp., *Trichoderma* sp., *Trichophyton* sp. It has been reported that extracts of mushrooms are also effective against fungal infections caused by *Candida* yeast (the most common cause of fungal infections worldwide). In addition, bioactive compounds isolated from several Basidiomycetes and one genus of Ascomycetes (*Ophiocordyceps sinensis*) reported anti-viral activities against the Influenza virus, Human Immunodeficiency Virus (HIV), Herpes Virus, Hepatitis Virus, West Nile Virus, Orth poxviruses, Poliovirus and Epstein-Barr virus.

CONSENT FOR PUBLICATION

Not applicable.

CONFLICT OF INTEREST

The author declares no conflict of interest, financial or otherwise.

ACKNOWLEDGEMENTS

Declared none.

REFERENCES

[1] Sun H, Zhao CG, Tong X, Qi YP. A lectin with mycelia differentiation and antiphytovirus activities from the edible mushroom Agrocybe aegerita. J Biochem Mol Biol 2003; 36(2): 214-22. [PMID: 12689522]

[2] Rosa LH, Machado KMG, Jacob CC, Capelari M, Rosa CA, Zani CL. Screening of Brazilian basidiomycetes for antimicrobial activity. Mem Inst Oswaldo Cruz 2003; 98(7): 967-74. [http://dx.doi.org/10.1590/S0074-02762003000700019] [PMID: 14762527]

[3] Rosa LH, Cota BB, Machado KMG, Rosa CA, Zani CL. Anti-fungal and other biological activities from Oudemansiellacanarii (Basidiomycota). World J Microbiol Biotechnol 2005; 21(6-7): 983-7. [http://dx.doi.org/10.1007/s11274-004-7553-7]

[4] Gao Y, Tang W, Gao H, *et al.* Antimicrobial activity of the medicinal mushroom Ganoderma. Food Rev Int 2005; 21(2): 211-29. [http://dx.doi.org/10.1081/FRI-200051893]

[5] Lima CUJO, Gris EF, Karnikowski MGO. Antimicrobial properties of the mushroom *Agaricus blazei* – integrative review. Rev Bras Farmacogn 2016; 26(6): 780-6. [http://dx.doi.org/10.1016/j.bjp.2016.05.013]

[6] Shen HS, Shao S, Chen JC, Zhou T. Antimicrobials from mushrooms for assuring food safety. Compr Rev Food Sci Food Saf 2017; 16(2): 316-29. [http://dx.doi.org/10.1111/1541-4337.12255] [PMID: 33371536]

[7] Soković M, Ćirić A, Glamočlija J, Stojković D. 7. Soković M, Ćirić A, Glamočlija J, Stojković D. The Bioactive Properties of Mushrooms. U: Ferreira, ICFR, Morales, P. and Barros, L.(Eds). Wild Plants, Mushroom and Nuts. Functional Food Properties and Applications. 2017; 4: 83-122.

[8] Smolskaitė L, Venskutonis PR, Talou T. Comprehensive evaluation of antioxidant and antimicrobial properties of different mushroom species. Lebensm Wiss Technol 2015; 60(1): 462-71. [http://dx.doi.org/10.1016/j.lwt.2014.08.007]

[9] Alves M, Ferreira I, Dias J, Teixeira V, Martins A, Pintado M. A review on antimicrobial activity of mushroom (Basidiomycetes) extracts and isolated compounds. Planta Med 2012; 78(16): 1707-18. [http://dx.doi.org/10.1055/s-0032-1315370] [PMID: 23023950]

[10] Alves MJ, Ferreira ICFR, Martins A, Pintado M. Antimicrobial activity of wild mushroom extracts against clinical isolates resistant to different antibiotics. J Appl Microbiol 2012; 113(2): 466-75. [http://dx.doi.org/10.1111/j.1365-2672.2012.05347.x] [PMID: 22621239]

[11] Bernardshaw S, Johnson E, Hetland G. An extract of the mushroom *Agaricus blazei* Murill administered orally protects against systemic Streptococcus pneumoniae infection in mice. Scand J Immunol 2005; 62(4): 393-8. [http://dx.doi.org/10.1111/j.1365-3083.2005.01667.x] [PMID: 16253127]

[12] Cui C, Xu H, Yang H, *et al.* Antibacterial Activity of Fruiting Body Extracts from Culinary-Medicinal Winter Mushroom, Flammulina velutipes (Agaricomycetes) against Oral Pathogen Streptococcus mutans. Int J Med Mushrooms 2020; 22(2): 115-24. [http://dx.doi.org/10.1615/IntJMedMushrooms.2020033335] [PMID: 32479000]

[13] Hirasawa M, Shouji N, Neta T, Fukushima K, Takada K. Three kinds of antibacterial substances from Lentinus edodes (Berk.) Sing. (Shiitake, an edible mushroom). Int J Antimicrob Agents 1999; 11(2): 151-7. [http://dx.doi.org/10.1016/S0924-8579(98)00084-3] [PMID: 10221419]

[14] Li S, Dai S, Shah NP. Sulfonation and antioxidative evaluation of polysaccharides from Pleurotus mushroom and Streptococcus thermophilus bacteria: a review. Compr Rev Food Sci Food Saf 2017; 16(2): 282-94. [http://dx.doi.org/10.1111/1541-4337.12252] [PMID: 33371533]

[15] Beelman RB, Royse DJ, Chikthimmah N. Bioactive components in button mushroom Agaricusbisporus (J. Lge) Imbach (Agaricomycetideae) of nutritional, medicinal, and biological importance. Int J Med Mushrooms 2003; 5(4): 18. [http://dx.doi.org/10.1615/InterJMedicMush.v5.i4.10]

[16] Mazzutti S, Ferreira SRS, Riehl CAS, Smania A Jr, Smania FA, Martínez J. Supercritical fluid extraction of Agaricus brasiliensis: Antioxidant and antimicrobial activities. J Supercrit Fluids 2012; 70: 48-56. [http://dx.doi.org/10.1016/j.supflu.2012.06.010]

[17] Berg A, Dörfelt H, Kiet TT, Schlegel B, Gräfe U. Agrocybolacton, a new bioactive metabolite from Agrocybe sp. HKI 0259. J Antibiot (Tokyo) 2002; 55(9): 818-20. [http://dx.doi.org/10.7164/antibiotics.55.818] [PMID: 12458772]

[18] Petrović J, Glamočlija J, Stojković D, *et al.* Bioactive composition, antimicrobial activities and the influence of Agrocybe aegerita (Brig.) Sing on certain quorum-sensing-regulated functions and biofilm formation by Pseudomonas aeruginosa. Food Funct 2014; 5(12): 3296-303. [http://dx.doi.org/10.1039/C4FO00819G] [PMID: 25367459]

[19] Liu XT, Winkler A, Schwan W, Volk T, Rott M, Monte A. Antibacterial compounds from mushrooms I: a lanostane-type triterpene and prenylphenol derivatives from Jahnoporus hirtus and Albatrellus flettii and their activities against Bacillus cereus and Enterococcus faecalis. Planta Med 2010; 76(2): 182-5. [http://dx.doi.org/10.1055/s-0029-1186001] [PMID: 19644795]

[20] Drehmel DC, Chilton WS. Characterization and toxicity of Amanita cokeri extract. J Chem Ecol 2002; 28(2): 333-41.
[http://dx.doi.org/10.1023/A:1017986108720] [PMID: 11925071]

[21] Johansson M, Sterner O, Labischinski H, Anke T. Coprinol, a new antibiotic cuparane from a Coprinus species. Z Naturforsch C J Biosci 2001; 56(1-2): 31-4.
[http://dx.doi.org/10.1515/znc-2001-1-205] [PMID: 11302209]

[22] Zahid S, Udenigwe CC, Ata A, Eze MO, Segstro EP, Holloway P. New bioactive natural products from ***Coprinus micaceus***. Nat Prod Res 2006; 20(14): 1283-9.
[http://dx.doi.org/10.1080/14786410601101829] [PMID: 17393652]

[23] Paterson RR, Lima N. Biomedical effects of mushrooms with emphasis on pure compounds. Biomed J 2014; 37(6): 357-68.
[http://dx.doi.org/10.4103/2319-4170.143502] [PMID: 25355390]

[24] Matijašević D, Pantić M, Rašković B, *et al.* The anti-bacterial activity of Coriolus versicolor methanol extract and its effect on ultrastructural changes of Staphylococcus aureus and Salmonella Enteritidis. Front Microbiol 2016; 7: 1226.
[http://dx.doi.org/10.3389/fmicb.2016.01226] [PMID: 27540376]

[25] Nicholas GM, Blunt JW, Munro MHG. Cortamidine oxide, a novel disulfide metabolite from the New Zealand basidiomycete (mushroom) *Cortinarius* species. J Nat Prod 2001; 64(3): 341-4.
[http://dx.doi.org/10.1021/np000408+] [PMID: 11277751]

[26] Saito M, Kuwahara S. Enantioselective total synthesis of enokipodins A-D, antimicrobial sesquiterpenes produced by the mushroom, *Flammulina velutipes.* Biosci Biotechnol Biochem 2005; 69(2): 374-81.
[http://dx.doi.org/10.1271/bbb.69.374] [PMID: 15725664]

[27] Shah SR, Ukaegbu CI, Hamid HA, Alara OR. Evaluation of antioxidant and antibacterial activities of the stems of *Flammulina velutipes* and *Hypsizygus tessellatus* (white and brown var.) extracted with different solvents. J Food Meas Charact 2018; 12(3): 1947-61.
[http://dx.doi.org/10.1007/s11694-018-9810-8]

[28] Girometta C. Antimicrobial properties of *Fomitopsis officinalis* in the light of its bioactive metabolites: a review. Mycology 2019; 10(1): 32-9.
[http://dx.doi.org/10.1080/21501203.2018.1536680] [PMID: 30834150]

[29] Popova M, Trusheva B, Gyosheva M, Tsvetkova I, Bankova V. Antibacterial triterpenes from the threatened wood-decay fungus Fomitopsis rosea. Fitoterapia 2009; 80(5): 263-6.
[http://dx.doi.org/10.1016/j.fitote.2009.02.008] [PMID: 19264113]

[30] Smania EFA, Delle Monache F, Yunes RA, Paulert R, Smania Junior A. Antimicrobial activity of methyl australate from Ganoderma australe. Rev Bras Farmacogn 2007; 17(1): 14-6.
[http://dx.doi.org/10.1590/S0102-695X2007000100004]

[31] Isaka M, Chinthanom P, Kongthong S, Srichomthong K, Choeyklin R. Lanostane triterpenes from cultures of the Basidiomycete Ganoderma orbiforme BCC 22324. Phytochemistry 2013; 87: 133-9.
[http://dx.doi.org/10.1016/j.phytochem.2012.11.022] [PMID: 23280041]

[32] Mothana RAA, Jansen R, Jülich WD, Lindequist U. Ganomycins A and B, new antimicrobial farnesyl hydroquinones from the basidiomycete Ganoderma pfeifferi. J Nat Prod 2000; 63(3): 416-8.
[http://dx.doi.org/10.1021/np990381y] [PMID: 10757736]

[33] Sato N, Ma CM, Komatsu K, Hattori M. Triterpene-farnesyl hydroquinone conjugates from Ganoderma sinense. J Nat Prod 2009; 72(5): 958-61.
[http://dx.doi.org/10.1021/np800687t] [PMID: 19331378]

[34] Moradali MF, Mostafavi H, Ghods S, Hejaroude GA. Investigation of anti-microbial fatty acids from medicinal artist conk mushroom Ganoderma applanatum (Pers.) Pat.(Aphyllophoromycetideae) by TLC and spectroscopic detection. Int J Med Mushrooms 2008; 10(2): 149-54.

[http://dx.doi.org/10.1615/IntJMedMushr.v10.i2.50]

[35] Barros L, Baptista P, Estevinho LM, Ferreira ICFR. Effect of fruiting body maturity stage on chemical composition and antimicrobial activity of Lactarius sp. mushrooms. J Agric Food Chem 2007; 55(21): 8766-71.
[http://dx.doi.org/10.1021/jf071435+] [PMID: 17927152]

[36] Bender S, Dumitrache-Anghel CN, Backhaus J, *et al.* A case for caution in assessing the antibiotic activity of extracts of culinary-medicinal Shiitake mushroom [Lentinusedodes (Berk.) Singer] (Agaricomycetideae). Int J Med Mushrooms 2003; 5(1): 6.
[http://dx.doi.org/10.1615/IntJMedMushr.v5.i1.40]

[37] Rao JR, Millar BC, Moore JE, Moore JE. Antimicrobial properties of shiitake mushrooms (Lentinula edodes). Int J Antimicrob Agents 2009; 33(6): 591-2.
[http://dx.doi.org/10.1016/j.ijantimicag.2008.10.018] [PMID: 19118982]

[38] Ishikawa NK, Kasuya MCM, Vanetti MCD. Anti-bacterial activity of Lentinula edodes grown in liquid medium. Braz J Microbiol 2001; 32(3): 206-10.
[http://dx.doi.org/10.1590/S1517-83822001000300008]

[39] Hatvani N. Antibacterial effect of the culture fluid of Lentinus edodes mycelium grown in submerged liquid culture. Int J Antimicrob Agents 2001; 17(1): 71-4.
[http://dx.doi.org/10.1016/S0924-8579(00)00311-3] [PMID: 11137653]

[40] Barros L, Baptista P, Estevinho L, Ferreira I. Bioactive properties of the medicinal mushroom Leucopaxillus giganteus mycelium obtained in the presence of different nitrogen sources. Food Chem 2007; 105(1): 179-86.
[http://dx.doi.org/10.1016/j.foodchem.2007.03.063]

[41] Stamets P. Novel anti-virals from mushrooms. HerbalGram 2001; 51: 24-7.

[42] Abraham WR. Bioactive sesquiterpenes produced by fungi: are they useful for humans as well? Curr Med Chem 2001; 8(6): 583-606.
[http://dx.doi.org/10.2174/0929867013373147] [PMID: 11281843]

[43] Okamoto K, Narayama S, Katsuo A, Shigematsu I, Yanase H. Biosynthesis of p-anisaldehyde by the white-rot basidiomycete Pleurotus ostreatus. J Biosci Bioeng 2002; 93(2): 207-10.
[http://dx.doi.org/10.1016/S1389-1723(02)80015-9] [PMID: 16233188]

[44] Göçenoğlu A, Pazarlioglu N. Cinnabarinic acid: Enhanced production from Pycnoporuscinnabarinus, characterization, structural and functional properties. J Biol Chem 2014; 42: 281-90.

[45] Centko RM, Ramón-García S, Taylor T, *et al.* Ramariolides A-D, antimycobacterial butenolides isolated from the mushroom Ramaria cystidiophora. J Nat Prod 2012; 75(12): 2178-82.
[http://dx.doi.org/10.1021/np3006277] [PMID: 23205944]

[46] Duan YC, Feng J, Bai N, Li GH, Zhang KQ, Zhao PJ. Four novel antibacterial sesquiterpene-α-amino acid quaternary ammonium hybrids from the mycelium of mushroom Stereum hirsutum. Fitoterapia 2018; 128: 213-7.
[http://dx.doi.org/10.1016/j.fitote.2018.05.026] [PMID: 29792906]

[47] Teplyakova TV, Kosogova TA. Anti-viral effect of Agaricomycetes mushrooms. Int J Med Mushrooms 2016; 18(5): 375-86.
[http://dx.doi.org/10.1615/IntJMedMushrooms.v18.i5.10] [PMID: 27649599]

[48] Santoyo S, Ramírez-Anguiano AC, Aldars-García L, Reglero G, Soler-Rivas C. Antiviral activities of Boletus edulis, Pleurotusostreatus and Lentinusedodes extracts and polysaccharide fractions against Herpes simplex virus type 1. J Food Nutr Res 2012; 51(4): 225-35.

[49] Kovalenko OG, Polishchuk ON, Wasser SP. Virus Resistance induced by glucuronoxylomannan isolated from submerged cultivated yeast-like cell biomass of medicinal yellow brain mushroom Tremella mesenterica Ritz.: Fr.(Heterobasidiomycetes) in hypersensitive host plants. Int J Med Mushrooms 2009; 11(2): 199-205.

[http://dx.doi.org/10.1615/IntJMedMushr.v11.i2.90]

[50] Minari M, Rincão V, Soares A, Ricardo N, Nozawa C, Linhares R. Antiviral properties of polysaccharides from Agaricus brasiliensis in the replication of bovine herpesvirus 1. Acta Virol 2011; 55(3): 255-9.
[http://dx.doi.org/10.4149/av_2011_03_255] [PMID: 21978159]

[51] de Sousa Cardozo FTG, Camelini CM, Mascarello A, *et al.* Antiherpetic activity of a sulfated polysaccharide from Agaricus brasiliensis mycelia. Antiviral Res 2011; 92(1): 108-14.
[http://dx.doi.org/10.1016/j.antiviral.2011.07.009] [PMID: 21787804]

[52] Wang H, Ng TB. Isolation and characterization of velutin, a novel low-molecular-weight ribosome-inactivating protein from winter mushroom (Flammulina velutipes) fruiting bodies. Life Sci 2001; 68(18): 2151-8.
[http://dx.doi.org/10.1016/S0024-3205(01)01023-2] [PMID: 11324720]

[53] Zjawiony JK. Biologically active compounds from Aphyllophorales (polypore) fungi. Journalof natural products. 2004 Feb 27 2004; 67(2): 300-10.

[54] Lindequist U, Niedermeyer THJ, Jülich WD. The pharmacological potential of mushrooms. Evid Based Complement Alternat Med 2005; 2(3): 285-99.
[http://dx.doi.org/10.1093/ecam/neh107] [PMID: 16136207]

[55] Okull DO, Beelman RB, Gourama H. Antifungal activity of 10-oxo-trans-8-decenoic acid and 1-octe--3-ol against Penicillium expansum in potato dextrose agar medium. J Food Prot 2003; 66(8): 1503-5.
[http://dx.doi.org/10.4315/0362-028X-66.8.1503] [PMID: 12929847]

[56] Barseghyan GS, Barazani A, Wasser SP. Medicinal mushrooms with anti-phytopathogenic and insecticidal properties. Mushroom Biotechnology 2016; 137-53.
[http://dx.doi.org/10.1016/B978-0-12-802794-3.00008-4]

[57] Ngai PHK, Zhao Z, Ng TB. Agrocybin, an antifungal peptide from the edible mushroom Agrocybe cylindracea. Peptides 2005; 26(2): 191-6.
[http://dx.doi.org/10.1016/j.peptides.2004.09.011] [PMID: 15629530]

[58] Kim YS, Lee IK, Seok SJ, Yun BS. Chemical constituents of the fruiting bodies of Clitocybenebularis and their anti-fungal activity. Mycobiology 2008; 36(2): 110-3.
[http://dx.doi.org/10.4489/MYCO.2008.36.2.110] [PMID: 23990744]

[59] Reis FS, Barros L, Calhelha RC, *et al.* The methanolic extract of Cordyceps militaris (L.) Link fruiting body shows antioxidant, antibacterial, antifungal and antihuman tumor cell lines properties. Food Chem Toxicol 2013; 62: 91-8.
[http://dx.doi.org/10.1016/j.fct.2013.08.033] [PMID: 23994083]

[60] Wang Y, Bao L, Yang X, *et al.* Bioactive sesquiterpenoids from the solid culture of the edible mushroom Flammulina velutipes growing on cooked rice. Food Chem 2012; 132(3): 1346-53.
[http://dx.doi.org/10.1016/j.foodchem.2011.11.117] [PMID: 29243621]

[61] Heleno SA, Barros L, Martins A, Queiroz MJRP, Santos-Buelga C, Ferreira ICFR. Phenolic, polysaccharidic, and lipidic fractions of mushrooms from northeastern Portugal: chemical compounds with antioxidant properties. J Agric Food Chem 2012; 60(18): 4634-40.
[http://dx.doi.org/10.1021/jf300739m] [PMID: 22515547]

[62] Smania EFA, Delle Monache F, Smania A Jr, Yunes RA, Cuneo RS. Antifungal activity of sterols and triterpenes isolated from Ganoderma annulare. Fitoterapia 2003; 74(4): 375-7.
[http://dx.doi.org/10.1016/S0367-326X(03)00064-9] [PMID: 12781809]

[63] Wang H, Ng TB. Ganodermin, an antifungal protein from fruiting bodies of the medicinal mushroom Ganoderma lucidum. Peptides 2006; 27(1): 27-30.
[http://dx.doi.org/10.1016/j.peptides.2005.06.009] [PMID: 16039755]

[64] Gilardoni G, Clericuzio M, Tosi S, Zanoni G, Vidari G. Antifungal acylcyclopentenediones from fruiting bodies of Hygrophorus chrysodon. J Nat Prod 2007; 70(1): 137-9.

[http://dx.doi.org/10.1021/np060512c] [PMID: 17253868]

[65] Lam SK, Ng TB. Hypsin, a novel thermostable ribosome-inactivating protein with antifungal and antiproliferative activities from fruiting bodies of the edible mushroom Hypsizigus marmoreus. Biochem Biophys Res Commun 2001; 285(4): 1071-5. [http://dx.doi.org/10.1006/bbrc.2001.5279] [PMID: 11467862]

[66] Luo DQ, Wang F, Bian XY, Liu JK. Rufuslactone, a new antifungal sesquiterpene from the fruiting bodies of the basidiomycete Lactarius rufus. J Antibiot (Tokyo) 2005; 58(7): 456-9. [http://dx.doi.org/10.1038/ja.2005.60] [PMID: 16161484]

[67] Pfister JR. Isolation and bioactivity of 2-aminoquinoline from Leucopaxillusalbissimus. J Nat Prod 1988; 51(5): 969-70. [http://dx.doi.org/10.1021/np50059a027] [PMID: 21401193]

[68] Aqueveque P, Anke T, Anke H, Sterner O, Becerra J, Silva M. Favolon B, a new triterpenoid isolated from the Chilean Mycena sp. strain 96180. J Antibiot (Tokyo) 2005; 58(1): 61-4. [http://dx.doi.org/10.1038/ja.2005.7] [PMID: 15813182]

[69] Ngai PHK, Ng TB. A napin-like polypeptide from dwarf Chinese white cabbage seeds with translation-inhibitory, trypsin-inhibitory, and antibacterial activities. Peptides 2004; 25(2): 171-6. [http://dx.doi.org/10.1016/j.peptides.2003.12.012] [PMID: 15062997]

[70] Chu KT, Xia L, Ng TB. Pleurostrin, an antifungal peptide from the oyster mushroom. Peptides 2005; 26(11): 2098-103. [http://dx.doi.org/10.1016/j.peptides.2005.04.010] [PMID: 15941607]

[71] Wang H, Ng TB. Eryngin, a novel antifungal peptide from fruiting bodies of the edible mushroom Pleurotus eryngii. Peptides 2004; 25(1): 1-5. [http://dx.doi.org/10.1016/j.peptides.2003.11.014] [PMID: 15003349]

[72] Quang DN, Bach DD, Hashimoto T, Asakawa Y. Chemical constituents of the Vietnamese inedible mushroom *Xylaria intracolorata*. Nat Prod Res 2006; 20(4): 317-21. [http://dx.doi.org/10.1080/14786410600650354] [PMID: 16644525]

[73] Faccin LC, Benati F, Rincão VP, *et al.* Antiviral activity of aqueous and ethanol extracts and of an isolated polysaccharide from Agaricus brasiliensis against poliovirus type 1. Lett Appl Microbiol 2007; 45(1): 24-8. [http://dx.doi.org/10.1111/j.1472-765X.2007.02153.x] [PMID: 17594456]

[74] Nguyen TL, Chen J, Hu Y, *et al. In vitro* antiviral activity of sulfated Auricularia auricula polysaccharides. Carbohydr Polym 2012; 90(3): 1254-8. [http://dx.doi.org/10.1016/j.carbpol.2012.06.060] [PMID: 22939338]

[75] Mothana RAA, Awadh Ali NA, Jansen R, Wegner U, Mentel R, Lindequist U. Antiviral lanostanoid triterpenes from the fungus Ganoderma pfeifferi. Fitoterapia 2003; 74(1-2): 177-80. [http://dx.doi.org/10.1016/S0367-326X(02)00305-2] [PMID: 12628419]

[76] Awadh Ali NA, Mothana RAA, Lesnau A, Pilgrim H, Lindequist U. Antiviral activity of Inonotus hispidus. Fitoterapia 2003; 74(5): 483-5. [http://dx.doi.org/10.1016/S0367-326X(03)00119-9] [PMID: 12837367]

[77] Zhao X, Hu Y, Wang D, *et al.* Optimization of sulfated modification conditions of tremella polysaccharide and effects of modifiers on cellular infectivity of NDV. Int J Biol Macromol 2011; 49(1): 44-9. [http://dx.doi.org/10.1016/j.ijbiomac.2011.03.010] [PMID: 21439998]

[78] Kaur P, Kumar Pandey D, Gupta RC, Dey A. Simultaneous microwave assisted extraction and HPTLC quantification of mangiferin, amarogentin, and swertiamarin in Swertia species from Western Himalayas. Ind Crops Prod 2019; 132: 449-59. [http://dx.doi.org/10.1016/j.indcrop.2019.02.055]

[79] Kaur P, Gupta RC, Dey A, Kumar Pandey D. Simultaneous quantification of oleanolic acid, ursolic

acid, betulinic acid and lupeol in different populations of five Swertia species by using HPTLC-densitometry: Comparison of different extraction methods and solvent selection. Ind Crops Prod 2019; 130: 537-46.
[http://dx.doi.org/10.1016/j.indcrop.2018.12.089]

[80] Pandey DK, Kaur P. Optimization of extraction parameters of pentacyclic triterpenoids from Swertia chirata stem using response surface methodology. 3 Biotech. 2018 Mar 1 2018; 8(3): 152.

[81] Umaña M, Eim V, Garau C, Rosselló C, Simal S. Ultrasound-assisted extraction of ergosterol and antioxidant components from mushroom by-products and the attainment of a β-glucan rich residue. Food Chem 2020; 332: 127390.
[http://dx.doi.org/10.1016/j.foodchem.2020.127390] [PMID: 32603922]

[82] Maeng JH, Muhammad Shahbaz H, Ameer K, Jo Y, Kwon JH. Optimization of microwave□assisted extraction of bioactive compounds from Coriolus versicolor mushroom using response surface methodology. J Food Process Eng 2017; 40(2): e12421.
[http://dx.doi.org/10.1111/jfpe.12421]

[83] Morales D, Gil-Ramirez A, Smiderle FR, Piris AJ, Ruiz-Rodriguez A, Soler-Rivas C. Vitamin D-enriched extracts obtained from shiitake mushrooms (Lentinula edodes) by supercritical fluid extraction and UV-irradiation. Innov Food Sci Emerg Technol 2017; 41: 330-6.
[http://dx.doi.org/10.1016/j.ifset.2017.04.008]

[84] Roselló-Soto E, Parniakov O, Deng Q, *et al.* Application of non-conventional extraction methods: Toward a sustainable and green production of valuable compounds from mushrooms. Food Eng Rev 2016; 8(2): 214-34.
[http://dx.doi.org/10.1007/s12393-015-9131-1]

[85] Sun Y, Zhang M, Fang Z. Efficient physical extraction of active constituents from edible fungi and their potential bioactivities: a review. Trends Food Sci Technol 2019.

[86] Pan Y, Dong S, Hao Y, *et al.* Ultrasonic-assisted extraction process of crude polysaccharides from Yunzhi mushroom and its effect on hydroxyproline and glycosaminoglycan levels. Carbohydr Polym 2010; 81(1): 93-6.
[http://dx.doi.org/10.1016/j.carbpol.2010.01.060]

[87] Luo Q, Yang XH, Cheng YX. Unprecedented tetranorsesquiterpenoids from Ganoderma applanatum mushrooms. Phytochem Lett 2019; 33: 81-3.
[http://dx.doi.org/10.1016/j.phytol.2019.06.002]

[88] Vidović S, Mujić I, Zeković Z, Lepojević Ž, Milošević S, Jokić S. Extraction of fatty acids from Boletus edulis by subcritical and supercritical carbon dioxide. J Am Oil Chem Soc 2011; 88(8): 1189-96.
[http://dx.doi.org/10.1007/s11746-011-1772-5]

[89] Tian Y, Zeng H, Xu Z, *et al.* Ultrasonic-assisted extraction and antioxidant activity of polysaccharides recovered from white button mushroom (Agaricus bisporus). Carbohydr Polym 2012; 88(2): 522-9.
[http://dx.doi.org/10.1016/j.carbpol.2011.12.042]

[90] Qiao DL, Zhao F, Huang HZ, Fan CC, Han L. Ultrasonic-assisted extraction and *in vitro* antioxidant activity of polysaccharides from Agaricusbisporus. Zhong yaocai= Zhongyaocai=. Zhong Yao Cai 2012; 35(8): 1340-7.
[PMID: 23320369]

[91] Heleno SA, Prieto MA, Barros L, Rodrigues A, Barreiro MF, Ferreira ICFR. Optimization of microwave-assisted extraction of ergosterol from Agaricus bisporus L. by-products using response surface methodology. Food Bioprod Process 2016; 100: 25-35.
[http://dx.doi.org/10.1016/j.fbp.2016.06.006]

[92] Parniakov O, Lebovka NI, Van Hecke E, Vorobiev E. Pulsed electric field assisted pressure extraction and solvent extraction from mushroom (Agaricusbisporus). Food Bioprocess Technol 2014; 7(1): 174-83.

[http://dx.doi.org/10.1007/s11947-013-1059-y]

[93] Li C, Mao X, Xu B. Pulsed electric field extraction enhanced anti-coagulant effect of fungal polysaccharide from Jew's ear *(Auricularia auricula)*. Phytochem Anal 2013; 24(1): 36-40. [http://dx.doi.org/10.1002/pca.2376] [PMID: 22689567]

[94] Unekwu HR, Audu JA, Makun MH, Chidi EE. Phytochemical screening and antioxidant activity of methanolic extract of selected wild edible Nigerian mushrooms. Asian Pac J Trop Dis 2014; 4: S153-7. [http://dx.doi.org/10.1016/S2222-1808(14)60431-X]

[95] Özyürek M, Bener M, Güçlü K, Apak R. Antioxidant/antiradical properties of microwave-assisted extracts of three wild edible mushrooms. Food Chem 2014; 157: 323-31. [http://dx.doi.org/10.1016/j.foodchem.2014.02.053] [PMID: 24679787]

[96] Chen R, Jin C, Li H, *et al.* Ultrahigh pressure extraction of polysaccharides from Cordyceps militaris and evaluation of antioxidant activity. Separ Purif Tech 2014; 134: 90-9. [http://dx.doi.org/10.1016/j.seppur.2014.07.017]

[97] Cheung YC, Siu KC, Wu JY. Kinetic models for ultrasound-assisted extraction of water-soluble components and polysaccharides from medicinal fungi. Food Bioprocess Technol 2013; 6(10): 2659-65. [http://dx.doi.org/10.1007/s11947-012-0929-z]

[98] Smiderle FR, Morales D, Gil-Ramírez A, *et al.* Evaluation of microwave-assisted and pressurized liquid extractions to obtain β-d-glucans from mushrooms. Carbohydr Polym 2017; 156: 165-74. [http://dx.doi.org/10.1016/j.carbpol.2016.09.029] [PMID: 27842810]

[99] Li L, Guo HJ, Zhu LY, Zheng L, Liu X. A supercritical-CO_2 extract of Ganoderma lucidum spores inhibits cholangiocarcinoma cell migration by reversing the epithelial–mesenchymal transition. Phytomedicine 2016; 23(5): 491-7. [http://dx.doi.org/10.1016/j.phymed.2016.02.019] [PMID: 27064008]

[100] Yin YG, Cui YR, Wang T. Study on extraction of polysaccharide from Inonotus obliquus by high intensity pulsed electric fields. Nongye Jixie Xuebao 2008; 39: 89-92.

[101] Ke L. Optimization of Ultrasonic Extraction of Polysaccharides from *L entinus Edodes* Based on Enzymatic Treatment. J Food Process Preserv 2015; 39(3): 254-9. [http://dx.doi.org/10.1111/jfpp.12228]

[102] Kitzberger CSG, Smânia A Jr, Pedrosa RC, Ferreira SRS. Antioxidant and antimicrobial activities of shiitake (Lentinula edodes) extracts obtained by organic solvents and supercritical fluids. J Food Eng 2007; 80(2): 631-8. [http://dx.doi.org/10.1016/j.jfoodeng.2006.06.013]

[103] Asgharpour F, Moghadamnia AA, Alizadeh Y, Kazemi S. Chemical Composition and antibacterial activity of hexane extract of Lycoperdon Pyriforme. S Afr J Bot 2020; 131: 195-9. [http://dx.doi.org/10.1016/j.sajb.2020.01.044]

[104] Fontes Vieira PA, Gontijo DC, Vieira BC, *et al.* Antioxidant activities, total phenolics and metal contents in Pleurotus ostreatus mushrooms enriched with iron, zinc or lithium. Lebensm Wiss Technol 2013; 54(2): 421-5. [http://dx.doi.org/10.1016/j.lwt.2013.06.016]

[105] Yan JK, Wang YY, Ma HL, Wang ZB. Ultrasonic effects on the degradation kinetics, preliminary characterization and antioxidant activities of polysaccharides from Phellinus linteus mycelia. Ultrason Sonochem 2016; 29: 251-7. [http://dx.doi.org/10.1016/j.ultsonch.2015.10.005] [PMID: 26585005]

[106] You Q, Yin X, Zhao Y, Jiang Z. Extraction and purification of polysaccharides from pine medicinal mushroom, Tricholoma matsutake (higher Basidiomycetes) fruit bodies. Int J Med Mushrooms 2014; 16(2): 149-60.
[http://dx.doi.org/10.1615/IntJMedMushr.v16.i2.50] [PMID: 24941036]

CHAPTER 10

Nano-particles for Microbial Growth and Drug Delivery

Pranjali Chole[1], Camil Rex[1], Sabia Imran[1], Kasi Gopinath[2] and **Lokesh Ravi[1,*]**

[1] *Department of Botany, St Joseph's University, Bengaluru, India*

[2] *School of Materials and Energy, Southwest University in Chongqing, BeiBei, China*

Abstract: The development of drug resistance in microorganisms has become one of the greatest global health challenges, as microorganisms tend to adapt to organic drugs *via* several mechanisms. Multi-drug resistance (MDR) in microorganisms not only increases the mortality rate of humans, but clinicians are also running out of options to treat MDR infections. A solution to this problem could be found in inorganic chemistry, where metal elements are converted in to nanoparticles that function as both drug and drug delivery agents to control microbial growth and overcome the resistance imposed on organic drugs. Nanoparticles have a high surface area to volume ratio, making them highly reactive with selective types of molecules such as bacterial/fungal cell walls. This makes nanoparticles an effective alternative to traditional chemical drugs. The development of resistance in the case of nanoparticles is almost nil. Nanoparticles of various elements have proven to be effective anti-microbial agents with several other pharmaceutical activities. Nanoparticles are also effective drug delivery agents that increase the bioavailability of drugs, enhance bioactivity, and increase drug flux into and through skin and biofilms. This chapter provides a compilation of various types of organic and inorganic nanoparticles, with their bioactivity, mode of action, synthesis, side effects, and mode of administration. Different types of nanoparticle-based drug delivery systems are summarised in this chapter, along with a summary of their organ-specific drug delivery. This report can provide a detailed understanding of nanoparticles in anti-microbial applications and aid in R&D to yield future nanomedicine.

Keywords: Antibacterial NPs, Antifungal NPs, Antimicrobial Nanoparticles, Antiparasitic NPs, Antiviral NPs, Chemical Synthesis, Drug Delivery, Drug-Resistant Pathogens, Gold Nanoparticles, Green Synthesis, Mechanism of NPs, Medicinal Applications, Metalic NPs, Nanoparticle Drug Delivery System, Nanoparticle Synthesis, Nanoparticles, Nanovehicles, Organic NPs, Pharmaceutical NPs, Silver Nanoparticles.

* **Corresponding author Lokesh Ravi:** Department of Botany, St Joseph's University, Bengaluru, India; E-mail: lokesh.ravi@sjc.ac.in

M. Amin-ul Mannan & Gaurav Kumar (Eds.)

INTRODUCTION TO NANOPARTICLES

The term 'Nano' refers to 'one billionth', of something. American theoretical physicist Richard Phillips Feynman introduced the concept of nanotechnology in his lecture titled 'There's plenty of room at the bottom' in 1959. The term 'nanotechnology' was proposed by professor Taniguchi (Tokyo University of Science), and he is the one who introduced a 'top-down approach' [1]. By definition, Nanoparticles are "solid particles or particulate dispersions with a size of 10-100NM" [2]. Nanoparticles plays a significant role in diverse factors, even from ancient times, such as in the making of the 'Lycurgus cup' in Rome and in the making of porcelain in the ancient Chinese dynasty [3]. The relationship between humans and Nanoparticles is inevitable because of their use in various fields, especially in medicine. The definition of nanoparticles, according to the American Society for Testing and Materials (ASTM) "particles with lengths that range from 1 to 100 nanometres in two or three dimensions" [4]. Nanoparticles are also called Artificial atoms [3]. Nanoparticles differ from bulk materials because of their distinctive physicochemical characteristics, such as extremely small size, immense surface area to mass ratio, and highly reactive nature [5]. The function of the nanoparticles in vivo is directly affected by the shape of the nanoparticles [6]. There are enormous chances of changes in physical properties when microparticles are transformed into nanoparticles. In the process of particles getting smaller, there is enhancing supremacy in the behaviour of atoms on the surface area of nanoparticles when compared to the atoms which are present inside the nanoparticles; this is due to the rise in the proportion of surface area to volume. These properties help to get the better of some constraints which are found in other therapeutic techniques [5]. Nanoparticles reveal their quantum mechanical property once they become sufficiently smaller. The transparent nature of nanoparticles is due to their dimension which is lower than the critical wavelength of light [3]. Nanotechnology is the field that involves the manipulation and creation of materials either by reducing or refining or by ratcheting up from a single group of atoms to produce materials on the nanometer scale [1]. One of the global problems concerning public health is infectious diseases; these infectious diseases can be intra or extracellular, diseases that are related to medical devices or biofilm mediated and these diseases paves way for millions of mortality. The use of nanoparticles is a hopeful approach against these infectious diseases.

Types of Nanoparticles

Organic Nanoparticles

Liposomes

Liposomes are microscopic, concentric, bilayered, and consist of spherical vesicles with the presence of aqueous chambers adjoining phospholipids [7]. The lipid bilayer of the liposomes can store drugs as the aqueous phase can be encapsulated. The diameter of the liposomes is about 400 nanometres to 2.5 micrometres [8]. A Tiny spherical system that is synthesized from non-Poisonous phospholipids and cholesterol is called Liposome. In the production process of liposomes, they can be engineered to have different properties based on the lipid of our interest [9] because liposomes are natural, have structural versatility, biocompatibility, non-immunogenicity, amphiphilic nature, and their ability to target specific delivery processes [7]. It can present in the bloodstream for a longer duration. Recent researches reveal that Liposomes are used as a successful drug carrier in cancer treatment without injuring the normal cells [9].

Dendrimers

The "word Dendrimers" is derived from the Greek ("Dendron"- Trees and "meros"- part). In the beginning, it was called "cascade molecules" by Vögtle et al. These molecules are named dendrimers by Tomalia et al. Dendrimers are radially symmetric, nano-sized molecules and possess monodisperse structures which have tree-like branches or arms. Fritz Vogtle was the first person to discover these hyperbranched molecules in 1978 [10]. Dendrimers are distinct multivalent molecules that possess a branched structure of nm size. Dendrimers have well-defined molecular architecture. The core of dendrimers starts with the group of atoms or a central atom, and from this, 'dendrons' (branches of other atoms) will grow through various chemical reactions [10]. The drug entrapment effectiveness is dictated by the exterior surface of the macromolecule, which has 3 various domains such as terminal functional groups, branches, and a central core. The terminal groups which are present in lagers numbers are responsible for the interactivity of the dendrimers with surfaces, solvents, and other molecules. Dendrimer' is not a compound; it is an architectural motif. Dendrimers are well known for their compact shape and amalgamation of many functional groups [10]. The binding properties, high solubility, and reactivity of the dendrimers are because of the presence of extremely active functionalities on their surface [11].

Carbon-based NPs

Two main carbon-based nanoparticles are represented by carbon nanotubes and Fullerenes. An Allotropic form of C is used in the Fullerenes [12]. Fullerene is inert and safe. H. Kroto, R. Smally, and R.Curl discovered the Fullerenes in the year 1985, and they were honoured with Nobel Prize (1966). Other than producing in the laboratory, Fullerenes also exist in nature (outer space). These are nothing but the closed polyhedra of C atoms. Different number of C atoms are present in discovered fullerenes. C-60 (n=60) is one of the famous fullerenes. Every fullerene possesses and hexagonal rings ((n/2)-10) [13]. It can transmit light based on intensity. They are employed in the environmental remediation process for the efficient absorption of gases. It is also used as a support medium for diverse organic and inorganic catalysts [12]. Carbon nanotubes (CNT) are made of graphene nanofoil. The diameter of the single-layered will be less than 0.7 nm, and for the multilayered, it is about 100nm, and the length will vary, respectively. To form nanotubes honeycomb lattice of C atoms will be wounded into hollow cylinders [14]. Carbon nanotubes are classified into MWCNT (multi-wall carbon nanotubes) and SWCNT (single-wall carbon nanotubes). The unique thermal, electrical and physical properties of carbon nanotubes make them more efficient in industrial and bio medicinal fields [15]. Carbon nanotubes possess high tensile strength, electrical and thermal conductivity, and flexibility [14].

Polymeric Nanoparticles

Polymeric nanoparticles possess a matrix structure composed of biocompatible and biodegradable polymers of natural or synthetic origin. Mostly used natural polymers are chitosan, alginates, *etc.*, and among the synthetic polymers the polylactide–polyglycolide copolymers, polylactide, polyacrylates, andpolycaprolactone [7]. The advantage of polymeric nanoparticles is that they can be fabricated in a wide variety and various sizes. Different kinds of therapeutic agents, such as proteins, DNA, and various low molecular weight compounds, can be carried by biodegradable polymeric nanoparticles. In the field of pharmacology, polymeric nanoparticles are utilized as drug carriers for vaccination, tumours, bone healing, and diabetics [9]. The extensive use of polymeric nanoparticles in the field of pharmaceutics and medicine is because of their controllable properties, subcellular size, and biocompatibility with various tissues and cells [16, 17]. Polymeric micelles are also non-biodegradable polymeric nanoparticles. It can securely deliver drugs to the target because physiologically, they are highly stable in the biological environment [9].

Metal Nanoparticles

Nanosized metals are used to call "metal nanoparticles," and their size is about 1 – 100nm. The major difference between bulk metal and metallic nanoparticles is in the bulk matrix because electron delocalization possesses high electrical and thermal conductivity, but in the metal nanoparticles, those delocalization is absent, which results in properties that are entirely different from bulk metals [18].

In the bio-medicinal field, metallic nanoparticles of the size of 5 to 16 nm are regarded as most hopeful for practical applications [19].

Because of their high specificity of interaction and chemical activity, metal nanoparticles have been paid great attention over the last decade. When compared to bulk equivalents, metal nanoparticles have an increased surface area-to-volume ratio, and this is the main reason for their catalytic applications. Some important features of metal nanoparticles include providing LDOS, which reveals a transition between molecular and metallic states, raising the number of kinks, edges and corners. One of the Advantages of metallic nanoparticles is that for a particular reaction, we can amplify the activity by giving needed surface properties and size. Increased availability of the surface system is also achieved by reducing the metal cluster or reducing the nanoparticle size. In therapeutics, metal nanoparticles pave a significant way in diagnostic applications as these metallic nanoparticles can be bound to the single strands of deoxyribonucleic acid nondestructively [16]. Some of the important metal nanoparticles are discussed as follows.

Gold Nanoparticles

Gold nanoparticles are also known as colloidal gold, which is nothing but the colloids of gold particles that are in nanometre size [20]. Gold nanoparticles play a vital role in catalysis. Gold was regarded as a poor catalyst and chemically inert metal. Gold can be a highly reactive catalyst in many reactions, such as hydrogenation, oxidation, and other related reaction, when its size is less than 10 mm and dispersed particle on the appropriate support. In the aqueous solution, because of their LSDD, the gold nanospheres with a diameter of 10nm will have a great absorption maximum (around 520 nm) [20]. Several reviews on gold nanoparticles reveal that the gold nanoparticles with the size of 1-10 nm are greatly dispersed on various supports. In immunochemical studies, Protein interaction can be identified using Gold nanoparticles [21]. The Au nanoparticles are used mainly in the oxidation of Cobalt [18]. Aminoglycosidic antibiotics such as neomycin, gentamycin, and streptomycin can be detected using Au nanoparticles [21].

Silver Nanoparticles

The particle size of the Ag nanoparticles is between 1 and 100 nm [20]. Silver is well known for its anti-microbial activity. Because of their anti-microbial property, silver nanoparticles are widely employed in drug delivery, health care products, optical sensors, pharmaceutical industries, orthopaedics, cancer treatment, and also in the food industry [22]. Silver nanoparticles are widely employed in various catalytic activities. Ag nanoparticles are crucial in the selective oxidation of alkenes and alkanes for the production of various industrial products, which include aldehydes and epoxides. In the heterogeneous oxidation of ethylene, silver nanoparticles supported on CaCo3 and Al_2O_3 are selective and highly active [18]. Various factors that, include the morphology of the particle, agglomeration, reducing agent, dissolution rate, surface chemistry, ion release efficacy, and size, are the crucial factors that determine the cytotoxicity of Ag nanoparticles [22]. Nowadays, silver sulfadiazine is extensively replaced by silver nanoparticles for treating wounds [20]. In addition to the field of medicine, silver nanoparticles are used in the manufacturing of sunscreen lotion, water treatment, and textile industries [21]. Controlled structures of Ag nanoparticles, such as uniform morphology, function, and size, are crucial for the usage of silver nanoparticles in the field of biomedicine.

Zinc Oxide Nanoparticles

Among the trace elements, zinc is inevitable in humans because it is involved in the activation of various enzymes, such as alcohol dehydrogenase and carbonic anhydrase [23]. Zinc oxide nanoparticles have the ability to impede the activation of nuclear factor-kappa B cells (NF-kB), which leads to the minimization of the inflammatory cytokine's mRNA expression [23]. Morphologically zinc oxide nanoparticles can be synthesized in various forms, such as Nanospheres, Nano cages, hexagonal, nanowires, nano boxes, nanorings, nanotubes, tripods, and tetrapods [23]. Due to their increased catalytic activity, nano dimensions, and great surface area, they can be applied to catalytic reactions [19, 24]. These nanoparticles are highly active against various pathogens, especially gram-positive bacteria [23]. On comparing to all other metal oxide nanoparticles, zinc oxide nanoparticles show maximum toxicity toward microorganisms [23]. By coating the Zinc oxide nanoparticles with a dissolvable polymeric material, it can be used to treat ulcers, microbial infections, treating wounds and also as a drug carrier to treat cancer [23]. Zinc nanoparticles are also used in ointments, cosmetic creams, *etc.* because it has a great ability to absorb radiation such as infrared waves, radio frequencies, microwaves and ultraviolet radiations [19].

Titanium Dioxide Nanoparticles

Titanium dioxide (TiO2) is naturally present in brookite, anatase, and rutile. It is white powdery, odourless, and noncombustible. It is used in capsules and toothpaste as a food additive [25].Tio2 nanoparticles have the property of anti-corrosive, photocatalysis, and great stability. It plays a great role in the making of sunscreen lotion because of the property that is transparent when its particle size is at the nanoscale (less than 100 nm) and it can absorb and reflect ultraviolet light [25].

Magnetic Nanoparticles

The major applications of Magnetic nanoparticles are in the making of recording tapes, biomedical materials, Ferrofluids, and also as a catalyst [26]. Magnetic nanoparticles provide a promising approach in biomedicine. One of the notable properties of magnetic nanoparticles is their controllable size (from very few to tens of nm) which is equal to or lesser to a virus, a cell, a gene, or a protein. This shows their ability to get closer to the biological system [27]. It can be used to bind or interact with a biological entity when it is covered with biological molecules provided that it should be 'tagged ' in a controllable manner [27]. Various medical applications of the magnetic nanoparticles are especially because of their response to the magnetic force, this enables the magnetic nanoparticles in the bioseparation and drug targeting [28]. It is also used as a contrast agent for heating mediators to treat hyperthermia and MRI [28]. Positively charged Magnetite cationic liposomes (MCL) (cationic magnetic particles) are used to introduce magnetite nanoparticles into cell targets as their cell surface is negatively charged [28]. Magnetic nanoparticles such as maghemite and magnetite are biocompatible. These nanoparticles are vitally involved in the sorting and manipulation of stem cells, DNA analysis, drug delivery, MRI, and gene therapy [21].

Ceramic Nanoparticles

Ceramic nanoparticles can be engineered with the preferred porosity and size. It is an inorganic system that can be used as a drug vehicle. In cancer therapy, biocompatible ceramic Nps such as alumina, silica, and titania are used. In photodynamic therapy against cancer, using silica-based doped nanoparticles is a successful tactic in drug delivery [9]. Nanomaterials of silica have a broad range of usage in Cosmetics, food, mechanical polishing, drug delivery, and varnishes. It is said to be (GRAS) "generally regarded as safe" material. In recent years it shows great progress in cancer therapy, Biosensors, enzyme immobilization, DNA delivery, and biomarkers [25].

Quantum Dots

Quantum dots are nano-sized particles that possess nearly 100 - 1000 atoms. These are very tiny semiconductor crystals [29]. Quantum dots are nothing but colloidal nanocrystals which plays a great role in various field because of their chemical and physical properties [30]. The radii of colloidal semiconducting quantum dots are about 2-10 nanometers in diameter, which is more or less equal to the Bohr radius. (QY) Quantum yield and the narrow spectral band of the QDs are because of their quantum confinement effect. QDs comprise III–V elements like InAs or II-VI groups such as CDS, ZnS, *etc.* Quantum dots possess distinctive electrical and optical properties. QDs have excellent luminescence emission properties. The quantum confinement effects make the QDs different from bulk materials. The use of Quantum dots as a luminescent probes for biomedical analysis is a great advantage in the field of medicine [31]. Quantum dots are semiconductor nanocrystals with higher fluorescence and longer durability, which pave the way for it visual technologies to relocating in-vivo imaging and living organism diagnostics [25]. To target specific receptors, biological events, and cellular structures, fluorescent Quantum dots can be combined with bioactive moieties such as antibodies [25]. It is widely used in transistors, medical imaging, laser diodes, display, and solar cells [29].

SYNTHESIS OF NANOPARTICLES

Nanoparticles synthesis has been followed by several approaches. The synthesis of nanoparticles is categorized mainly three different methods as followed;

Green Synthesis

This method is also known as a biological method. In this method, nanoparticle synthesis plays an important role to control the reducing ability of enzymes, and biological molecules of microbial cells. This method of nanoparticle synthesis is environmentally friendly. Metal nanoparticles are biosynthesized by using different bacteria, plants, and fungi. *E. coli*, *Staphylococcus aureus*, *Enterobacter cloacae*, *Pseudomonas aeruginosa*, and some *lactobacilli species* have shown the ability to produce silver nanoparticles successfully. These biologically synthesized Ag NPs show higher stability than chemically synthesized ones.

The presence of enzymes and different phytochemicals in plant extracts helps to reduce silver ions into nano-size material. In leaf extracts of black tea and green tea biomolecules like phenol, flavonoids and polyphenols help to stabilize the Ag NPs by reducing Ag ions [32]. The green synthesis method has proved beneficial for renewable sources as well.

Chemical Synthesis

Reducing chemical organic and inorganic agents commonly synthesise silver nanoparticles. This synthesis process requires three components mainly; reducing agents, stabilizing agents, and metal precursors. Other different chemical methods for the synthesis of nanoparticles are the electrochemical synthetic method, UV-initiated photoreduction, photo-induced reduction, irradiation methods, and microemulsion method. As compared to the physical method, the chemical synthesis method gives a higher yield.

Physical Synthesis

In this method of synthesis of nanoparticles, temperature, concentration, and size play an important role. Generally, high temperature plays an important role, but it shows disadvantages like required space and time as well as harmful effects on the environment, with the temperature of the heater surface, the concentration of nanoparticles and size var [32].

ANTIBACTERIAL NANOPARTICLES

The antibacterial agents mainly show two kinds of activities; either they inhibit the growth of bacteria or kill the bacteria without harming the surrounding tissue. To fight against infectious diseases is the main adjective of these antibacterial agents. Antibacterial agents have increased in medicine, various industries, food packaging and water disinfection. The wide and extensive use of such organic compounds shows toxic disadvantages to the environment, including humans [33]. The continuous use of antibacterial agents has challenged the formation of bacterial resistance in the research community. This resistance is evolutionary.

To overcome this threat, antibacterial agents like metal nanoparticles were developed by scientists. Several laboratory studies proved that these metal nanoparticles showed strong antibacterial activities against these resistant agents.

A single time, antibacterial nanoparticles target multiple biomolecules and decrease the resistant potential of these resistant bacteria by avoiding the development of resistant strains [34]. Antibacterial nanoparticles include metals and metal oxides like silver (Ag), iron (Fe), copper oxide (CuO), zinc oxide (ZnO), titanium, *etc.*

ANTIBACTERIAL APPROACHES

Bacterial Cell Wall

The bacterial cell wall and membrane play important functions to provide shape, rigidity, and strength to the bacterial cell. It protects the bacterial cell from mechanical damage. According to components and structure, bacterial cell walls are differentiated into two types; 1. Gram-positive bacteria, 2. Gram-negative bacteria.

The cell wall of Gram-positive bacteria is made up of peptidoglycan and is attached to teichoic acids. This is the unique feature of a Gram-positive bacterial cell wall. While the Gram-negative cell wall shows a chemically as well as physically more complex structure. It is made up of lipopolysaccharides, lipoproteins, and phospholipids, allowing only entry to macromolecules. Also, Gram-negative bacterial membrane shows resistance to hydrophobic compounds like detergents. The cell walls of Gram-positive bacteria have high negative charges, which are helpful to attract nanoparticles. Hence, it shows that nanoparticles show higher activity against Gram-positive bacteria than against Gram-negative. Some recent studies showed that, in *E.coli*

A phospholipid is the head group of the lipopolysaccharide membrane, which interconnect with ε-poly-L-lysine by forming electrostatic attraction and damaging the cell membrane [33].

Role Of Biofilm Formation

Biofilm is a complex microbial community that is formed by adhesion with a solid surface and covers the bacterial cell community by secretion of a matrix, i.e., extra-polysaccharide and proteins [33]. The formation of biofilm is a significant problem. It protects pathogenic bacteria against antibiotics which is the main cause of the production of chronic infections [35]. For biofilms, the metabolism of bacteria is an important activity. For example, for the growth and development of *streptococcus*, the metabolism of D-alanine is essential. A recent study proved that in bacterial biofilms, nanoparticles can act as ion channels as well [36]. By using an external magnetic field, Mg nanoparticles show the ability to penetrate biofilm.

Growth Rate Control

The rate of bacterial growth is another important factor that can affect the tolerance of bacteria. It is scientifically studied that; slow-growing bacteria are less receptive than fast-growing bacteria to nanoparticles. Consequently, the particular strain is highly responsible for the antibacterial effect.

Other than this, the different physicochemical properties of nanoparticles, like size, surface morphology, charge as well as crystal structure, are helpful elements that regulate the nanoparticle's action on the bacterial cell [33]. Some more, bacterial strain, environmental conditions, and exposure time are the major factors that also affect nanoparticle antibacterial activity.

ANTIBACTERIAL MECHANISM OF NANOPARTICLES

Nanoparticles can change bacterial metabolic activity. This capacity of nanoparticles leads to the cure of various bacterial diseases by eliminating those bacteria. Hence, in the field of medicine, the use of nanoparticles has increased with an increase in the study of potential ability as well as the antibacterial mechanism of these nanoparticles.

When bacterial cells come in contact with nanoparticles, they attain the antibacterial function of the bacterial cell. Some important antibacterial mechanisms of nanoparticles are discussed below.

Release Of Metal Ion From Nanoparticles

The electrochemical potential present in the solution leads to the dissolution of nanoparticles. The anti-microbial activity is mainly depending on the number of ions released. It means the toxicity is directly proportional to the concentration of nanoparticles [33]. When the nanoparticle gets interconnected with the bacterial cell wall, it gives rise to a specific localization source for ions and releases ions continuously, which results in to increase in toxicity to the cell. The large concentration of these ions helps to penetrate the bacterial cell wall. After penetration, they interact with functional groups like amino (-NH) and carboxyl (-COOH), which are present in nucleic acid and proteins. They inhibit the bacteria by affecting physiological processes and inhibiting the activity of enzymes as well. The effect of the metal ion depends on pH during metal oxide suspension of the anti-microbial process, which is present in lipid vesicles [34]. For anti-microbial substances, heavy metal ions act as bearers indirectly.

Oxidative Stress

During basic metabolism production of Reactive Oxygen Species (ROS) takes a highly reactive place. Reducing oxygen molecules of different nanoparticles forms different kinds of ROS. Hydroxyl radical (-OH), hydrogen peroxide(H_2O_2), superoxide radical (O^{2-}), and singlet oxygen (O2) are four types of ROS that show different activities from each other [36]. All different four types of ROS produced by copper oxide. The catalase, as well as superoxide radicals, show the ability to acute microbial death while H2O2 and O2- lead to a reduction of less acute stress.

Oxygen vacancies that are present in the crystal, restructuring, and defect sites are mainly responsible for the production of ROS. In case of a change in the permeability of the bacterial cell membrane and to damage it, high stress of ROS plays a key factor. Even recent scientific studies proved that ROS acts as an important factor for the interlink between bacterial cells and DNA [36]. ROS is also important to increase the level of oxidative protein in gene expression, which is an important mechanism of apoptosis in bacterial cells. ROS can attack and depresses the physiological activities of the protein.

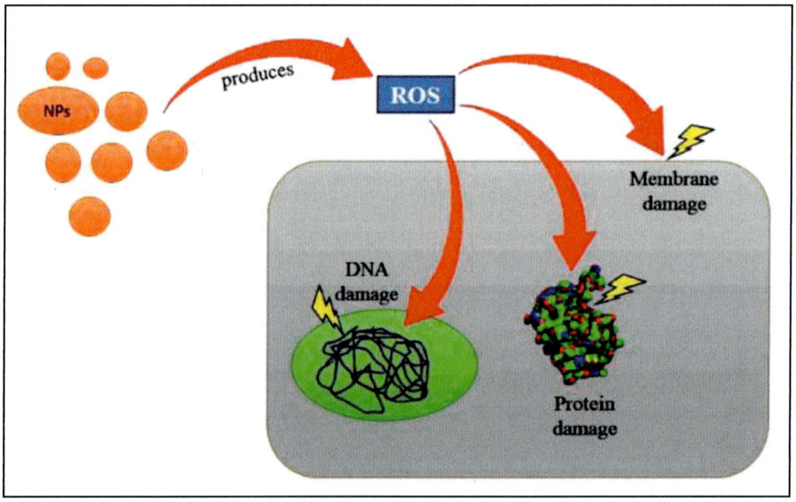

Fig. (1). Graphical representation of ROS mechanism of antibacterial NPs.

When a cell undergoes stress like UV light, DNA damage occurs. ROS production leads to an increase in the toxicity level of the cell and can cause death or the destruction of the cell. In the presence of light, the important anti-microbial substance (OH) is produced by zinc oxide. For example, when zinc oxide is stimulated by light, it produces the main anti-microbial substance, i.e., OH, in a minimal amount. The graphical representation of the mechanism is shown in Fig. (**1**).

NANOPARTICLES AGAINST DRUG-RESISTANT BACTERIA

Multi-drug resistant (MDR) bacteria are a globally increasing challenge. Developing new formulations by using nanotechnology has offered some prospects in recent days. Nanoparticles can act as natural anti-microbial compounds as well as a carrier for antibiotics. Table. **1** shows the list of nanoparticles that have been proven to be effective against drug-resistant bacterial pathogens.

Table 1. List of antibacterial nanoparticles against various drug-resistant pathogenic bacteria

Targeted bacteria	Types of NPs	Resistance antibiotic type	Mechanism of antibacterial action	References
S.aureus	Ag NPs	Methicillin-resistant	Combine with antibiotics	Yael N, *et al2017*
E.coli, S.aureus, K.pneumoniae	Ag NPs	Erythromycin resistant	Damaging cell surface as well as loss of chain integrity	Yael N, *et al2017*
E.coli	Ag NPs	MDR	Generating ROS	Yael N, *et al2017*
P.aeruginosa	Ag NPs	Ofloxacin resistant	Multi-drug efflux pump	Yael N, *et al*2017
S.aureus	Au NPs	Methicillin-resistant	Photothermal therapy by generating ROS	SSN Fernando, *et al 2018*
P.aeruginosa	Au NPs	Formation of biofilm	Interact with surface of cell	SSN Fernando, *et al 2018*
E.coli, S.aureus, Enterobacter aerogenes	Au NPs	Cefotaxime resistant	Bacterial cell disruption, damage of DNA	SSN Fernando, *et al 2018*
S.aureus	ZnO NPs	Methicillin resistant	Inhibition of enzyme	SSN Fernando, *et al 2018*
E.coli	ZnO NPs	MDR	Generation of ROS, bacterial cell wall disruption	SSN Fernando, *et al 2018*
K.pneumoniae	ZnO NPs	Ampicillin resistant	Generation of ROS, bacterial cell wall disruption	SSN Fernando, *et al 2018*
P.aeruginosa	Cu NPs	MDR	Cu hydrosol generation	SSN Fernando, *et al 2018*

(Table 1) cont.....

Targeted bacteria	Types of NPs	Resistance antibiotic type	Mechanism of antibacterial action	References
P.aeruginosa	Cu NPs	Formation of biofilm	Penetrate cell wall & damage nucleic acid	SSN Fernando, *et al 2018*
E.coli, S.aureus	CuO NPs	MDR	ROS generation	Yael N, *et al 2017*
E.coli	TiO_2NPs	MDR	Generation of ROS & cell wall disruption	Yael N, *et al2017*
S.aureus	Graphene	Methicillin-resistant	Combine antibiotics with NIR exposure.	Yael N, *et al2017*

ANTI-FUNGAL NANOPARTICLES

Fungi are unique, diverse group of organisms and ubiquitous. The pathogenic group of fungi causes several infections in human beings. Several studies described and detected continuous emergence and re-emergence of fungal infection. An increase in fungal infections is causing healthcare concerns worldwide. Developed research derived different anti-fungal nanoparticles to successfully overcome this problem, which kills fungal pathogens through different mechanisms.

Nanoparticles like Ag, ZnO and TiO_2 show anti-fungal activity. A recent study proved that biosynthesized AgNPs show superior anti-fungal activity than chemically synthesized ones. It shows anti-fungal activity against *Alternariaalternata*, *Rhizopusstolonifer*, *Aspergillus niger* and *Phomopsisasparagi*. Chitosan is a demanding material for the preparation of nanoparticles. It shows great anti-fungal activity, biodegradable in nature as well as shows non-toxicity.

ANTI-FUNGAL MECHANISM OF NPS

There is not exactly only one mechanism studied in anti-fungal nanoparticles as more study is going on. Different nanoparticles show different mechanisms based on their production, toxicity against the pathogen, size, and growth as well. It is scientifically proved that nanoparticles that are biosynthesized, disrupt cellular integrity, generation of superoxide radicals, and produce an osmotic imbalance in the pathogen. When fungal pathogens like *Alternariabrassicicola*treated with biosynthesized nanoparticles and scanned in an electron-microscope, it is observed that the osmotic imbalance and membrane disintegrate, the main cause for the death of the fungal cell.

Chitosan nanoparticles propose three main mechanisms to inhibit the fungal activity of pathogens. In the first mechanism, chitosan targets the plasma membrane of pathogenic fungi. Negatively charged components like phospholipids present in the membrane interact with the positively charged chitosan. This activity leads to an increase in the permeability of the membrane and the death of the cell by cellular leakage. In the second mechanism, chitosan penetrates the wall of the fungal cell and binds with DNA. This binding leads to the inhibition of the synthesis of mRNA, which results in an effect on the production of proteins and enzymes. While in the third mechanism, chelating agents present in the chitosan-formed nanoparticles bind trace elements of the pathogen and cause the unavailability of nutrients and inhibit the growth of the fungal pathogen.

ANTI-PARASITIC ACTIVITIES

Nanoparticles are ultra-fine particles with a larger surface area to volume ratio that can reduce the incipient melting temperature of nanoparticles. A recent approach in nano-technology and nanoparticle drug delivery to promote disease diagnostics against pathogens like parasites, fungi, bacteria, and viruses is trending and beneficent. Natural phytochemicals present in the plant-like alpha-pinene, eugenol, alpha-cubebene, cyperene, methyl eugenol, alpha-gurjunene, beta-caryophyllene, trans-alpha-bergamotene, beta-farnesene, gamma-muurolene, alpha-curcumene, selinene, beta-bisabolene and delta-morphine are known to have anti-parasitic efficacy against parasitic pathogens like Giardia lambia, Toxoplasma gondii, Trichomornasvaginalis which can be utilized in synthesizing organic nanoparticles. Furthermore, parasites are the class of pathogens that are more dangerous than bacteria to livestock and humans, and some of the anti-parasitic drugs, like Leishmania amphotericin B, ivermectin and praziquantel, have been facing side-effects due to weak transmembrane, intracellular transferability, poor absorption, insolubility and discharged with excrement which needs to be prevented with the enhanced therapeutic activity of anti-parasitic drugs with the aid of organic nanoparticles. Also, the anti-parasitic nanoparticles make it very cost-effective against parasites which decreases the mortality rate in the endemic areas that promote drug therapy.

ANTI-PARASITIC MECHANISM OF NPS

Acanthamoeba is a species of parasite that is responsible for a fatal disease named granulomatous amebic encephalitis (GAE). It also causes skin infection, Pneumonitis, and nasopharyngeal infection. Poly (DL-lactide-co-glycolide) (PLGA) microspheres were synthesized as delivery systems for the natural anti-parasitic phytochemicals like Betulinic acid BA-(PLGA), Periglaucine A PGA-

(PLGA), to be encapsulated on it, thus, enhancing the therapeutic efficiency of the anti-parasitic drugs. Poly (DL-lactide-co-glycolide) (PLGA) is a copolymer that is used in nano-technology for an efficient drug delivery system.

The PLGA nanoparticles encapsulated with anti-parasitic drugs were successfully synthesized by single emergent solvent evaporation with different concentrations of nano-particles observed at 25 microgram/ml, 50 micrograms/ml, and 100 micrograms/ml. The prepared PGA-(PLGA) and BA-(PLGA) showed maximum absorbance at 630 and 210 nanometers with sizes approximately of 500 nm and 100 nm when observed under a microscope that showed inhibition in the viability of Acanthamoeba triangularis 74.9%, 59.9%, 49.9% and 71.2%, 52.2%, 88% respectively.

Mechanism

Porphyrin is a group of heterocyclic compounds having greater catalytic, luminescence properties and has shown greater potential concerning medical interventions. These pharmaceutical features and the prominent nature of porphyrin and its derivatives like chlorins, benzoporphyrins, purpurins, texaphyrins, phthalocyanines, and protoporphyrin IX with nano-carriers can enhance the bioavailability and drug-targeted delivery. Structural modification in porphyrin macrocycles by changing the peripheral substituents can enable anti-parasitic activity. Considering this, the modification in the beta position of the porphyrin ring displays potent anti-parasitic activity against cutaneous Leishmaniasis or prepared meso-substituted porphyrins, including pyrrolidinone through Buchwald–Hartwig amination reactions to assess anti-parasitic activity against *Leishmaniadonovani*, *Trypanosomabrucei,* and *Plasmodium species.*

Providing nanoformulations like encapsulation in liposomes, micelles, polymeric particles, cyclodextrin, and dendrimers on the modified porphyrins can increase the therapeutic efficacy of the drug.

NPS AGAINST DRUG RESISTANT PARASITES

Chitosan Nanoparticles

Chitosan nanoparticles are widely used as an antibacterial agent because of their mass and known concentration. Using the ionic gelation method, we can use different concentrations and molecular weights to check anti-parasitic activity against Toxoplasma gondii RH strains. Acid-base titration and the viscometric method were used to characterize the deacetylation of chitosan. Furthermore, therapeutic anti-parasitic efficacy was checked with various concentrations and molecular weights using scanning electron microscopy, bioassay in mice, and many others.

Gold Nanoparticles

Gold nanoparticles are metal nanoparticles possessing special features of very small size and high surface area to volume ratio, which has an effective efficacy against microsporidian parasite Heterosporissaurida that affects the economy of lizardfish aquaculture. Gold nanoparticles were synthesized by the chemical reduction of tetrachloroauric acid as a metal precursor. Furthermore, it was observed that EK-1 cell spores produced were reduced according to the concentration of gold nanoparticles.

Silver Nanoparticles

Silver nanoparticles are metal nanoparticles possessing special physical and chemical properties. The green synthesis method with the extract of phoenix dactylifera seeds and Nabka leaves was used to synthesize nanoparticles. Silver Nitrate was used as a metal precursor at a particular concentration in the process. Later, Transmission electron microscopy was used to characterize the shape and morphology of the silver nanoparticle, and it was identified a reduction in the immune system of NF-kB and TGF-Beta in hepatic tissues, which is required for the treatment of Toxoplasmosis.

ANTI-VIRAL NANOPARTICLES

Virus nanoparticles are virus-like particles that have prominent formulations against virus heterogeneity and diversity, which further promotes nanotechnology that improves biomedical applications when it comes to drug delivery system, permeability, therapeutic efficacy and disease diagnosis. Recent viruses like Coronavirus and Nipah virus have wider heterogeneity and versatility, making them of higher antibiotic resistance. It is further important to come up with different technology through bio nanomaterials to combat these viruses.

ANTI-VIRAL MECHANISM OF NANOPARTICLES

Mechanism 1

Herpes simplex virus type 1 (HSV-1) is a virus enveloped with linear double-stranded DNA that causes various diseases like encephalitis, keratoconjunctivitis, cold sores, and genital herpes in humans, which calls upon a mechanism of effective anti-viral agents. Rhubarb is a rhizome of a Chinese plant that consists of essential natural chemical agents like aloe-emodin, rhein, emodin and chrysophanol, which has anti-viral effects but with limited bioavailability and hydrophobicity. Species of Rhubarb plant named R. tanguticum was grounded into small powder-like and later subjected to gel and water grinding to obtain a size of lesser than 50 nanometers, followed by cell breaking using a high-pressure homogenizer to obtain the dry nanoparticles. Furthermore, the active component analysis in the plant was performed by high-performance liquid chromatography and morphology was obtained by Transmission electron microscopy.

Mechanism 2

Recent aerosols caused by pathogens are deadly and dangerous to humans. Researchers have found some respiratory infection outbreaks from lethal viruses like middle east respiratory syndrome, swine flu virus, and acute respiratory coronavirus that have adverse health impacts. Air filters were coated with silver nanoparticles which were synthesized using a lab-made spark discharge method. Changes and evolution in the silver nanoparticles were confirmed by field emission scanning electron microscope- energy dispersive X-ray and shown to have anti-viral activity towards some bacteriophage.

DRUG-DELIVERY SYSTEM

Present-day drugs that are available for injection or oral administration lack optimum formulations to some extent. To enhance the efficiency in the product containing nucleic acids or proteins and against undesirable degradation, there is a need for an innovative drug carrier system [17]. Nanotechnology can act at the molecular level paves the way for manipulating and controlling the structures, which greatly impact the pharmaceutical sciences and medical technology. Nanoparticle-based drug delivery is one of the innovative and efficient drug delivery systems mainly because of their large surface area and tiny size [17]. This allows the drug nanoparticles to manifest great solubility, ability to get into the pulmonary system, crossing of BBB (Blood-brain barrier), intense bioavailability, and can be absorbed through the compact junctions of cells of the endothelium of the skin. The advances in nano-biotechnology enable accurate drug release into extremely specified targets [9]. Circulation stability, Biocompatibility, and Biodegradability are the characteristic features of active systems with increased efficacy and no side effects. Older drug delivery techniques are reported to have low efficacy as they depend on nanoparticles with no propulsion capability [37]. But in the last decade, active drug delivery vehicles with self-propulsion ability in aqueous media, which is achieved by a multitude of approaches such as magnetic field, light, or catalytic decomposition, have been developed. Modulation of cell targeting, motor speed regulation, and internalization phenomena, thereby providing enhanced control of the system [37].

TYPES OF NANOTECHNOLOGY-BASED DRUG DELIVERY SYSTEMS

Nanobots

Notable traits such as Navigation, self-propulsive velocity, cell targeting, towing, and penetration of tissues are necessary for the nanoparticles to have a targeted delivery of drugs at the desired site of action [38]. Wireless Nanobots have been widely used for drug delivery systems in recent times. Since it is a wireless transfer system, the data or particles to be transferred or lodged are coded [39]. Then, using the frequency of 10 GHz to THz, the data is modulated. Modulated data get communicated using wireless antennas and gets decoded at the site of action. These easier techniques of transferring data using wireless networks help to solve various biological problems. They are also used in the diagnosis of various tumours, which adds more value to these techniques [39]. Various fields of medicine, such as Neurology, Genetics, Gastrology, and Immunology, use nano communication to treat the diseases which they deal with. Nanobot bit, Nanobot gateway, and Nanobot node are the types of bots which is commonly used to achieve effective nano communication.

Nanoghosts

Bone marrow stem cells are used to produce Nanoghosts by removing cytoplasmic content using homogenization, thereby creating ghost cells. PEGs (polyethylene glycol), at a minimum concentration, don't affect the nature of NGS ability to target [40]. The natural vesicles, such as exosomes and a few other vesicles which are originally reconstructed from RBCs aren't equipped with target capabilities. To make the target capabilities feasible and to stabilize a technology that produces such NGs are processed. As a result of this, the term Nanoghosts is born with an inbuilt targeting mechanism and differs from conventional nano-vesicles in various factors of processing, production, and prospects. Immune Evasiveness despite surface-based target delivery system, different NGs from other nanovesicles. These immune responses are widely used in various treatments, such as malignancies, immune disorders, and degenerative disorders [41].To prevent drug leakage and enhance drug loading capacity, it is designed as NGs surrounding NPs such as gold nanocages, gelatin, mouse melanoma, *etc.*, thereby favouring cancer-specific immune response [42]. Hypogenecity and cancer-target response, including chemotaxis and interaction with the surface, are the major factors in choosing MSC (Mesenchymal stem cells)as the Nanoghost producer. Even though it is tumour-specific, it is not species-specific. sTRAIL is a soluble form of tumour necrosis factor involved self-death-inducing ligand, which has shown considerable anticancerous impact when associated with the target MSC. The biological half-life of a very short period, enormous potential, and selectivity are the factors that supported sTRAIL selection [43].

Nanoclews

Nanoclews (NC) are bioinspired drug delivery system that shows anti-cancer properties. Rolling Circle Amplification is used to synthesize long-chain ssDNA from which Nanoclews are assembled. Decoration by electrostatic interactions of DNase into nanoclews is facilitated by the encapsulation of negatively charged DNase into positively charged polymeric nanogel. In an acidic environment, the shedding of the shell made of polymers results in self- degradation and release of doxorubicin which shows enhanced therapeutic efficacy [44]. DNA NC can be efficiently used as a delivery vehicle for CRISP Cas9 protein. To facilitate an endosomal escape, it is coated with a polyethyleneimine, from where it is transferred to the nucleus by peptides of nucleus localizing ability. DNA NCs, as a delivery vehicle, help in uniform particle formations and insertion or deletion at the targeted sites without any alteration in the cell viability [45]. To access membrane-targeted delivery in plasma, two functional components are involved: liposome shell useful in destroying PLA2 and two DNA NCIs forming the core. DNA NCs hybridization presents TRAIL to death receptors and reduces endocytosis, thereby increasing the anticancer efficacy [46]. Electrochemiluminescence is used for the detection of mi-RNA21 using techniques such as Toehold-mediated strand displacement (TMSD). It is a kind of technique that possesses no enzyme amplification with good specificity and sensitivity. Nanoclews, when loaded with $Ru(phen)_3^{2+}$, NCs act as signal tags. These tags are used in the TMSD technique of biosensors for the detection of mi-RNA21 [22].

Nanoneedles

The efficient transfer of various bioactive reagents to too many cells is carried out by nanoneedles. They possess other properties like the transduction of optical properties and sensitizing electrical activities. Microfabrication of metal-assisted chemical etch results in the production of nanoneedles. However, they degrade over time due to increased porosity thinning and finally losing their morphology [47]. Delivery of drug molecules with very low cytotoxicity into the cytoplasm is made possible by nanoneedles. Apart from this, they allow the cytoplasmic entry of other nanoscale drugs, thereby increasing their ability. Even though there is no exertion of force on the needle during the process, membrane puncture is made possible due to its sharp edge morphology [48]. Despite being efficient in the process of the controlled rate of drug delivery, nanoneedles are also used as optical biosensors and electrochemical probes. This paves the way for understanding cellular physiology and the environment in the effect of nanoparticles. Povidone is the commonest polymer used to produce nanoneedles [49]. The efficiency of nanoneedles as a drug delivery device is due to cell fixation, decreased fluidity in the membranes, and the sealing time of long duration. These notable properties during the process facilitate more time, thereby, more cargo enters the cell without delay [50].

Nanoclusters

Challenge in developing sophisticated drug delivery nanoplatforms paved the way for the simpler and more efficient nanoclusters made by assembling lanthanide dope particles with anti-cancer peptides and one iNGRof cyclic nature to attack the cancer cells. LDC- iNGR, along with PBI and BIM, can undergo inhibition of the cancer cell viability in a manner dependent on p-53 [51]. They differ from other nanoparticles due to their properties, namely biosensors, large photons excitation with molecular properties, and long-time fluorescence, which enables bioimaging both *in-vivo* & *in-vitro* procedures. Other than this, they possess pH-dependent reactions, photostability, and good colloidal in water medium [52]. Gold Nanoclusters have created a wide range of attention because of their various properties in drug delivery mechanisms. They also do not contain heavy metals with toxicity which differs them from organic dyes. For multimodal therapeutics, GNC (Glod nanoclusters) along with reduced graphene oxide, namely GNC-RGO, is used, which could inhibit the Hep-2 gene and thereby possess anticancer properties in association with doxorubicin [53]. Another combination of gold nanoclusters with DOX (Doxorubicin) in the bovine serum albumin (BSA) particles shows an enhanced imaging technique, where the clusters show the uptake of BSA particles by the cancer cells and DOX shows the intracellular drug release. Sometimes nanoclusters, in combination with anti-cancer drugs of conventional nature, are useful in radiational therapy [54].

Liposomes

Natural or synthetic phospholipids make up the liposome bilayer. The net characters of the phospholipid constituents determine the chemical and physical properties of the liposome, such as charge density, steric hindrance, and permeability. Several methods are involved in loading drugs into liposomes, including the pH gradient method, a saturation of soluble drugs with an aqueous solution of liposomes, utilizing solvent exchange mechanisms, and utilizing lipophilic drugs. Targeting a desired tissue by the liposomes is achieved using both active and passive strategies. This is possible because of the manipulating ability of the liposomes that molecules can be added to the exterior surface of the lipid bilayer. Very less particle size of the liposomes allows the MPS system to swiftly clear them [8].

Nanobubbles

Nanobubbles are widely used as contrast agents in ultrasound techniques, because of their size, stabilization with polymers, surfactants and lipids, and confinement to the vessels. They have the potential to move into the parenchyma of a tumour and specific targeting of biomarkers located in the surface of the cell [55]. Gold nanoparticles combined with Doxil an encapsulated drug, give plasmonic nanobubbles (PNB). These PNB facilitates the release of the drug into the cytoplasm and enhances therapeutic efficacy against various conditions such as the total duration of the treatment, cells that are resistant against drugs, toxicity, and drug dosage. Intracellularly they attack individual cells while extracellularly the drug is removed and a new PNB is generated [56]. In drug loading, liposome, emulsion, and coumarin are a few control agents used for nanobubbles drug delivery. These cell tests with the control agents show nanobubbles as a promising drug vector in therapeutic systems [57]. Cell Permeable Peptide (lysine or arginine-rich particles) with Campothecin Conjugate, when loaded into the nanobubble shows enhanced efficacy. This combination allows nanobubbles to release a large amount of CPP (Cell permeable peptides) and thereby contributing to increased efficiency. They also show cytotoxic activity against HeLa cells of Xenograft nature with good cell uptake compared to the conventional injection groups [58].

Exosomes

Exosomes are natural vehicles or vesicles of nanometres in size useful in drug delivery systems. They consist of lipid bilayers with proteins of adhesive nature. Since it is found naturally, it is not attacked or eliminated by the body's immune system and therefore it shows increased drug therapeutic efficacy. Few alterations in the exosomes with saponin permeabilization show increased loading capacity, controlled release of the drug, and prevention against destruction. They are widely used in neuronal disorders [59]. Exosomes contain various proteins that are often used as markers for identifying them. Few exosomes extracted from dendritic cells reveals that it comprises miRNA of a different variety. Other than this exosomes of myoblasts and astrocytes contains mitochondrial DNA. They also facilitate the delivery of short interfering RNA, which has been a great challenge due to its large size, rapid degradation and negative charge [60]. The delivery of anti-inflammatory drugs such as Curcumin is more stable and it ensures high concentration for a long time in the blood with exosomes. As exosomes are target specific, they direct curcumin towards the inflamed cells and thus enhance the anti-inflammatory effect with minimal or no side effects. Also. curcumin with exosomes has shown protection against septic shock induced by lipopolysaccharides [61]. Nowadays, Mesenchymal stem cells are shown to be the largest producer of exosomes. These mesenchymal stem cells being immunosuppressive, exosomes also shows immunosuppressive action. Since the qualities of exosomes derived from MSC are uncompromised, they are the perfect source for production [62].

Injectable Nanoparticle Generator

Due to the limitation that only a small amount of drug gets deposited in the tumour, the drug efficacy is too low. To encounter this, iNPG (Injectable Nanoparticle Generator) gives a solution by overcoming all the barriers. It is a micrometre in size and can be loaded with therapeutic chemicals. Doxorubicin and poly- glutamic acid in combination with iNPG give iNPG-pDOX that shows increased efficiency against breast cancer of metastatic condition [63]. iNPG shows various advantages in the drug delivery system such as drug concentration maintenance for a long time, site/target-specific delivery, chance to revaluate drugs of poor pharmacokinetics. Nanomaterials with inorganic layers have the advantage of increased ion exchange, adjustable shape, size, and structure. Even though it has many advantages, the question of toxicity due to nanoparticles is still in research [64]. The various techniques models used to overcome the problem of decreased efficacy with nanoparticles are lithography, emulsification with analysis on physicochemical characteristics, imprinting, properties of the drug such as release and its degradation mechanism, micro moulding. The composition of nanoparticles varies from macromolecules to micro molecules other than nanotubes and dendrimers. These particles prevent degradation during delivery and thereby show increased efficacy [65].

Nano Terminator

The concept of nanoterminator is derived from the body's natural mechanism of response to infections, popularly called apoptosis – cell suicide. The nanoparticles or nanorobots designed to deliver drugs may sometimes show dysfunction and may cause life-threatening events to the patient. To avoid these consequences, nanoterminators have been designed to kill themselves whenever there is a dysfunction in the delivery mechanism or post-delivery circumstances. Even though it is in no way the same as apoptosis, it results in outputs that are similar to apoptosis [66]. The nano terminators are designed with a polymeric endoskeleton as a base. This polymeric endoskeleton is a conjugated polymer that consists of hight bright particles. When the body's natural killer cells are coated over the endoskeleton, which shows good biocompatibility and reasonable quantum yield. They also play a role in the modulation of tight junctions and provide a pathway called the green pathway, which enables them to cross barriers in the blood-brain cascade. Apart from this, they are useful in skull imaging by the deposition in the glioblastoma cells [67]. The major application of nano terminators is seen in the field of dentistry, especially during root canal treatment. Anaesthesia given during the treatment using nanorobots/nanoparticles to the pulp cavity may cause catastrophic effects by encountering the vital centres such as the conduction system of the heart and the respiratory centre in the brain. They are used as a bottom-up approach in dentistry [68]. Apart from this few nano terminators are used in aeromedicine to tackle emergencies [69].

Niosomes

Similar to transferosomes, ethosomes, and liposomes, Niosomes are surfactants of non-ionic nature, and they are usually multilamellar, being vesicular in structure. Mainly it comprises two components, namely additives such as cholesterol and a few molecules, which are ionic and non-ionic surfactants being the basement to vesicular structure [70]. The uniqueness of niosomes lies in their structure which enables them to encapsulate substances such as water and fat-attracting substances. Because of its unique nature, it can entrap a wide range of proteins, vaccines, and drugs easily. Various amphiphiles under niosomes are amides, ethers, and esters of alkyl nature. Other than cholesterol, diacetyl phosphate is being used as an additive that facilitates the imparting of negative charges [71]. Niosomes are used as mucoadhesive to deliver solutions of timolol maleate, which are coated over Carbopol using a technique called reverse-phase evaporation. Timolol maleate is the first line of drug for glaucoma of open-angle type. To increase its bioavailability at the ocular site, delivery using niosomes is being followed. Unlike conventional techniques, they reduce systemic absorption and side effects [72]. Niosomes are widely used in the treatment of various organisms which possess the ability to cause systemic illness to serious mortality conditions. They are used in the treatment of bacterial, fungal, leishmaniasis, and viral diseases. Also, they are used in cosmetics, tubercular diseases, inflammatory conditions, diabetics, gene and protein delivery, relaxation of muscle, contraception, anaesthesia, vitamin deposition, immunization, and in the diagnosis of various diseases [73].

Carbon Nanotubes

Carbon nanotubes are new forms of drug delivery system wrapped by chitosan, enabling more biocompatibility and water solubility for the delivery system. It is considered special over other nanoparticles, for its features like increased time of circulation, enhanced carrying capacity, unloading depending on pH, and incomparable cell permeability [74]. Usually, nanotubes are single-walled. On the other hand, electrical sensitivity is increased using multi-walled nanotubes as an additive. The mechanism is made possible by the dissolution of polymer networks consisting of polyethene oxide. Thus the electrical voltage of carbon nanotubes enables the release of a comparatively large amount of drug during delivery [75]. Due to their mechanical and chemical properties, they are not easily degraded in the body. This paves the way for the use of nanotubes as nanovectors in the cells. Cisplatin, a chemotherapy drug based on platinum, is combined with single-walled carbon nanotubes. This combination is used against cancer cells in the prostate [76].To deliver drugs against the cancer cells located in the lymphatics, the size of the nanoparticles has to be adjusted. To make this possible, a magnet is placed inside the nanotubes to control the drug composition externally and guide them to the exact site of action i.e. lymphatic cells. Along with these magnet placed nanotubes, folic acid is also embedded to enable higher selectivity [77].

APPLICATIONS OF DRUG DELIVERY NANOPARTICLES

Drug Delivery System (DDS) is defined as a system by which drugs are delivered by subcellular organs, cells, organs, and tissues for absorption through a variety of drug carriers and controlled drug release. In the prevailing situation, nanoparticles as drug carriers have shown great potential with regard to the encapsulation of drugs in nanoparticles, nanospheres, and nanocapsules which reduces the side effects and increases the therapeutic index. It is also essential to draw attention to the enhanced techniques in drug delivery to improve the modern drug delivery system.

- Recent progress and perspective of drug delivery in D-alpha-tocopheryl polyethylene glycol succinate (TGPS) approved by the food and drug administration (FDA) for efficient biocompatibility, drug permeation can inhibit the activity of ATP-dependent P-glycoprotein and act as a potent excipient for overcoming multi-drug resistance (MDR) in the tumour.

- The ideal goal achieved through controlled drug delivery concerns relevant drug formulations at desired kinetics to treat patients effectively with convenience and compliance.

• Limitations in bioavailability and poor water solubility have decreased the therapeutic efficacy of the naturally available potential plant product, In respect to this, recent studies have attempted to address this problem with the aid of nano-carriers proved by the FDA.

• Learning and preventing the pathophysiology of a chronic wound is important because of the increased incidence of obesity and diabetes. Recent advances in pharmaceutical and pathological science have introduced the drug delivery system mechanism to improve the tissue regeneration rate and expedite the halted physiological process.

CONCLUSION

Nowadays, Nanoparticles are majorly employed in various applications. It becomes an irreplaceable part of our day-to-day life. Various challenges in the field of medicine are solved by utilizing the advancements in nanomedicine. In the process of developing drugs, diagnosing various diseases and drug delivery, nanoparticles play an indispensable role as they are safe, customizable, can overcome drug resistance, and can cross barriers. The biological substances which are less soluble and absorbable can be converted into hopeful deliverable substances utilizing nanoparticles. Compared to the organic drug delivery system, the nanoparticle-based drug delivery system is more precise, enhanced bioavailability, drugs can be released sustainably, and possesses the advantage of targeting a wide range of cells in the biological system.

CONSENT FOR PUBLICATION

Not applicable.

CONFLICT OF INTEREST

The author declares no conflict of interest, financial or otherwise.

ACKNOWLEDGEMENTS

Declared none.

REFERENCES

[1] Benelmekki M, Benelmekki M. An introduction to nanoparticles and nanotechnology. 2015; 1-14. [http://dx.doi.org/10.1088/978-1-6270-5469-0ch1]

[2] Rai A, Comune M, Ferreira L. Nanoparticle-based drug delivery systems: Promising approaches against bacterial infections. Antibact Drug Discov to Combat MDR Nat Compd Nanotechnol Nov Synth Sources 2019; 57: 605-33. [http://dx.doi.org/10.1007/978-981-13-9871-1_27]

[3] Holister P, Weener J-W, Vas CR, Harper T. Holister P, Weener J-W, Vas CR, Harper T, Cientifica. Nanoparticles: Technology White Paper No.3. Científica 2003; 11

[4] Lewinski N, Colvin V, Drezek R. Cytotoxicity of nanopartides 2008; Vol. 4 [http://dx.doi.org/10.1002/smll.200700595]

[5] Zhang L, Gu FX, Chan JM. AZ Wang RL and OF. Therapeutic, Nanoparticles in Medicine: Applications and Developments. Educ Policy Anal Arch 2007; 8: 761-9. [http://dx.doi.org/10.1038/sj.clp]

[6] Doane TL, Burda C. The unique role of nanoparticles in nanomedicine: imaging, drug delivery and therapy. Chem Soc Rev 2012; 41(7): 2885-911. [http://dx.doi.org/10.1039/c2cs15260f] [PMID: 22286540]

[7] Pund S, Joshi A. Nanoarchitectures for Neglected Tropical Protozoal Diseases: Challenges and State of the Art Nano- Microscale Drug Deliv Syst Des Fabr 2017; 439-80. [http://dx.doi.org/10.1016/B978-0-323-52727-9.00023-6]

[8] Malam Y, Loizidou M, Seifalian AM. Liposomes and nanoparticles: nanosized vehicles for drug delivery in cancer. Trends Pharmacol Sci 2009; 30(11): 592-9. [http://dx.doi.org/10.1016/j.tips.2009.08.004] [PMID: 19837467]

[9] Yih TC, Al-Fandi M. Engineered nanoparticles as precise drug delivery systems. J Cell Biochem 2006; 97(6): 1184-90. [http://dx.doi.org/10.1002/jcb.20796] [PMID: 16440317]

[10] Abbasi E, Aval SF, Akbarzadeh A, *et al.* Dendrimers: synthesis, applications, and properties. Nanoscale Res Lett 2014; 9(1): 247. [http://dx.doi.org/10.1186/1556-276X-9-247] [PMID: 24994950]

[11] Nimesh S, Kulkarni MS, Karva GS, Giram PS, Gugulkar RR. Dendrimers. Int J Pharm Sci Rev Res 2013; 33: 259-85. [http://dx.doi.org/10.1533/9781908818645.259]

[12] Khan I, Saeed K, Khan I. Nanoparticles: Properties, applications and toxicities. Arab J Chem 2019; 12(7): 908-31. [http://dx.doi.org/10.1016/j.arabjc.2017.05.011]

[13] Obodovskiy I. Radiation-Chemical Nanotechnologies 2019; 351-9. [http://dx.doi.org/10.1016/B978-0-444-63979-0.00027-6]

[14] Ealias AM, Saravanakumar MP. A review on the classification, characterisation, synthesis of nanoparticles and their application. IOP Conf Ser Mater Sci Eng. [http://dx.doi.org/10.1088/1757-899X/263/3/032019]

[15] Khalili Fard J, Jafari S, Eghbal MA. A review of molecular mechanisms involved in toxicity of nanoparticles. Adv Pharm Bull 2015; 5(4): 447-54. [http://dx.doi.org/10.15171/apb.2015.061] [PMID: 26819915]

[16] Thakkar KN, Mhatre SS, Parikh RY. Biological synthesis of metallic nanoparticles. Nanomedicine 2010; 6(2): 257-62. [http://dx.doi.org/10.1016/j.nano.2009.07.002] [PMID: 19616126]

[17] Rizvi SAA, Saleh AM. Applications of nanoparticle systems in drug delivery technology. Saudi Pharm J 2018; 26(1): 64-70. [http://dx.doi.org/10.1016/j.jsps.2017.10.012] [PMID: 29379334]

[18] Campelo JM, Luna D, Luque R, Marinas JM, Romero AA. Sustainable preparation of supported metal nanoparticles and their applications in catalysis. ChemSusChem 2009; 2(1): 18-45. [http://dx.doi.org/10.1002/cssc.200800227] [PMID: 19142903]

[19] Bogutska K, Sklyarov Y, Prylutskyy Y. Zinc and zinc nanoparticles: biological role and application in biomedicine. Ukr Bioorg Acta 2013; 1: 9-16.

[20] Mody V, Siwale R, Singh A, Mody H. Introduction to metallic nanoparticles. J Pharm Bioallied Sci 2010; 2(4): 282-9.
[http://dx.doi.org/10.4103/0975-7406.72127] [PMID: 21180459]

[21] Hasan S. Hasan S. A Review on Nanoparticles: Their Synthesis and Types. Res J Recent Sci Res J Recent Sci Uttar Pradesh (Lucknow Campus) 2014; 4: 1-3.

[22] Zhang Y, Xu G, Lian G, *et al.* Electrochemiluminescence biosensor for miRNA-21 based on toehold-mediated strand displacement amplification with $Ru(phen)_3^{2+}$ loaded DNA nanoclews as signal tags. Biosens Bioelectron 2020; 147111789
[http://dx.doi.org/10.1016/j.bios.2019.111789] [PMID: 31655383]

[23] Siddiqi KS. ur Rahman A, Tajuddin, Husen A. Properties of Zinc Oxide Nanoparticles and Their Activity Against Microbes. Nanoscale Res Lett 2018; 13
[http://dx.doi.org/10.1186/s11671-018-2532-3] [PMID: 29740719]

[24] Li S, Silvers SJ, Samy El-Shall M. Preparation, characterization and optical properties of zinc oxide nanoparticles. Proc MRS 1996; 452: 389-94.
[http://dx.doi.org/10.1557/PROC-452-389]

[25] Webster TJ. Safety of Nanoparticles From Manufacturing to Medical Applications. 2009.
[http://dx.doi.org/10.1007/978-0-387-78608-7]

[26] Patron L, Marinescu G, Mindru I. Magnetic Nanomaterials 2008.
[http://dx.doi.org/10.1201/NOE0849396397.ch173]

[27] Bárcena C, Sra AK, Gao J. Applications of magnetic nanoparticles in biomedicine. Nanoscale Magn Mater Appl 2009; 167: 591-626.
[http://dx.doi.org/10.1007/978-0-387-85600-1_20]

[28] Ito A, Shinkai M, Honda H, Kobayashi T. Medical application of functionalized magnetic nanoparticles. J Biosci Bioeng 2005; 100(1): 1-11.
[http://dx.doi.org/10.1263/jbb.100.1] [PMID: 16233845]

[29] Sumanth Kumar D, Jai Kumar B, Mahesh HM. Quantum Nanostructures (QDs): An Overview 2018.
[http://dx.doi.org/10.1016/B978-0-08-101975-7.00003-8]

[30] Jha M. Current trends in industrial scale synthesis of quantum dots and its application in electronics 2018.
[http://dx.doi.org/10.1016/B978-0-12-813351-4.00023-7]

[31] Kara HEŞ, Ertaş N. Quantum Dots for Pharmaceutical and Biomedical Analysis 2017.
[http://dx.doi.org/10.5772/intechopen.70034]

[32] 2018.

[33] Hajipour MJ, Fromm KM, Akbar Ashkarran A, *et al.* Antibacterial properties of nanoparticles. Trends Biotechnol 2012; 30(10): 499-511.
[http://dx.doi.org/10.1016/j.tibtech.2012.06.004] [PMID: 22884769]

[34] Yael N. Slavin1, Jason Asnis2 UOH and HB. Metal nanoparticles: understanding the mechanisms behind antibacterial activity. J Nanobiotechnology 2017.

[35] Bjarnsholt T. The role of bacterial biofilms in chronic infections 2013.
[http://dx.doi.org/10.1111/apm.12099]

[36] Wang L. 1,* Chen Hu, 2 and LS. The anti-microbial activity of nanoparticles: present situation and prospects for the future. Int J Nanomedicine 2017.

[37] Hortelão AC, Patiño T, Perez-Jiménez A, Blanco À, Sánchez S. Enzyme-Powered Nanobots Enhance Anticancer Drug Delivery. Adv Funct Mater 2018; 28(25)1705086
[http://dx.doi.org/10.1002/adfm.201705086]

[38] Andhari SS, Wavhale RD, Dhobale KD, *et al.* Self-Propelling Targeted Magneto-Nanobots for Deep

Tumor Penetration and pH-Responsive Intracellular Drug Delivery. Sci Rep 2020; 10(1): 4703. [http://dx.doi.org/10.1038/s41598-020-61586-y] [PMID: 32170128]

[39] Krishnaswamy D, Ramanathan R, Qamar A. Collaborative wireless nanobots for tumor discovery and drug delivery 2012; 6203-8. [http://dx.doi.org/10.1109/ICC.2012.6364972]

[40] Oieni J, Levy L, Khait NL, Yosef L, Schoen B, Fliman M, *et al.* Nano-Ghosts: Biomimetic Membranal Vesicles, Technology and Characterization. Methods 2019; ••• [http://dx.doi.org/10.1016/j.ymeth.2019.11.013] [PMID: 31794834]

[41] Kaneti L, Bronshtein T, Malkah Dayan N, *et al.* Nanoghosts as a Novel Natural Nonviral Gene Delivery Platform Safely Targeting Multiple Cancers. Nano Lett 2016; 16(3): 1574-82. [http://dx.doi.org/10.1021/acs.nanolett.5b04237] [PMID: 26901695]

[42] Krishnamurthy S, Gnanasammandhan MK, Xie C, Huang K, Cui MY, Chan JM. Monocyte cell membrane-derived nanoghosts for targeted cancer therapy. Nanoscale 2016; 8(13): 6981-5. [http://dx.doi.org/10.1039/C5NR07588B] [PMID: 26975904]

[43] Toledano Furman NE, Lupu-Haber Y, Bronshtein T, *et al.* Reconstructed stem cell nanoghosts: a natural tumor targeting platform. Nano Lett 2013; 13(7): 3248-55. [http://dx.doi.org/10.1021/nl401376w] [PMID: 23786263]

[44] Sun W, Jiang T, Lu Y, Reiff M, Mo R, Gu Z. Cocoon-like self-degradable DNA nanoclew for anticancer drug delivery. J Am Chem Soc 2014; 136(42): 14722-5. [http://dx.doi.org/10.1021/ja5088024] [PMID: 25336272]

[45] Sun W, Ji W, Hall JM, *et al.* Self-assembled DNA nanoclews for the efficient delivery of CRISPR-Cas9 for genome editing. Angew Chem Int Ed 2015; 54(41): 12029-33. [http://dx.doi.org/10.1002/anie.201506030] [PMID: 26310292]

[46] Amaral G, Bushee J, Cordani UG. DNA Nanclew-Mediated Drug Delivery. J Petrol 2013; 369: 1689-99. [http://dx.doi.org/10.1017/CBO9781107415324.004]

[47] Chiappini C, De Rosa E, Martinez JO, *et al.* Biodegradable silicon nanoneedles delivering nucleic acids intracellularly induce localized in vivo neovascularization. Nat Mater 2015; 14(5): 532-9. [http://dx.doi.org/10.1038/nmat4249] [PMID: 25822693]

[48] Kolhar P, Doshi N, Mitragotri S. Polymer nanoneedle-mediated intracellular drug delivery. Small 2011; 7(14): 2094-100. [http://dx.doi.org/10.1002/smll.201100497] [PMID: 21695782]

[49] Shende P, Sardesai M, Gaud RS. Micro to nanoneedles: a trend of modernized transepidermal drug delivery system. Artif Cells Nanomed Biotechnol 2018; 46(1): 19-25. [http://dx.doi.org/10.1080/21691401.2017.1304409] [PMID: 28355887]

[50] Liu Z, Nie J, Miao B, *et al.* Self□Powered Intracellular Drug Delivery by a Biomechanical Energy□Driven Triboelectric Nanogenerator. Adv Mater 2019; 31(12)1807795 [http://dx.doi.org/10.1002/adma.201807795] [PMID: 30721538]

[51] Yan J, He W, Yan S, *et al.* Self-Assembled Peptide–Lanthanide Nanoclusters for Safe Tumor Therapy: Overcoming and Utilizing Biological Barriers to Peptide Drug Delivery. ACS Nano 2018; 12(2): 2017-26. [http://dx.doi.org/10.1021/acsnano.8b00081] [PMID: 29376322]

[52] Yahia-Ammar A, Sierra D, Mérola F, Hildebrandt N, Le Guével X. Self-Assembled Gold Nanoclusters for Bright Fluorescence Imaging and Enhanced Drug Delivery. ACS Nano 2016; 10(2): 2591-9. [http://dx.doi.org/10.1021/acsnano.5b07596] [PMID: 26845515]

[53] Wang C, Li J, Amatore C, Chen Y, Jiang H, Wang XM. Gold nanoclusters and graphene nanocomposites for drug delivery and imaging of cancer cells. Angew Chem Int Ed 2011; 50(49):

11644-8.
[http://dx.doi.org/10.1002/anie.201105573] [PMID: 21990208]

[54] Khandelia R, Bhandari S, Pan UN, Ghosh SS, Chattopadhyay A. Gold Nanocluster Embedded Albumin Nanoparticles for Two-Photon Imaging of Cancer Cells Accompanying Drug Delivery. Small 2015; 11(33): 4075-81.
[http://dx.doi.org/10.1002/smll.201500216] [PMID: 25939342]

[55] Hernandez C, Gulati S, Fioravanti G, Stewart PL, Exner AA. Cryo-EM Visualization of Lipid and Polymer-Stabilized Perfluorocarbon Gas Nanobubbles - A Step Towards Nanobubble Mediated Drug Delivery. Sci Rep 2017; 7(1): 13517.
[http://dx.doi.org/10.1038/s41598-017-13741-1] [PMID: 29044154]

[56] Lukianova-Hleb EY, Belyanin A, Kashinath S, Wu X, Lapotko DO. Plasmonic nanobubble-enhanced endosomal escape processes for selective and guided intracellular delivery of chemotherapy to drug-resistant cancer cells. Biomaterials 2012; 33(6): 1821-6.
[http://dx.doi.org/10.1016/j.biomaterials.2011.11.015] [PMID: 22137124]

[57] Wang Y, Li X, Zhou Y, Huang P, Xu Y. Preparation of nanobubbles for ultrasound imaging and intracelluar drug delivery. Int J Pharm 2010; 384(1-2): 148-53.
[http://dx.doi.org/10.1016/j.ijpharm.2009.09.027] [PMID: 19781609]

[58] Xie X, Lin W, Liu H, *et al.* Ultrasound-responsive nanobubbles contained with peptide–camptothecin conjugates for targeted drug delivery. Drug Deliv 2016; 23(8): 2756-64.
[http://dx.doi.org/10.3109/10717544.2015.1077289] [PMID: 26289216]

[59] Haney MJ, Klyachko NL, Zhao Y, *et al.* Exosomes as drug delivery vehicles for Parkinson's disease therapy. J Control Release 2015; 207: 18-30.
[http://dx.doi.org/10.1016/j.jconrel.2015.03.033] [PMID: 25836593]

[60] Schiffelers R, Kooijmans S, Vader , van Dommelen , Van Solinge . Exosome mimetics: a novel class of drug delivery systems. Int J Nanomedicine 2012; 7: 1525-41.
[http://dx.doi.org/10.2147/IJN.S29661] [PMID: 22619510]

[61] Sun D, Zhuang X, Xiang X, *et al.* A novel nanoparticle drug delivery system: the anti-inflammatory activity of curcumin is enhanced when encapsulated in exosomes. Mol Ther 2010; 18(9): 1606-14.
[http://dx.doi.org/10.1038/mt.2010.105] [PMID: 20571541]

[62] Yeo RWY, Lai RC, Zhang B, *et al.* Mesenchymal stem cell: An efficient mass producer of exosomes for drug delivery. Adv Drug Deliv Rev 2013; 65(3): 336-41.
[http://dx.doi.org/10.1016/j.addr.2012.07.001] [PMID: 22780955]

[63] Xu R, Zhang G, Mai J, *et al.* An injectable nanoparticle generator enhances delivery of cancer therapeutics. Nat Biotechnol 2016; 34(4): 414-8.
[http://dx.doi.org/10.1038/nbt.3506] [PMID: 26974511]

[64] Du Y, Chen B. Combination of drugs and carriers in drug delivery technology and its development. Drug Des Devel Ther 2019; 13: 1401-8.
[http://dx.doi.org/10.2147/DDDT.S198056] [PMID: 31118575]

[65] Aryal S, Park H, Leary JF. 2019.

[66] Chaughule RS. Dental applications of nanotechnology 2018.
[http://dx.doi.org/10.1007/978-3-319-97634-1]

[67] Deng G, Peng X, Sun Z, Zheng W, Yu J, Du L, *et al.* Natural Killer Cell Inspired AIE Nanoterminator for Blood-Brain-Barrier Crossing *via* Tight-Junction Modulation and NIR-II Gliomas Theranostics 2020.
[http://dx.doi.org/10.26434/chemrxiv.12504404.v1]

[68] Dwivedi S, Dwivedi CD, Chandra A, Sharma PN. Nanotechnology boon or bane for restorative dentistry. RE:view 2013; 2: 1-5.

[69] Voronka NR, Hoyt RP, Technologist C, Scientist C. Endbody Plasma Contactors 2005.

[70] Rajera R, Nagpal K, Singh SK, Mishra DN. Niosomes: a controlled and novel drug delivery system. Biol Pharm Bull 2011; 34(7): 945-53. [http://dx.doi.org/10.1248/bpb.34.945] [PMID: 21719996]

[71] Moghassemi S, Hadjizadeh A. Nano-niosomes as nanoscale drug delivery systems: An illustrated review. J Control Release 2014; 185: 22-36. [http://dx.doi.org/10.1016/j.jconrel.2014.04.015] [PMID: 24747765]

[72] Aggarwal D, Kaur IP. Improved pharmacodynamics of timolol maleate from a mucoadhesive niosomal ophthalmic drug delivery system. Int J Pharm 2005; 290(1-2): 155-9. [http://dx.doi.org/10.1016/j.ijpharm.2004.10.026] [PMID: 15664141]

[73] Sankhyan A, Pawar P. Recent trends in niosome as vesicular drug delivery system. J Appl Pharm Sci 2012; 2: 20-32. [http://dx.doi.org/10.7324/JAPS.2012.2625]

[74] Ji Z, Lin G, Lu Q, *et al.* Targeted therapy of SMMC-7721 liver cancer in vitro and in vivo with carbon nanotubes based drug delivery system. J Colloid Interface Sci 2012; 365(1): 143-9. [http://dx.doi.org/10.1016/j.jcis.2011.09.013] [PMID: 21974923]

[75] Im JS, Bai BC, Lee YS. The effect of carbon nanotubes on drug delivery in an electro-sensitive transdermal drug delivery system. Biomaterials 2010; 31(6): 1414-9. [http://dx.doi.org/10.1016/j.biomaterials.2009.11.004] [PMID: 19931904]

[76] Tripisciano C, Kraemer K, Taylor A, Borowiak-Palen E. Single-wall carbon nanotubes based anticancer drug delivery system. Chem Phys Lett 2009; 478(4-6): 200-5. [http://dx.doi.org/10.1016/j.cplett.2009.07.071]

[77] Yang F, Fu DL, Long J, Ni QX. Magnetic lymphatic targeting drug delivery system using carbon nanotubes. Med Hypotheses 2008; 70(4): 765-7. [http://dx.doi.org/10.1016/j.mehy.2007.07.045] [PMID: 17910909]

CHAPTER 11

Current Approaches to Antimicrobial Formulations and their Delivery

Kaushita Banerjee[1] and **Harishkumar Madhyastha**[2,*]

[1] *Department of Biomedical Sciences, School of Biosciences and Technology, Vellore Institute of Technology, Vellore-632014, India*

[2] *Department of Cardiovascular Physiology, Faculty of Medicine, University of Miyazaki, Miyazaki, Japan*

Abstract: With the escalating concerns about antimicrobial resistance and the intractable nature of microbial infections, there is a demand for the expansion and development of alternative stratagems for treating microbial diseases. At present, the advent of antimicrobial resistance amidst microbial pathogens, especially the 'drug-resistant' ones, has led to poor clinical consequences, thus, shooting up healthcare outlays and mortality. Moreover, the formation of biofilms-like assemblies by microorganisms and their surface association mechanisms have led to secondary infections in immunocompromised individuals and further muddled the prophylaxis. Such microbial resistance is primarily attributed to the inapt and undue use of antimicrobials in humans/animals and the unregulated administration of these drug formulations. Therefore, there is an urgent need to propose and imbibe various modern, multifaceted antimicrobial formulation approaches to prevent the fatal consequences of antibiotic resistance and enhance the effectiveness of microbial growth control. Currently, several new-age antimicrobial formulation therapies are being explored and have shown promising results as efficacious preventatives, diagnostics, and drug carriers in comparison to conventional antibiotic therapy being used. In this chapter, we highlight the different categories of new-age antimicrobial formulation therapies currently in use, their molecular mechanism of microbial targeted delivery, their effectiveness over the traditional therapies, the challenges in their development and the future outcome of these contemporary formulations.

Keywords: Antimicrobial formulation, Antimicrobial resistance, Biofilms, Contemporary formulation therapies, Diagnostics, Drug carriers, Microbial infections, Microbial growth control, Secondary infections, Targeted delivery.

* **Corresponding author Harishkumar Madhyastha:** Department of Applied Physiology, Faculty of Medicine, University of Miyazaki, Miyazaki, Japan; E-mail: hkumar@med.miyazaki-u.ac.jp

M. Amin-ul Mannan & Gaurav Kumar (Eds.)

INTRODUCTION

The aggregation of all micron-sized organisms that are diverse, abundant, omnipresent and invisible to the naked eye comprises the 'microbial biosphere'. Not limited to only bacteria, fungi or viruses, the biosphere encompasses an array of multicellular life forms that manifests the basis of a global ecological niche [1, 2]. These microorganisms dwell at all possible biological nooks and are most beneficial to the host, with a handful of these being pathogenic in nature triggering damage, disease, and even fatal consequences giving rise to infection. To curb the growth of such pathogens and combat the infection, natural or synthetic chemical substances known as 'Antimicrobial Agents' have been devised. Antimicrobials have further been customized and amplified as per the end requirements to enhance target-specific antimicrobial activity [3, 4]. These broad-spectrum compounds, categorized into antiseptics, antibiotics, anti-fungals, antihelminthics, antiprotozoals, *etc.*, target an array of microbial pathogens. Nevertheless, the usage of antimicrobial agents for prophylaxis of infectious disease has incited an evolutionary response within the microbial genome, making them resistant to administered antimicrobials [5]. 'Antimicrobial Resistance' is one of the prime public well-being challenges, and fighting this threat requires a combined global approach across diverse health care sectors. The antimicrobial resistance (AMR) predicament has increased both the occurrence of infectious ailments and the sky-high treatment costs in the human populace, causing many infections to recur post-treatment and sometimes become untreatable [6, 7].

Among many, the four crucial factors that regulate this predicament are the increasing rate of resistant phenotypes among microorganisms, the uncontrolled shoot in population allowing easy microbial accessibility, the interaction with the host transmission on a large scale to cause the overexploitation of antimicrobials and compensatory resistance progression. AMR dates back to the 19th century when penicillin was first used for human trials. Initially, easy regulatory antibiotic discovery programmes and the financial recompense that followed a led to the over and uncensorious usage of antibiotics without considering its societal implications [8 - 11]. Post the 'golden antibiotic era', therapeutic firms faced hindrances while screening and developing new antibiotics, specifically against gram-negative strains with high reflux-efflux cellular mechanisms, which required inflated doses of antibiotics that ultimately led to toxic effects. Gradually, these firms restricted antibiotic production because of declining market growth and profits. This major setback by the pharma and curative sectors brought challenges in the scientific and research communities and health care arm [12, 13]. This ultimately caused an empty therapeutic conduit for newer discoveries and set the alarm for the urgent production of newer classes of antimicrobials. With time as the pattern of resistance has clinically transformed, the problem has been addressed either by chemical/functional modification of antimicrobial classes with restricted cross-resistance to existing antibiotics or by the invention of a newer generation of such drugs [14 - 16].

In recent times, newer and advanced screening methods for natural and modified products are evolving, but with the baggage of antimicrobial resistance that deteriorates the long-term effectivity of antimicrobial compounds. Different approaches to assess the microbial resistance progression have been employed, which involve detecting the spontaneous mutational alterations by zone inhibition assays, whole genome/targeted sequencing of mutants to detect the mutation loci and the extent of resistance caused by such mutations. However, the resistance evolution process is quite intricate and difficult to accomplish, as observed in the case of older antibiotics, which though clinically competent, induce high levels of *in vitro* resistance in bacterium like Rifampicin [17 - 19]. Uncritical and unreasonable usage of antibiotics has resulted in an unparalleled stir within human civilization as it has been a challenge to treat infections because of the microbe's inherent ability to become resistant to antimicrobials. Such agents target one of the following mechanisms: meddle with cell wall synthesis, DNA and RNA synthesis, lyse the bacterial outer sheath, impede protein synthesis or specific metabolic pathways, *etc.* Several drugs with a specific mode of activity against infectious pathogens have been used to treat a variety of microbial (bacterial, fungal, viral, protozoal) infections to a considerable extent and have also shown promising results. However, the clinical treatment available is not completely effective in eradicating microbial resistance as it comes with limitations of infection relapse, resistance to the drugs administered or additional secondary fatal infections [20].

Table **1** summarizes some of the different microbial infections and the associated disease/symptoms it causes, their mechanism of resistance and the mode of action of the treatment drugs. Consequently, the rising concern of such antimicrobial resistance along with the intractable nature of secondary infections, demands the need for advancements in building up alternative strategies to treat such infectious ailments. Hence the need of the hour is to switch over to alternative antimicrobial formulations and therapies that could tackle microbial resistance to a substantial extent. Current antimicrobial formulations have distinctive physicochemical characteristics that can be fine-tuned to fit the requirements to combat microbial infections [21]. Advanced antimicrobial formulations include immunomodulators, adjuvants, vaccines, plant-based antimicrobials, phage cocktails, liposomal and polymeric agents, nanoemulsions, drug and antibiotic conjugated systems, *etc.*, that play either a 'static' or 'cidal' role to specifically target the molecular and the biochemical functioning of a microorganism. Each of these antimicrobials has a specific mode of inhibition action on the microbial population, and hence we discuss some of the aspects that would and give a clearer portrayal of the emergent path on antimicrobial resistance and the development of newer high-tech antimicrobial formulations.

Table 1. Different microbial infections, their associated disease/symptoms, the mechanism of resistance and mode of action of the treatment drugs

Infection	Microbe	Associated Symptoms/Infection	Mode of Resistance	Mode of Action of Treatment Drug
Bacterial	*E. coli*	Urinary tract infection, septicaemia, gastroenteritis, bowel necrosis	Efflux pump system, hydrolysation; Acetylation	Beta-lactams (biosynthesis of peptidoglycan); Fluoroquinolones (replication of DNA) [2, 4, 6]
	E. faecalis	Endocarditis, Septicaemia, Urinary tract infection, meningitis	Phosphorylation; Efflux pump system; Glycosylation	Aminoglycosides (genetic translation); Beta-lactams (biosynthesis of peptidoglycan); Macrolides (genetic translation) [2, 4, 6]
	K. pneumoniae	Pneumonia, bronchitis, meningitis	Efflux pump system; Acetylation; Monooxygenation	Aminoglycosides (genetic translation); Fluoroquinolones (replication of DNA); Tetracyclines (genetic translation) [2, 4, 6 - 8]
	H. pylori	Peptic ulcers, gastritis, nausea	Efflux pump system, hydrolysation; Acetylation	Beta-lactam (biosynthesis of peptidoglycan); Fluoroquinolones (Replication of DNA) [20]
	M. tuberculosis	Pulmonary tuberculosis, meningitis	Efflux pump system, hydrolysation, target alteration	Beta-lactam (biosynthesis of peptidoglycan) [6 - 8, 10]

(Table 1) cont.....

Infection	Microbe	Associated Symptoms/Infection	Mode of Resistance	Mode of Action of Treatment Drug
Fungal	*C. albicans*	Oral and vaginal candidiasis	Altered drug expression, up-regulated drug transporters, drug-induced stress	Azoles (biosynthesis of ergosterol); Polyenes (biosynthesis of ergosterol); Echinocandins (biosynthesis of beta-(1,3--glucan synthesis) [4, 9, 10]
	C. glabrata	urinary tract infection, tonsils and oral infection	Altered expression of drug target, phenotypic switching, biofilm formation, drug-induced stress	Azoles (biosynthesis of ergosterol); Echinocandins (biosynthesis of beta-(1,3--glucan synthesis) [4, 9, 10]
	C. neoformans	Pulmonary infection, central nervous system infection	Chromosomal alterations, hindrance of ABC transporters	Azoles (biosynthesis of ergosterol); 5-fluorocytosine (loss of permease activity) [4, 9, 10]
	A. fumigatus	chronic pulmonary and invasive aspergillosis	stress-response pathway mutations; Mutations in 1,3-bet--glucan synthase gene	Azoles (biosynthesis of ergosterol); Echinocandins (biosynthesis of beta-(1,3--glucan synthesis) [4, 9, 10]
	S. cerevisiae	Gastrointestinal and urinary tract infection	Mutations in 1,3-bet--glucan synthase gene	Azoles (biosynthesis of ergosterol) [9]
Protozoan	*P. falciparum/ P.vivax*	Malaria and multi-drug *Plasmodium* infections	Chromosomal and gene mutations	Quinine (heme detoxification blocking); Sulfadoxine (dihydropteroate synthetase inhibition); Artemisinin (free radical and reactive species generation) [9, 10]
	Leishmania sp.	Leishmaniasis	Laboratory derived mutations	Amphotericin B (cell apoptosis); Sodium Stibogluconate (triggering of thiol levels, hindrance of ABC transporters) [9, 10]
	Trypanosoma sp.	Human African trypanosomiasis	P2 aminopurine transporter and laboratory-derived mutations	Pentamidine (DNA/RNA synthesis inhibition); Fexinidazole (DNA synthesis inhibition) [9, 10]

ANTIMICROBIAL FORMULATIONS AND THEIR TYPES

Plant Antimicrobials

Such antimicrobial agents exhibit the enormous capability to fight a wide arena of diseases, such as bacterial, fungal, protozoal and viral, that are deprived of any identified side effects. Plant-based antimicrobials are essentially secondary metabolites that comprise quinines, lectins, alkaloids, peptides, flavonoids, tannins, coumarins and essential oils. These bioactive secondary metabolites from plants can serve as potential inhibitory mediators to microorganisms, which could be a potent alternate and efficacious approach to interrupt microbial resistance [22]. Research on the discovery of novel and innovative molecules is being explored that could cover a wide range of gram-positive and gram-negative bacteria. Medicinal herbs have shown great potential since the ancient ages in the form of Ayurveda, Unani, Homeopathy, *etc.* It has been utilized to treat some of the extremely difficult diseases like extensively drug-resistant tuberculosis, human immunodeficiency virus, H1N1 virus, encephalitis and hepatitis [22, 23]. Plant-based drugs have an immeasurable repertoire of aromatic components that target some of the irreversible indications of liver functioning, kidney damage, brain-blood barrier function, *etc.* The ethanolic and aqueous extracts of *Phyllanthus* can seamlessly bind to the surface antigen and hinder the enzymatic activity of DNA polymerases, thus exhibiting anti-hepatitis B activity [24]. *Allium* and *Acalypha* species and their distilled extracts have antituberculosis activity [25].

Plant bio actives individually also contain several organic and inorganic constituents that exhibit microbial potency and have been incorporated and developed into tailor-made formulations of gels, sols, emulsions, films, scaffolds, and ointments to target specific routes of antimicrobial infections. Conjugated cyclic dienes (also known as quinones) have antimicrobial action against *Pseudomonas aeruginosa* [26]. Certain polycyclic aromatic hydrocarbons like Hypericin can inhibit the growth of methicillin-resistant and methicillin-sensitive *Staphylococcus* [27]. One of the most potent pain killers and stimulants, morphine and caffeine, are psychoactive molecules that work on the suppression of neurotransmitters in the central nervous system, thus reducing the intense stimuli of pain. These alkaloids are also potential antimicrobial candidates for most of the *Staphylococcus* and *Streptococcus* contagion. Berberine, an alkaloid present in *Rhizoma coptidis*, inserts itself within the DNA to disintegrate the structural membrane leading to its permeability in *Streptococcus agalactiae* [28]. Hasubanan alkaloids such as Oxoepistephamiersine extracted from the root tubers of Stephania are effective *Staphylococcus aureus*, *Microsporum gypseum*, *Trichophyton rubrum*, *etc.* [29]. Similarly, flavonoids and flavones that occur extensively in fruits, honey, seeds and vegetables are a class of phytochemicals that are vital anti-fungals. A combination of various flavonoids possesses anti-inflammatory, antiviral, antiallergic and antimicrobial activities, and has been widely explored as herbal formulations in folk medicine for ages. Tetrahydroxyflavone, Sophoretin, Birutan and some other flavones are active against *Botrytis cinerea* and *Aspergillus flavus* [30, 31]. Ozcelik *et al.* described the antimicrobial effectivity of diverse flavonoids isolated from *Galium* and *Viscum* species against prolonged spectrum beta-lactamase-producing *Klebsiella pneumoniae* [32]. Another crucial antimicrobial constituent expansively being exploited for oral prophylaxis is 'Coumarins'. These phenolic counterparts of benzene and pyrone ring molecules are administered as an anticoagulant for curing blood clots, thrombosis and pulmonary blockage. Coumarins present in plants like *Angelica Lucida* L. have inhibitory mechanisms against cariogenic *S. mutans* and *S. viridans* [33]. Another type of secondary metabolite is essential oils containing isoprene and terpenoids and a variety of functional groups that have been long investigated in traditional remedies and have antibacterial, anticancer, and further pharmacological functions. These aromatic oils have a characteristic odour due to the presence of terpenes and phenols in them. Most of the oils have either been used per se or in the form of formulation products for a range of antimicrobial quandaries. Some examples of essential oil with antimicrobial potential include ginger, garlic, tea tree, cinnamon, clove, myrrh, lemongrass, cedarwood, lavender, rosemary, sage, *etc.* [34, 35].

Antimicrobial Peptides

Peptides, in general, hold much potential as a key component in advanced supramolecular assemblies. These are vital for vaccine preparation, microbial chemo and immunotherapy, food nutrition maintenance, transplantation, dental material design, diabetic concoction, and various other significant applications. Now that infectious diseases have taken a toll on human health, expanding investigations on antimicrobial peptides have shown path-breaking results [36, 37]. These are essentially the host defense peptides and belong to the innate immunity of the body, and are derived with or without the aid of ribosomes [38]. Some of the common antimicrobial peptides are polymyxin B, bacitracin, vancomycin (non-ribosomal) and nisin (ribosomal). Antimicrobial peptides may have diverse structural and functional residues that possess potent antiviral, antiparasitic and immune modulation properties. Structurally, these are linear with amphipathic and hydrophobic residues, and miniature beta-folded protein sheets seldom ringed conformation with exclusive amino acid chains. These can also sometimes form bundles of alpha-helical rods within the lipid bilayer by the process of self-assembly or could bear cationic/anionic charges [39]. Alamethicin is a negatively charged peptide that constitutes the hexamer clusters of helices that passes through the lipid layer surrounded by a hydrophilic sheath. Another hydrophobically helical transmembrane peptide bearing a positive charge is gramicidin [40].

Defensins, a minor cysteine-containing positively charged protein, are found in prokaryotic and eukaryotic organisms exhibiting antimicrobial and immunomodulatory potential. These defend the immune system against various infections like the stimulation of antimicrobial peptides, bacterial wound colonization, confine virulence factors and *Shigella* infections by enhancing beta-defensin-1 expression levels and also recruit white blood cells to enhance the cell-mediated defence system [41, 42]. Though much has been spoken about the antimicrobial efficacy of these host defence peptides, their clinical possibility outdoes its treatment as antimicrobial agents; it exhibits a host immune cellular immune response to impede the invasion of pathogens in most mammals [43]. For instance, a non-bacteriostatic antimicrobial peptide effector quickens the healing of wounds in mice models. The cell permeating quality of peptides exhibits high healing potency against cancer and accelerates tumour cell death by restricting new blood vessel formation, thus holding a great possibility as an anti-carcinoma drug. The tumoricidal activity of magainin-2 against human small-cell lung cancer cell lines [44] and a wide range of hematopoietic cell lines has been reported [45]. Research and investigation on different peptides and their formulations are in the pipeline, and among them, a ringed undecapeptide traversed with amino acid moieties, cyclosporin A, is an excellent immunosuppressing agent. Synthesized from a fungus, *T. inflatum,* it shows great effectivity in arthritic inflammation, psoriasis, immune-suppressing activity in inhibiting the rejection in transplantation procedure [46 - 48]. Cyclosporin A exhibits antagonism to calcineurin, a calcium-dependent phosphatase enzyme that triggers the transcription factor, which in turn activates the interleukin -2 expression to repress the proliferation of T-helper cells [49, 50]. However, there are limitations. The therapeutic potency of cyclosporin A is short-term, as it is hepatotoxic, cytotoxic and neurotoxic to a considerable extent [51]. Hence, newer formulations based on such peptides are leading the market in their target specificity, less toxicity, substantial stability and bioavailability with high efficacy *in vivo.* Formulating peptides with lipids and emulsifiers can confer stability to the overall system and has an easy interrelation with the hydrophobic residues of the peptide. Also, polymeric particles are common vehicles for antimicrobial peptide delivery, and most of the polymers are prepared by spraying and drying forms large phospholipidic crystals [52]. Polyethylene glycol-chitosan layers surrounding the lipopeptide combined cyclosporin A form well-distributed particles in an aqueous solution, sometimes giving rise to nanometric, polymeric micellar systems. One such example is the ethylene glycol-poly lactide nanoparticles [53]. Cyclosporin-conjugated nanoparticles are usually small dimensional structures that are stable colloids in isotonic solutions [54]. These injectable colloid solutions can balance and overpower the T-helper cell proliferation and production of inflammatory cytokines *in vivo* [55]. One disadvantage of cyclosporin is its solubility. Therefore, the formulations prepared are often oleaginous with low bioavailability due to their high affinity on lipids than the aqueous residues and thus become unsuitable for industrial scale-up processing. Antimicrobial peptides are like cyclic cyclosporins, and most of them have been used to develop formulations in combination with lipids, micelles, nanoparticles, nanocapsules, nanogels, nano-emulsions *etc.*, for sustained delivery to the infected intracellular targets. The peptide and vancomycin drug-based formulations can destroy the methicillin-resistant *Staphylococcus* upon systemic administration and can become even more beneficial when the drug formulation is conglomerated with phosphocholine and cholesterol [56]. Additionally, peptide formulations administered orally can prevent proteolytic destruction when combined with polymers. Encapsulation of a drug becomes easier when a polymer capsule is used as it enables ready absorption and controlled dispersion across the membrane-like in case of pulmonary infections. Antimicrobial lipopeptide formulations with sulphated polymyxin B and non-ionic surfactants polyoxyethylene and sorbitan esters show a decreased toxic activity against *P. aeruginosa* [57]. Lately, a sustainable nasal chitosan-poly lactic acid based-nanoparticles holding exendin-4 is being explored for type 2 diabetes [58]. Thus, these antimicrobial peptides undeniably possess properties that confer the developed new formulations with low toxicity, high bioavailability and extreme stability. Inventive formulations engraved with nanotechnology are unmatchable in their mode of action for antimicrobial chemotherapy, including cancer, diabetes, transplants, cell diffusion *etc.*

Immunomodulatory Agents as Antimicrobials

Immunomodulators are extremely efficient in enhancing the host defences to fight various infections in the form of vaccines or clinical therapy. Vaccines lately are made of adjuvants that help in uplifting the immune response to an antigenic determinant. A novel vaccine necessarily requires an adjuvanted molecule but currently also utilizes different immunomodulators like cytokine, toll-like receptors, peptides, *etc.*, that have antimicrobial properties and help in boosting the immune response. Adjuvants were first observed to increase the effectiveness of vaccines when aluminium salts were added to the vaccines. It is from there that adjuvants form a vital component in vaccine preparation. Adjuvanted molecules, even though non-immunogenic, assist immunogenic response to vaccine components, lessen the vaccination dose, and enhance immunological memory [59]. The first vaccines that were commercialized were the live-attenuated and inactivated ones derived from organisms like viruses and bacteria, but if scrutinized thoroughly, these lacked immunomodulatory activity and exhibited a humoral immune response that was unsuccessful in declining the infections of intracellular pathogenic organisms. It can be rightly said that 'Immunomodulatory Agents' are the future of the antimicrobial resistance developed by antibiotics. New and advanced therapeutically active composites are highly effective against diverse broad-spectrum drug-resistant organisms and, in turn, augment host immunity. At present, approaches of modifiable host immune responses by immunomodulators are being developed into clinical therapeutics and are significant counterparts between the adjuvanted molecules used and those that have healing effectiveness, perhaps reflecting the prerequisite of initiation of a suitable innate immune response [60]. Immunomodulators can suitably be used for oral, intranasal, and parenteral delivery of antigenic epitopes to induce the immune response of mucosal sites [61], and their connotation with intracellular lipid bilayers enables their entrapment within the cytosol present in antigen producing cells. Both antibody and cytokine-mediated immunity against a range of antigenic determinants as well as T helper type-1 and 2 pathways, are actuated by these immune-stimulatory carriers [62]. Besides the presentation of major histopathology complex-I (MHC-I) protein on antigen-processing cell surfaces with its exposure to the transmembrane glycoprotein CD8, immunomodulators also assist the cross-presentation of external antigens in the endogenous pathway, proving their exclusivity which is not seen in most other nanocarriers [62, 63]. Investigation of vaccine delivery by these immunomodulatory adjuvants to oral and nasopharyngeal sites has also gained importance [64]. Studies to boost the mucosal and systemic adjuvant function of B-cell targeted fusion protein cholera toxin A-1 have shown that immunomodulators preserve this fusion protein and avert its enzymatic disruption in the digestive tract [65]. Different classes of immunomodulators, such as cytokines, exhibit antiviral immune responses and have been used for treating chronic hepatitis infections and could be a possible candidate for SARS-Coronavirus therapy. Interferon-based formulations are suitable for regulating virus-mediated inflammation in multiple sclerosis. Another of its kind is the colony-stimulating factor (CSF), which can act as a potent antibacterial and anti-fungal therapeutic and could be the immune system supplementation for neutropenia [66]. Toll-like receptors are inflammation stimulators and possess antibacterial, anti-fungal and antiviral efficacy. It targets and activates the leucocytes to initiate antimicrobial response and is the promoter of cell-mediated immune responses [67].

Colloid-based Antimicrobials

Colloids are a part of the antimicrobial schema and have been explored in detail due to their small hydrodynamic size-to-volume ratio along with their inherent quality to display broad-spectrum antimicrobial action. Although it is not very clear how mechanistically these colloid particles inhibit microbial growth, a range of mechanisms has been proposed, with expansive research still ongoing. Presently colloidal particles such as zinc, titanium, magnesium, and silver have been developed into formulations by combining them with polymeric, ceramic, emulsions, lipids, oleogels, hydrogels, *etc.*, for antibacterial, anti-fungal, antiprotozoal and antiviral activities. Solutions of titanium and zinc are being studied as pharmaceutical formulations and are coherent towards several bacteria: *E. coli*, *S. aureus*, *P. aeruginosa* and *B. phytofirmans* [68]. Colloidal copper particles own incomparable biotic and physicochemical properties and are very prevalent to scientists developing innovative antimicrobial agents because of their cost compatibility [69]. Accordiny to Usman *et al.*, the antimicrobial action of copper-doped chitosan nanoparticles was assessed for its anaerobic antibacterial and anti-fungal potencies on different microorganisms *viz.,* methicillin-resistant *Staphylococcus*, *Salmonella choleraesuis*, *Candida albicans*, *Pseudomonas aeruginosa*, and *Bacillus subtilis* were reported by Usman and co-workers [70]. The silver ions in colloidal silver can exhibit different mechanistic approaches to the microorganism to inhibit its growth. It can act on the DNA and RNA to disrupt its function; can target the cellular sulfo-peptides on the plasma membrane to hinder their viability; can destabilize the plasma membrane proteins and block enzyme activity; can also leach cytoplasmic components and disrupt the respiratory enzymes and proton channel pump [71]. Silver nano-colloidal formulations are also active against gram negatives and oral pathogens such as *Streptococcus mutans*, *Streptococcus oralis*, and *Lactobacillus lactis*. Gold colloidal particles are important antimicrobial agents due to their photothermal action, non- toxic nature and high ability of functionalization. Tiwari *et al.* depicted the antibacterial and anti-fungal action of colloidal gold particles modified with 5-fluorouracil against *Staphylococcus aureus*, *Escherichia coli*, *Pseudomonas aeruginosa*, *Aspergillus fumigatus*, *Aspergillus niger* and *Micrococcus luteus* [72]. They also concluded that these modified nano-colloids were much more effective in gram negatives than in gram positives due to their high dispersibility and permeability into the cell wall of gram negatives.

MECHANISM OF ACTION OF ANTIMICROBIAL FORMULATIONS AND THEIR DELIVERY

The mechanistic approach that any antimicrobial component in a formulation adapts is unique and different for every microorganism. Antimicrobial agents act as selective impeding substances which block the enzymatic activity specific or non-specific to the prokaryotic cell. Based on the site of action and mechanism of activity, any antimicrobial formulation can be categorized into one of the four groups, *viz.,* impeders of the cell wall, protein and nucleic acid synthesis and cell membrane depolarizers in microorganisms [12, 17]. An overlap of one or more mechanisms is more efficacious in controlling the disease. However, the lack of proper supervision of antimicrobial agent-doped formulations has led to microbial resistance to these [6, 7, 11]. An outline of the chief mechanisms is discussed here. Several antimicrobial mediators target microbial cell wall synthesis. During the development of an anti-infective drug moiety, the cell wall is one of the most important therapeutic targets. The antimicrobial formulations that target the disruption of cell wall synthesis mainly do so by either obstructing the cytoplasmic path of peptidoglycan precursor synthesis, the passage of lipid-bound precursors across the cytosolic membrane, by positioning polysaccharide moieties into the cell wall or by chemically impeding the transpeptidase crosslinking [12, 15]. The modulatory effectiveness of an antimicrobial agent is reliant on the cell differentiation caused by the cell-bound antimicrobial agent. While some hinder mitosis, certain others affect the cell wall formation and lyse the cell through osmosis. Some antimicrobials contain bulky multifaceted glycosylated cyclic or polycyclic non-ribosomal short-chain amino acids that block murein synthesis in most organisms. Its activity is further enhanced when various glucose moieties are attached to it. These glycosylated amino acids bind to the D-Ala-D-Ala stem terminus of murein side-chain pentapeptide to inhibit trans-glycosylation and transpeptidation reaction in the prolonging murein chains disordering the cell wall [15, 16].

Certain antimicrobials inhibit nuclei acid metabolism leading to the disorientation of the DNA/RNA structure and its synthesis [15]. Their mechanistic approach is either intrusion with nucleotide sequences, nucleic acid moieties, or the whole biochemical pathway of the cell. Some plant antimicrobials with phenolic and flavonoids are potent purine, and pyrimidine inhibitors like the tea epicatechin gallate (ECG) based polyethylene glycol formulations [23]. Tea ECG at the same time with the bacteria contests with the active site of proteases, thus bocking the nucleotide precursor molecule [23]. Polyethylene glycol-folate-linked microemulsions have been anticipated to reduce the activity of the dihydrofolate reductase enzyme to ultimately block the folate metabolism within cells [27]. The development of an antimicrobial gel formulation with silver nanoparticles for topical administration exhibited anti-polymerase and anti-topoisomerase enzyme activity and selectively hindered the process of cell transcription [30]. Topoisomerase II and IV are the key enzymes for DNA supercoiling and antimicrobial agent doped formulations that stop their mechanism of action do so by affecting the DNA/RNA biology of the cell and generating certain mutations. This, in turn, causes disbalance in the efflux pump systems and leakage of the cell wall components in the whole [19]. As for protein synthesis, most of the antimicrobials interfere with the ribosomal subunit at various protein synthesis stages. Mostly, protein synthesis is hindered during the genetic code proofreading to give rise to misread mRNAs. Formulations with the antibiotic of the class aminoglycosides do so by upsetting the translocation step of the polypeptide formation and modifications in one of the 30S or 50S ribosomal units by certain transferases. They then cause the tRNA molecules to detach from the ribosomal compartments [38, 39]. Inhibition of polypeptide synthesis is a common mechanism shown by most antimicrobials that exhibit activity against a wide range of gram-positives and gram-negatives along with chlamydia, mycoplasmas, rickettsia, and protozoans [46, 48]. Undeniably, the foremost challenge lies in the efforts required to control microbial resistance to these antimicrobial agents with time. Thus, it is also important to highlight the resistance mechanisms employed by various microorganisms that need to be addressed for any therapeutic or prophylactic remedy development.

Mechanistic Approach of Resistance to Antimicrobials in Microorganisms

Microorganisms can adapt to various mechanisms and become resistant to antimicrobial drugs. Microbial resistance can either be ‘inherent’, where the microbe could be resistant to one or more antimicrobials or ‘acquired’ where the genetic constitution of the microbe may enable the production of enzymes to hinder the efficacy of drug/agent, may trigger efflux pump system inhibiting the drug to reach the specific intracellular site, may alter the drug activation position or shift to an alternative metabolic pathway that blocks the effect of the drug [73, 74]. This changeover and modification of microbial genetics may occur *via* gene transfer methods like conjugation, transduction, transfection, *etc.*, that integrate varied resistance genes into the host or plasmids. Therefore, antimicrobial agents do lead to a discerning burden for resistant strains of microorganisms. Modification of genotypic machinery in microbes can cause them to form multifaceted hindrances (multi-drug resistance), emerging into serious disquietude in healthcare sectors. The hindrances could be due to newer alterations in nucleotide, target proteins and upregulation/downregulation of membrane protein channels to clog the drug entry passage [75, 76]. Also, most antimicrobials considered for prophylaxis of infections follow one of the mechanistic features *viz.,* targeting the cell wall or protein, nucleotide synthesis or blocking a particular metabolic pathway. In all these situations, the microorganism-bearing resistance conversing transmutations are shortlisted for antimicrobial testing, which stops the vulnerable strains but may allow the resistant strains to multiply. The mechanisms of microbial resistance are discussed here.

Inactivation of Antimicrobials

A few strains of pathogenic organisms acquire resistance to beta-lactam moieties either by altering the chemical structure or enzymatic digestion of the antimicrobials [77]. The genes encoding for resistance to a drug is present in extrachromosomal plasmids in the case of bacteria, and most of the beta-lactam class of drugs lose their potency due to the action of the lactamase enzyme that disrupts the amide bonds by hydrolysation [78]. Gram-negative bacteria, however, inactivate the aminoglycoside group of the antibiotic by post-transitional modification of proteins and other biochemical processes [79].

Modification of the Target

A 'target site' is the binding site of any antimicrobial to change its normal function. When a target site is altered due to enzymatic action, a microbe may develop resistance. For example, the pathogenic strain of *Streptococcus* dodges the mechanistic behaviour of macrolides and lincosamides antibiotics by restricting their attachment to the 50S ribosomal unit and inhibiting protein synthesis by co-transcriptional variation of 23S rRNA constituent present in the ribosomal unit by methylating the amino group of adenine moiety [80]. Okuma and co-workers discussed the dissemination of a staphylococcal cassette chromosome mec, a mobile genetic element in *Staphylococcus* species possessing the mec-A gene, that contributes to the resistance of methicillin-resistant strains of this bacteria [44]. Another likeness of target site modification is the attachment and modification of the resistant strains from D-Ala-D-Ala to D-alanyl-D-serine/D-alanyl lactate, thus exhibiting resistance to the glycopeptide antibiotic Vancomycin [81]. Certain antibiotics, such as fusidic acid, linezolid, chloramphenicol derivatives, alter the site by the hindrance of protein synthesis in pathogenic bacterial strains [82]. Resistance developed in *Enterococcus* to certain protein synthesis-inhibiting antibacterial agents is likely due to the mutation in the 23S rRNA unit that correspondingly causes mutations in the 16S rRNA as well, thereby decreasing the target binding to the drug [83]. *Mycobacterium tuberculosis* and *Escherichia coli's* vulnerability towards streptomycin is due to the mutated ribosomal S12 protein gene. Likewise, the broad-spectrum bactericidal compound with a bicyclic core structure like quinolone obstructs the function of Type II and IV topoisomerases, and in both *E. coli* and *S. aureus,* the hindrance to the two topoisomerases occur due to changes in either the primary/secondary target sites or both simultaneously [84].

Drug Efflux Pumps in Resistance Development

Antibiotic resistance inherently developed in a bacterium is the result of the genes encoding for efflux pump system proteins that balance the functions with bacterial cells. Established reports on the efflux system-mediated resistance to antibiotics have been well-studied. Efflux pumps in bacteria consist of majorly non-drug specific protein residues that conveniently expel any unrelated substances, chemicals, or antimicrobials without disintegrating their structural composition. This expulsion in turn, leads to a lower ineffective concentration of the antimicrobial in the cells that fail to inhibit the growth of bacteria. In the case of gram-negative bacteria, where the transportation of the drug is through the outer lipopolysaccharide layer, the permeation of the drug across the membrane is quite difficult [85, 86].

Role of Plasmid-mediated Drug Efflux Pump Systems

The gaining of novel genomic material from other resistant microbes could be a cause for bacterial resistance. This transfer generally takes place *via* lateral genetic transfer between similar or dissimilar bacterial species with genetic material transfer mechanisms like transformation, conjugation and transduction. Occasionally this genetic transfer is facilitated by mobile transposable elements, the plasmids that carry and transmit resistant or pathogenic genes [87]. Enterococci-mediated pheromone-responsive plasmids bearing the mobile genetic element contribute to the antibiotic resistance of the bacteria. Plasmid-borne quinolone resistance genes (qnr), discovered by Martinez and co-workers depict that this 'qnr' gene encoded for quinolone resistance along with sheathing the bacterial DNA gyrase from disruption by this antibiotic [88]. Fig. (**1**) depicts the mechanistic approach of microbial resistance developed in an organism.

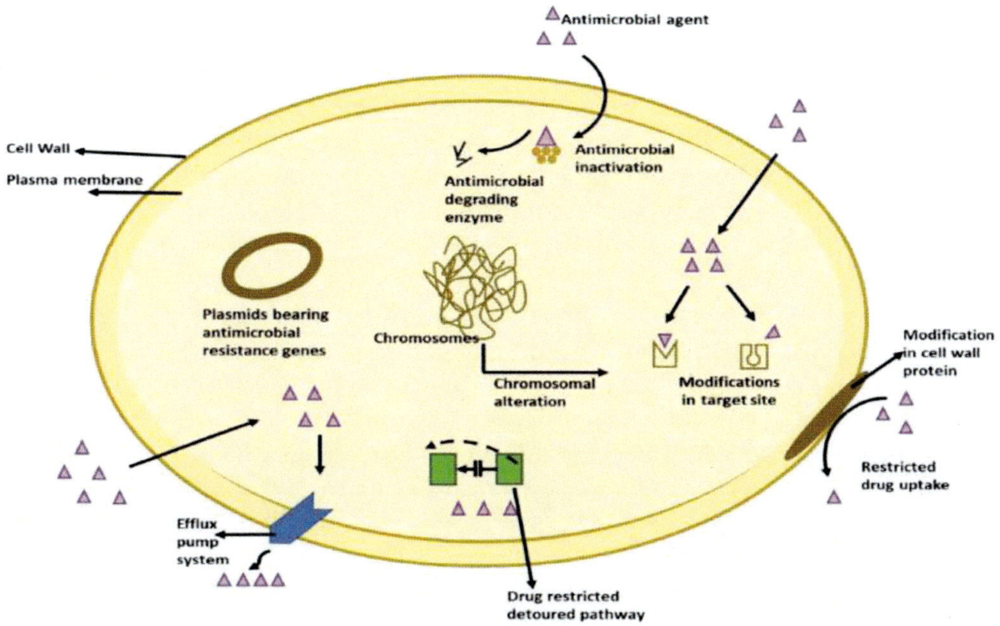

Fig. (1). Mechanistic approach of antimicrobial resistance in a microorganism [86].

In the event of fungal and protozoal machinery of resistance to one or more antimicrobials, the therapeutic choices are fewer because the number of anti-fungal and antiprotozoal drugs is scarce. Moreover, the clinical manifestation of these diseases is further conceded by the development of drug resistance that has been alarmingly rising in most of the medical and healthcare sectors. Obstinate multi-drug resistance is also a common entity in the case of fungal species like *Candida* and protozoan species like *Leishmania* and *Trypanosoma* [89]. In the case of protozoan diseases, the scarcity of vaccines is also a discerning problem. The molecular mechanism employed comprises reformed affinity to the drug, easy target availability, decreased efflux system conjugated intracellular drug levels and biofilm formation. These mechanisms accountable are frequently shared by resistant strains exhibiting integrally abridged vulnerability to the drug during treatment. Contemporary understanding of the resistance mediation and regulation by genotypic material and cellular factors gives a better idea of the emergence of resistance in these microorganisms. The most regular used anti-fungal agents are polyenes, azoles, allylamines, echinocandins, 5-fluorocytosine and antiprotozoal agents are chloroquine phosphates, metronidazole, sulfa drugs and diloxanide furoate [90, 91]. The mode of action of anti-fungal drugs encompasses (a) the restricted synthesis of ergosterol (azoles and allylamines), (b) binding of ergosterol to the plasmalemma to disintegrate the cell function (polyenes), (c) constrained synthesis of DNA and pyrimidine (5-fluorocytosine) and (d) restricted synthesis of beta-1,3-glucans (echinocandins).

It is important to state here that an unbridled and extensive usage of anti-fungal/antiprotozoal agents is the cause of drug resistance [92]. Also, the advent of attained resistance amid the widespread pathogen limits medication choices, and in turn, modifies patient supervision. A more elaborate and comprehensive knowledge of the mechanism-specific hindrances and their biochemical mode of action is the top priority need for developing improved therapeutic formulations, and diagnostics to be used for first-line therapy to overcome such microbial resistance [93]. A description of the mechanistic approach of resistance to different anti-fungal drugs is shown in Table **2**.

Table 2. Mechanistic approach of resistance to different anti-fungal drugs [93].

A. Azoles: These are plasma membrane targets. Resistance to azoles can arise by blocking of ergosterol biosynthesis via multifaceted mechanism like overexpression or modification of the drug target; up-regulation of the drug carriers; cellular alterations that diminishes drug toxicity or enable tolerance of drug impelled extracellular stress response.	***Candida Albicans***	***Cryptococcus Neoformans***	***Aspergillus Fumigatus***
B. Polyenes: For Polyenes, the resistance emerges via the exhaustion of the target ergosterol molecules attributable to loss-of-function mutations in the genes that cause the biosynthesis of ergosterol. **C. Echinocandins:** These drugs target the fungal cell wall and mutate the 1,3-beta-glucan synthase component FKS1; a gene that encodes for echinocandin target beta-D glucan synthase and causes mutations to its catalytic subunit. Cellular stress responses also regulate the resistance phenotypes of the fungi.	• Genetic mutations in various genes • Heterozygosity depletion • Mutations in ergosterol triggering genes implying cellular stress dependent resistance • Calcineurin, sgt1: stress response pathway mediators • Unbalanced structural deformity in chromosomal arms • Loss of function mutations in ergosterol biosynthesis • Fks1 gene mutations and stress response pathway mediators: Hsp90, calcineurin	• Mutations, abnormal expression, and chromosomal duplication in ergosterol triggering genes • Stress response pathways mediators: ATPase -------- ---------	• Gly54 and Cyp51A mutations and abnormal expression of Cyp51 A • Stress response pathway key mediators: Hsp90 • Abnormal expression of Afr1 --------- ---------

The major hardship with protozoan infections lies in the rise and uncontrolled spread of drug resistance without the availability of efficacious vaccines. Besides, the investigation and experimentation on drug resistance in protozoan pathogens are usually carried out under laboratory conditions that do not mimic the actual host-parasite infection site and association. The host health also impacts the extent of communication of infection. Hosts that can balance their defence system and control the parasitic infection might act as carriers of the infection for other individuals resulting in a high probability of re-infection sometimes. One of the most prevalent parasitic diseases that has become a major setback to public health is 'malaria'. *Plasmodium species viz., falciparum* and *vivax,* are accountable for the prime number of severe cases of the disease and also maximum drug-resistant infections. Most of the antimalarial drugs in use (chloroquine, quinine, artemisinin, sulfadoxine, *etc.*) chiefly inhibit the development of intra-erythrocytes, targeting either the food vacuole of the ring stage and trophozoites of the blood stage or blocking the enzymatic biosynthesis of folic acid in trophozoites. In both these cases, infection eradication becomes tough as drugs to completely stop the transmission and elimination of hepatic forms are lacking [90 - 93]. Sometimes the malarial drug resistance is also due to genotypic factors of the parasite, such as one or more genetic mutations in the genes of the parasite, making them resistant to drugs by impeding their thorough dispersion [94]. Leishmaniasis, another protozoan infection caused by the parasite *Leishmania donovani,* can exist in forms like visceral, dermal, cutaneous or mucocutaneous forms and is mostly transmitted *via* sand-fly species to mammalian hosts. Anti-leishmanial drugs used for treating leishmaniasis fail to completely omit the parasite, and a potential relapse of infection is expected, especially in T-cell compromised individuals. While pentavalent antimonial remains the gold standard for leishmaniasis treatment, second-and third-line drugs are gaining importance and are currently being investigated [95].

CHALLENGES IN DEVELOPING CONTEMPORARY ANTIMICROBIAL FORMULATIONS

Earlier, research on decoding a microbe's ability to obstruct a given chemical therapy for treatment and develop novel stratagems to inactivate some of the antibiotics was fundamental. In this day and age, the challenge of developing a new and more potent antimicrobial formulation against infection is extremely crucial. Regardless of the frontier technology adapted to regulate infectious diseases, the new phenotypic and genotypic adapted organisms, the predicament in antibiotic resistance, and the risk of biological attack calls for the necessity to enliven basic research and modernise expertise of anti-infective chemical and biological therapy [96]. Consequently, the mechanism of infection by any microorganism needs to be better scrutinized in terms of physiology, adaptability to a particular drug and development of interruptions to discover novel contemporary formulations and medicaments.

There are several hints that advanced technology-driven approaches are required to combat such a scenario. Firstly, the pattern of death rates from infectious diseases observed before and in the present: During the 19th century, the rate of death due to infections escalated predictably, with rates dropping in a controlled fashion in comparison to the 20th century, where due to the outspread of HIV and methicillin-resistant *Staphylococcus* strains, vancomycin-resistant enterococci, multidrug-resistant, *etc.* caused the death rate to be almost twice [97]. This doubling of mortality rate is also due to the immunocompromised individuals who are unable to recover from the infection. Secondly, the discovery time of a drug is unhurried. There has been a discovery gap of approximately 4 decades between the introduction of a new class of antibiotics: fluoroquinolones in 1962 and oxazolidinone in 2000 [98, 99]. Thirdly the current inclination of pharmaceuticals to leave their research on areas of antimicrobial therapeutics further narrows down the scientific expertise in antimicrobial drug finding and its expansion. Some of the strategic methods that could be considered for contriving new classes of antimicrobials and novel screening techniques for the pathogens are discussed here.

Inexorableness to Antimicrobial Resistance

Quinolones, a synthetic class of antibiotic, the susceptibility of bacterial resistance is low given the fact the number of times of exposure to the drug is less and hence the mechanism of evolved resistance might not be in the cards. Nevertheless, the pervasive and uninhibited efflux systems evolved, sheathing the microorganism from various toxic products that occur in nature, habitually delivering cross-protection to the microorganism. In turn, the genes encoding resistance elements are deep-rooted in the genomes, thus rendering continuous protection against such toxic products even if there is no prior exposure [100]. Besides the antimicrobial resistance, genotypic elements transferred between microorganisms by the help of plasmids, transposons, and integrins are collectively responsible for several resistance genes; that give rise to multi-drug resistant strains, some of which might be uncurable with the current cache of permitted medications [17].

Manageability of Resistance

Though antimicrobial resistance is unavoidable, collecting and managing the data on resistance among various geographical regions and healthcare sectors is needed. Also, managing the technical and administrative difficulties might help in monitoring the resistance and controlling the spread of infection within a population in a particular area. Collection of data on prior infections and subsequent resistance development might assist to target both the treatment of new infections and effort to curb the existing infection locally as well as universally. A surveillance network set up to monitor the data is also essential as it will enable us to understand the nature of resistance in a microorganism [101]. For example, the molecular studies carried out on a particular bacterial phenotype, when equated with the antibiotic usage data, might give an idea about the selection of the exact antibiotic needed to treat that specific disease and the discovery of newer diagnostics, respectively.

There is no doubt that the need of the hour demands novel and cutting-edge policies and molecules to target pathogenic organisms that show resistance to virtually all modern-day antibiotics. We now sit on a tie-ticking bomb where infections are much different from the pre-antibiotic age and sometimes are uncurable leading to casualty.

EFFECTIVENESS OF CONTEMPORARY ANTIMICROBIALS OVER THE TRADITIONAL THERAPIES

In 2007, a publication by GlaxoSmithKline revealed that the pharmaceutical company had screened 70 libraries to obtain around five lakh molecules as candidates for targeted drug research, out of which hardly a few have cleared the subsequent medical trials to be accredited (Payne *et al.*, 2007 [102]. This highlights the complexity of screening new antimicrobial agents.

Numerous drug candidates are formerly being assessed clinically, but none have cleared the approval for market endorsement. Hence the search for alternatives to traditional therapies is still in a nutshell. In this section, the emergence of some of the newer drug development stratagems, such as antimicrobial peptides, phage therapy, vaccines, potentiators of traditionally used antibiotics and antibacterial biomaterials, are discussed.

Antimicrobial Peptides

These are evolutionarily preserved residues of the innate immune system that are found in all animals [40]. These 'antimicrobials' have the potency to disintegrate microbial membranes, causing cell death. Peptides have been currently classified as immunomodulators [43]. Antimicrobial peptides are classified as broad-spectrum drugs active against most prokaryotic microorganisms but may cause toxicity in eukaryotes. A peptide isolated from frogs, called Magainin, is one of the extensively explored antimicrobial peptides [44, 45]. Other examples include bacteriocins (nisin), a broad-spectrum antimicrobial, exploited in food conservation [103]. Hence, peptides are being studied as potential topical drug candidates due to their broad spectrum of action, low toxicity and no induced resistance [38]. However, developing the peptides to complete clinical drugs can be tough as the production costs are sky-high and their stability *in vivo* is a problem [39]. Though These peptides are functionally being explored as synthetic mimics for therapeutics, surpassing the limitations. The ongoing research on antimicrobial peptides comprises pexiganan, a synthetically derived magainin used for topical administration in foot ulcers. The effectivity of synthetic peptide mimics, arylamide oligomers and lipopeptides has been evaluated *in vivo* in animal models [104 - 106]. Likewise, arylamide foldamers have also shown promising results when tested on mice models with *Staphylococcus* infections [106]. Another broad-spectrum peptidomimetic, the LTX-109, is also under Phase I/II evaluation for nasal decolonization of methicillin-resistant *S. aureus* [107]. Endogenous peptide like the human-derived peptide DPK-060 have also been investigated and has gone through successful phase II clinical trials for their administration topically against eczema [108]. Another bacterial lanobiotic and fungal plectasin are under pre-clinical evaluation for their efficiency against *Streptococcus pneumoniae* [109].

Phage Therapy

Bacteriophage-based therapeutics have been in global use since 1901, but with substantial hullabaloo. It then became underutilized in the Western world after antibiotics came into existence. In the year 2006, the FDA permitted bacteriophage therapy against *L. monocytogenes* contamination in poultry [110]. Since bacteriophages are derivatives of bacteria, they make good antimicrobial candidates because of their high specificity and cost-effectivity but might develop resistance with time owing to their problem with neutralization in host immune systems. A substitute hence could solve the problem where whole phage particles called lysins can be utilized, which are exceedingly powerful enzymes for destroying the cell wall of microbes [111]. A category of phage called the BioPhage-PA has been under clinical testing for phase I and II trials as a topical agent for the treatment of *Pseudomonas* ear infections [111]. Clinical trials for an aerosol variant for cystic fibrosis are also under investigation. BFC-1, a phage mixture of *P. aeruginosa* and *S. aureus,* is under assessment for treating burn wounds [112]. Phages have also been compatible with animal models and successful infections with *Burkholderia, Enterococcus, Escherichia*, *Klebsiella, Salmonella enterica* and *Vibrio vulnificus* [113, 114]. Bioengineered phage is also under current screening.

Vaccines

The first vaccines were attenuated and killed whole-cell ones. With time new techniques for having been employed for producing vaccines that are modified toxins and protein-conjugated polysaccharide types. The field of genomics has also thrown light on an apt screening of antigenic determinants for vaccine preparation. Bioinformatic analysis has led to the identification of precise surface-protein groups and also thrown light to “reverse vaccinology” and “ANTIGENome technology” [115, 116]. Screening of the whole genomic material of a pathogen is accomplished by peptide libraries where quick identification of any developed immunogenicity in humans upon pathogen exposure can be mapped. Vaccines could be the most cost-efficient option for an antibacterial medicament. However, the matter of safety in vaccine administration must be looked upon stringently. Recent research and development of vaccines are vast, with many of them undergoing pre-clinical identification. Much work has gone into the *P. aeruginosa* vaccine progress, and quite a few have appeared promising in early clinical investigation [117]. The vaccine IC43, the recombination of two fusion proteins of *P. aeruginosa* has reached phase II trials [118]. In the intervening time, three vaccines for curing Staph infections are under evaluation. Local use of a few vaccines based on outer membrane vesicles is also under efficacy testing.

Potentiators Replacing the Currently used Antibiotics

Potentiators to the existing antibiotics are substances that enhance sensitization to an antigenic determinant. These have been used clinically and their mode of action is either to act by adapting to resistance reversal mechanisms in inherent sensitive pathogens or initiate per se sensitization to the resistant pathogenic strains. Reversal of resistance has been observed in beta-lactam antibiotics by producing the enzyme lactamase or penicillin-binding proteins. Some of the beta-lactam inhibitors that work on reverse resistance mechanisms are clavulanic acid, tazobactam and sulbactam [119]. Gram-negative bacteria also exhibit such a mechanism by the overexpression of efflux systems. Bacterial efflux pump systems are also a target competitor for potentiators, though none of the efflux systems inhibitors are available commercially. Potentiators are effective and have been used clinically. However, its limitation remains in its resistance to beta-lactamase inhibitors [120]. Some of the beta-lactamase inhibitor potentiators are under development: NXL104, in combination with ceftazidime inhibitor, has shown inhibitory effects on *E. coli* and *K. pneumoniae* and is currently under phase II and II trials [121]. Another piperacillin-penem beta-lactamase inhibitor combination has been efficaciously investigated and showed promising results in murine models of local infection of *E. cloacae* [122]. Though no efflux pump system inhibitors have received approval, one potential contender MP-601,205 combined with quinolone antibiotics developed by the company Mpex Pharma has been explored for trials but suffered a tolerability problem. A recent approach to targeting the stress response regulators in *Pseudomonas* is still under preliminary study but has shown sensitivity to tobramycin [123].

Biomaterials Combined Antimicrobials

Microbial colonization might happen in medical implants and equipment, and the most common pathogens that dwell here are *Staphylococcus* species. Biofilms formed on medical devices also contribute to a high risk of infections. *Proteus mirabilis* is responsible for encrustation in catheters [124]. Substantial efforts have been made to decrease the number of medical device-related infections by systemic antibiotic administration, but the severity of such infections is difficult to cover, which might range from mild to fatal medical conditions in patients. The cost of prophylaxis for replacing the infected implants is also a major concern. Much exploration and work have been implemented in developing antimicrobial medical implants. It is convenient for bacteria to colonize in polymers used in urinary medical equipment *viz.,* catheters. Hence antimicrobial coatings on such equipment can be imbibed, which are bound to the coating covalently and restrict bacterial colonization. Similarly, implants for orthopaedics are made of alloys like stainless steel and titanium that though highly biocompatible with good mechanical properties, limit bacterial colonization and growth inhibition. Nowadays, silver has been expansively used in coatings as it possesses a broad-spectrum antibacterial activity toward both Gram-positives and Gram-negative bacteria [125]. Antibiotic and antiseptic engraved coatings are also currently available in the market. Triclosan-based coatings have shown good responses in both *in vitro* and *in vivo* studies and are under phase II clinical investigation. Fluorouracil-coated medical implants, due to their good biocompatibility, have been used as antimicrobial carriers [126]. A condensed tabular representation of some of the current therapies under development and use has been summarized in Table **3**.

Table 3. Some of the antimicrobial formulation therapies currently under development and use that have potentially enhanced the effectiveness of their treatment.

Drug/Formulation	Source/Active Drug Involved	Targeted Microorganism	Mechanism of Action
Antimicrobial Peptides			
Human β-defensin-3	Human defensin	Broad-spectrum of gram-positive and gram-negative bacteria	Promotes wound healing, triggers keratinocytes and cytokines, aids in proliferation and migration *via* phosphorylation of STAT and EGFR proteins [104].

(Table 3) cont.....

Drug/Formulation	Source/Active Drug Involved	Targeted Microorganism	Mechanism of Action
LTX-109 (Lytixar)	Synthetic antimicrobial peptidomimetic (SAMP)	Methicillin-resistant *S.aureus* /methicillin-susceptible *S.aureus*	Causes bacterial membrane disruption, treats most of the preliminary gram-positive skin infections, prevents the growth of methicillin-resistant strains of *Staphylococcus aureus* in wound infections topically [106].
Pexiganan (Locilex)	Analog of African frog peptide magainin	*S. aureus*, *S.epidermidis*, *S. saprophyticus,* Enterococci, β-haemolytic Streptococci	Treats diabetic foot infections when applied topically, reduces the hypoglycaemic state of diabetes, which induces oxidative stress on nerve cells [105].
HB1275	Lipohexapeptide with potent anti-fungal activity	Yeast and filamentous fungi *Trichophyton* infection	Treats dermatophytosis, immunosuppression of cell wall mannans restricts the growth of arthroconidia in keratinized tissue [127].
Phage Therapy			
BioPhage-PA	A concoction of six *P. aeruginosa* specific phage	Antibiotic resistant *P. aeruginosa*	Chronic otitis infection, mechanisms not established yet [111].
BFC-1	A phage mixture	*P. aeruginosa* and *S. aureus*	Treats burn wounds, known to fasten granulation and keratinocyte migration [112].
Vaccines			
JNJ-63871860 (ExPEC 4V)	Multivalent glycoprotein conjugate vaccine	Invasive extra-intestinal pathogenic *E. coli*	Treats urinary tract infections leading to adult bacteraemia, blocks the putative virulent genes. (currently under phase II trial) [127].

(Table 3) cont.....

Drug/Formulation	Source/Active Drug Involved	Targeted Microorganism	Mechanism of Action
Group B *Streptococcus* vaccine (GBS)	Polysaccharide vaccine of group B *Streptococcus*	*Streptococcus agalactiae*	Treats invasive neonatal GBS infections, Stimulates the production of functionally active antibodies that crosses the placenta to deliver defence against neonatal GBS infection. (currently under phase II trial) [128].
IC43	Recombination of two fusion proteins of *P. aeruginosa*	*P. aeruginosa*	Nosocomial infections in intensive care units (ICUs) (currently under phase III trial) [118].
Potentiators replacement to currently used antibiotics			
NXL104	Non-lactam inhibitor of β-lactamase	*Mycobacterium tuberculosis*	Inhibits BlaC, the only chromosomally encoded β-lactamase from *Mycobacterium tuberculosis* infections, by forming a carbamoyl adduct. (currently under phase III trial) [121].
SPR741	Polymyxin B derivative	Multidrug-resistant gram-negative bacteria	Disrupts the outer membrane of gram-negative bacteria
Biomaterials as Antimicrobials			
Dipalmitoylphosphatidylcholine (DPPC), cholesterol, and dimethylammonium ethane carbamoyl cholesterol (DC-chol) [Liposomal formulation]	Benzylpenicillin	*S. aureus*	Disrupts the cell membrane integrity [128].
Poly-lactide-co-glycolide (PLG) nanoparticle [Polymeric nanoparticle formulation]	Rifampicin, isoniazid, Pyrazinamide, Ethambutol.	*Mycobacterium tuberculosis*	Augmented absorption of the drug disrupts nucleic acid synthesis [128].
Poly (D,L-lactide) (PLA) Nanospheres	Arjunglucoside	*Leishmania donovani*	Condensed toxicity, blocks the *Leishmania* antigen TIMP-1 [128].

(Table 3) cont.....

Drug/Formulation	Source/Active Drug Involved	Targeted Microorganism	Mechanism of Action
Pegylated lysine based copolymeric dendrimer	Artemether	*Plasmodium falciparum*	Augments the drug stability, highly soluble and efficient delivery vehicles that inhibits the paracellular and endocytosis pathway in the malarial parasite [128].
Glyceryl tripalmitate and tyloxapol [Solid-lipid nanoparticle formulation]	Clotrimazole	Fungi, yeast	Hinders free radical formation when applied topically [128].

CONCLUSION

It is high time that we realise how vital it is to develop antimicrobial agents to fight microbial resistance. Quite a lot of reports are roving on the big pharmaceutical companies curbing down the basic research on the screening and development of newer therapeutic options for antimicrobial infections, thus prompting less approving monetary incentives compared to the available medicaments. Both academia and industry are trying to accomplish novel stratagems for newer antimicrobial medications. Though most of the current generation antimicrobial formulations are still under unapproved trials, in this acute hour of the emergence of microbial resistance, it is a taste of various categories of drugs that are asked for. The more diversity in the cache of pathogens, the more convenient it is to offer custom-made antimicrobial alternatives. The first-line therapy remains the broad-spectrum antibiotics, but an alternative high-tech-driven approach would aid in controlling most of the commensals and opportunistic ailments. Hence, precise diagnostics with novel antimicrobial machinery are looked upon to terminate this complexity in medications. Understanding the biological and metabolic modifications in a microorganism will better depict the direction of treatment, for instance, studies on latent bacteria. Nevertheless, this will need a more extensive investigation to clinically target a particular microorganism's physiological function. This can be done by exploring molecular aspects, for instance, gene knockouts, genome-scale technologies *etc.* More research on screening the resistance within common microbial flora and utilizing combinational therapies to treat them would be a beneficial step in growing patient amenability, the probability for resistance expansion, lessening drug costs and upsurge efficiency, eventually encompassing the life of these antimicrobial agents.

CONSENT OF PUBLICATION

Declared none.

CONFLICT OF INTEREST

The authors declare no conflict of interest.

ACKNOWLEDGEMENTS

The authors thank the management of Vellore Institute of Technology, Vellore and the University of Miyazaki, Japan.

REFERENCES

[1] Green JL, Holmes AJ, Westoby M, *et al.* Spatial scaling of microbial eukaryote diversity. Nature 2004; 432(7018): 747-50. [http://dx.doi.org/10.1038/nature03034] [PMID: 15592411]

[2] Whitman WB, Coleman DC, Wiebe WJ. Prokaryotes: The unseen majority. Proc Natl Acad Sci USA 1998; 95(12): 6578-83. [http://dx.doi.org/10.1073/pnas.95.12.6578] [PMID: 9618454]

[3] D'Costa VM, King CE, Kalan L, *et al.* Antibiotic resistance is ancient. Nature 2011; 477(7365): 457-61. [http://dx.doi.org/10.1038/nature10388] [PMID: 21881561]

[4] Hofer U. The cost of antimicrobial resistance. Nat Rev Microbiol 2019; 17(1): 3. [http://dx.doi.org/10.1038/s41579-018-0125-x] [PMID: 30467331]

[5] Karkey A, Thwaites GE, Baker S. The evolution of antimicrobial resistance in *Salmonella Typhi.* Curr Opin Gastroenterol 2018; 34(1): 25-30. [http://dx.doi.org/10.1097/MOG.0000000000000406] [PMID: 29059070]

[6] Rice LB. Antimicrobial stewardship and antimicrobial resistance. Med Clin North Am 2018; 102(5): 805-18. [http://dx.doi.org/10.1016/j.mcna.2018.04.004] [PMID: 30126572]

[7] Eliopoulos GM, Cosgrove SE, Carmeli Y. The impact of antimicrobial resistance on health and economic outcomes. Clin Infect Dis 2003; 36(11): 1433-7. [http://dx.doi.org/10.1086/375081] [PMID: 12766839]

[8] Harbarth S, Samore MH. Antimicrobial resistance determinants and future control. Emerg Infect Dis 2005; 11(6): 794-801. [http://dx.doi.org/10.3201/eid1106.050167] [PMID: 15963271]

[9] Russell AD. Similarities and differences in the responses of microorganisms to biocides. J Antimicrob Chemother 2003; 52(5): 750-63. [http://dx.doi.org/10.1093/jac/dkg422] [PMID: 14519671]

[10] Nweze JA, Mbaoji FN, Huang G, *et al.* Antibiotics development and the potentials of marine-derived compounds to stem the tide of multidrug-resistant pathogenic bacteria, fungi, and protozoa. Mar Drugs 2020; 18(3): 145. [http://dx.doi.org/10.3390/md18030145] [PMID: 32121196]

[11] McArthur AG, Tsang KK. Antimicrobial resistance surveillance in the genomic age. Ann N Y Acad Sci 2017; 1388(1): 78-91. [http://dx.doi.org/10.1111/nyas.13289] [PMID: 27875856]

[12] Juliano RL. Pharmaceutical innovation and public policy: The case for a new strategy for drug discovery and development. Sci Public Policy 2013; 40(3): 393-405. [http://dx.doi.org/10.1093/scipol/scs125]

[13] Abadie R. The exploitation of professional "guinea pigs" in the gig economy: The difficult road from consent to justice. Am J Bioeth 2019; 19(9): 37-9. [http://dx.doi.org/10.1080/15265161.2019.1630513] [PMID: 31543031]

[14] Davidovich C, Bashan A, Yonath A. Structural basis for cross-resistance to ribosomal PTC antibiotics. Proc Natl Acad Sci USA 2008; 105(52): 20665-70. [http://dx.doi.org/10.1073/pnas.0810826105] [PMID: 19098107]

[15] von Nussbaum F, Brands M, Hinzen B, Weigand S, Häbich D. Antibacterial natural products in medicinal chemistry--exodus or revival? Angew Chem Int Ed 2006; 45(31): 5072-129. [http://dx.doi.org/10.1002/anie.200600350] [PMID: 16881035]

[16] Wright GD. The antibiotic resistome: the nexus of chemical and genetic diversity. Nat Rev Microbiol 2007; 5(3): 175-86. [http://dx.doi.org/10.1038/nrmicro1614] [PMID: 17277795]

[17] Kim S, Lieberman TD, Kishony R. Alternating antibiotic treatments constrain evolutionary paths to multidrug resistance. Proc Natl Acad Sci USA 2014; 111(40): 14494-9. [http://dx.doi.org/10.1073/pnas.1409800111] [PMID: 25246554]

[18] Aslam B, Wang W, Arshad MI, *et al.* Antibiotic resistance: a rundown of a global crisis. Infect Drug Resist 2018; 11(11): 1645-58. [http://dx.doi.org/10.2147/IDR.S173867] [PMID: 30349322]

[19] Schroeder M, Brooks B, Brooks A. The complex relationship between virulence and antibiotic resistance. Genes (Basel) 2017; 8(1): 39. [http://dx.doi.org/10.3390/genes8010039] [PMID: 28106797]

[20] Romano M, Iovene MR, Russo MI, *et al.* Failure of first-line eradication treatment significantly increases prevalence of antimicrobial-resistant *Helicobacter pylori* clinical isolates. J Clin Pathol 2008; 61(10): 1112-5. [http://dx.doi.org/10.1136/jcp.2008.060392] [PMID: 18755715]

[21] Jones ML, Ganopolsky JG, Labbé A, Wahl C, Prakash S. Antimicrobial properties of nitric oxide and its application in antimicrobial formulations and medical devices. Appl Microbiol Biotechnol 2010; 88(2): 401-7. [http://dx.doi.org/10.1007/s00253-010-2733-x] [PMID: 20680266]

[22] Bush K. Antibacterial drug discovery in the 21st century. Clin Microbiol Infect 2004; 10(4) (Suppl. 4): 10-7. [http://dx.doi.org/10.1111/j.1465-0691.2004.1005.x] [PMID: 15522035]

[23] Abreu AC, McBain AJ, Simões M. Plants as sources of new antimicrobials and resistance-modifying agents. Nat Prod Rep 2012; 29(9): 1007-21. [http://dx.doi.org/10.1039/c2np20035j] [PMID: 22786554]

[24] Venkateswaran PS, Millman I, Blumberg BS. Effects of an extract from Phyllanthus niruri on hepatitis B and woodchuck hepatitis viruses: in vitro and in vivo studies. Proc Natl Acad Sci USA 1987; 84(1): 274-8. [http://dx.doi.org/10.1073/pnas.84.1.274] [PMID: 3467354]

[25] Gupta R, Thakur B, Singh P, *et al.* Anti-tuberculosis activity of selected medicinal plants against multi-drug resistant Mycobacterium tuberculosis isolates. Indian J Med Res 2010; 131(6): 809-13. [PMID: 20571171]

[26] Habbal O, Hasson SS, El-Hag AH, *et al.* Antibacterial activity of Lawsonia inermis Linn (Henna) against Pseudomonas aeruginosa. Asian Pac J Trop Biomed 2011; 1(3): 173-6. [http://dx.doi.org/10.1016/S2221-1691(11)60021-X] [PMID: 23569753]

[27] Dadgar T, Asmar M, Saifi A, *et al.* Antibacterial activity of certain Iranian medicinal plants against methicillin-resistant and sensitive *Staphylococcus aureus.* Asian J Plant Sci 2006; 5(5): 861-6. [http://dx.doi.org/10.3923/ajps.2006.861.866]

[28] Peng L, Kang S, Yin Z, *et al.* Antibacterial activity and mechanism of berberine against *Streptococcus agalactiae* . Int J Clin Exp Pathol 2015; 8(5): 5217-23. [PMID: 26191220]

[29] Semwal D, Rawat U. Antimicrobial hasubanalactam alkaloid from *Stephania glabra.* Planta Med 2009; 75(4): 378-80. [http://dx.doi.org/10.1055/s-0028-1112223] [PMID: 19148860]

[30] Fragopoulou E, Nomikos T, Karantonis HC, *et al.* Biological activity of acetylated phenolic compounds. J Agric Food Chem 2007; 55(1): 80-9. [http://dx.doi.org/10.1021/jf0627221] [PMID: 17199317]

[31] Tahara S, Ingham JL, Nakahara S, Mizutani J, Harborne JB. Fungitoxic dihydrofuranoisoflavones and related compounds in white lupin, *Lupinus albus.* Phytochemistry 1984; 23(9): 1889-900. [http://dx.doi.org/10.1016/S0031-9422(00)84936-5]

[32] Özçelik B, Orhan DD, Özgen S, Ergun F. Antimicrobial activity of flavonoids against extended-spectrum β-lactamase (ESβL)-producing Klebsiella pneumoniae. Trop J Pharm Res 2008; 7(4): 1151-7. [http://dx.doi.org/10.4314/tjpr.v7i4.14701]

[33] Widelski J, Popova M, Graikou K, Glowniak K, Chinou I. Coumarins from Angelica lucida L.--antibacterial activities. Molecules 2009; 14(8): 2729-34. [http://dx.doi.org/10.3390/molecules14082729] [PMID: 19701119]

[34] Ragasa CY, Ha HK, Hasika M, Maridable JB, Gaspillo PA, Rideout JA. Antimicrobial and cytotoxic terpenoids from Cymbopogon citratus Stapf. Philipp Sci 2008; (45): 111-22.

[35] González-Mas MC, Rambla JL, López-Gresa MP, Blázquez MA, Granell A. Volatile compounds in Citrus essential oils: a comprehensive review. Front Plant Sci 2019; 10: 12. [http://dx.doi.org/10.3389/fpls.2019.00012] [PMID: 30804951]

[36] Fernández L, Hancock REW. Adaptive and mutational resistance: role of porins and efflux pumps in drug resistance. Clin Microbiol Rev 2012; 25(4): 661-81. [http://dx.doi.org/10.1128/CMR.00043-12] [PMID: 23034325]

[37] de la Fuente-Núñez C, Reffuveille F, Fernández L, Hancock REW. Bacterial biofilm development as a multicellular adaptation: antibiotic resistance and new therapeutic strategies. Curr Opin Microbiol 2013; 16(5): 580-9. [http://dx.doi.org/10.1016/j.mib.2013.06.013] [PMID: 23880136]

[38] Papagianni M. Ribosomally synthesized peptides with antimicrobial properties: biosynthesis, structure, function, and applications. Biotechnol Adv 2003; 21(6): 465-99. [http://dx.doi.org/10.1016/S0734-9750(03)00077-6] [PMID: 14499150]

[39] Afacan NJ, Yeung AT, Pena OM, Hancock RE. Therapeutic potential of host defense peptides in antibiotic-resistant infections. Curr Pharm Des 2012; 18(6): 807-19. [http://dx.doi.org/10.2174/138161212799277617] [PMID: 22236127]

[40] Wiesner J, Vilcinskas A. Antimicrobial peptides: The ancient arm of the human immune system. Virulence 2010; 1(5): 440-64. [http://dx.doi.org/10.4161/viru.1.5.12983] [PMID: 21178486]

[41] Cole AM, Shi J, Ceccarelli A, Kim YH, Park A, Ganz T. Inhibition of neutrophil elastase prevents cathelicidin activation and impairs clearance of bacteria from wounds. Blood 2001; 97(1): 297-304. [http://dx.doi.org/10.1182/blood.V97.1.297] [PMID: 11133774]

[42] Xu D, Liao C, Zhang B, *et al.* Human enteric α-defensin 5 promotes Shigella infection by enhancing

bacterial adhesion and invasion. Immunity 2018; 48(6): 1233-1244.e6. [http://dx.doi.org/10.1016/j.immuni.2018.04.014] [PMID: 29858013]

[43] Islam D, Bandholtz L, Nilsson J, *et al.* Downregulation of bactericidal peptides in enteric infections: A novel immune escape mechanism with bacterial DNA as a potential regulator. Nat Med 2001; 7(2): 180-5. [http://dx.doi.org/10.1038/84627] [PMID: 11175848]

[44] Ohsaki Y, Gazdar AF, Chen HC, Johnson BE. Antitumor activity of magainin analogues against human lung cancer cell lines. Cancer Res 1992; 52(13): 3534-8. [PMID: 1319823]

[45] Lehmann J, Retz M, Sidhu SS, *et al.* Antitumor activity of the antimicrobial peptide magainin II against bladder cancer cell lines. Eur Urol 2006; 50(1): 141-7. [http://dx.doi.org/10.1016/j.eururo.2005.12.043] [PMID: 16476519]

[46] Thell K, Hellinger R, Schabbauer G, Gruber CW. Immunosuppressive peptides and their therapeutic applications. Drug Discov Today 2014; 19(5): 645-53. [http://dx.doi.org/10.1016/j.drudis.2013.12.002] [PMID: 24333193]

[47] Bernardi F, Gaggelli E, Molteni E, Porciatti E, Valensin D, Valensin G. 1H and 13C-NMR and molecular dynamics studies of cyclosporin a interacting with magnesium(II) or cerium(III) in acetonitrile. Conformational changes and cis-trans conversion of peptide bonds. Biophys J 2006; 90(4): 1350-61. [http://dx.doi.org/10.1529/biophysj.105.074245] [PMID: 16299069]

[48] Chen M, Kumar S, Anselmo AC, *et al.* Topical delivery of Cyclosporine A into the skin using SPACE-peptide. J Control Release 2015; 199(199): 190-7. [http://dx.doi.org/10.1016/j.jconrel.2014.11.015] [PMID: 25481447]

[49] Harrison CA, Bastan R, Peirce MJ, Munday MR, Peachell PT. Role of calcineurin in the regulation of human lung mast cell and basophil function by cyclosporine and FK506. Br J Pharmacol 2007; 150(4): 509-18. [http://dx.doi.org/10.1038/sj.bjp.0707002] [PMID: 17200674]

[50] Kim ST, Jang DJ, Kim JH, *et al.* Topical administration of cyclosporin A in a solid lipid nanoparticle formulation. Pharmazie 2009; 64(8): 510-4. [PMID: 19746839]

[51] Keenan B, Finol-Urdaneta RK, Hope A, *et al.* *N*-alkylisatin-based microtubule destabilizers bind to the colchicine site on tubulin and retain efficacy in drug resistant acute lymphoblastic leukemia cell lines with less in vitro neurotoxicity. Cancer Cell Int 2020; 20(1): 170. [http://dx.doi.org/10.1186/s12935-020-01251-6] [PMID: 32467666]

[52] Chen J, Wang F, Liu Q, Du J. Antibacterial polymeric nanostructures for biomedical applications. Chem Commun (Camb) 2014; 50(93): 14482-93. [http://dx.doi.org/10.1039/C4CC03001J] [PMID: 25110921]

[53] Dong Y, Feng SS. Methoxy poly(ethylene glycol)-poly(lactide) (MPEG-PLA) nanoparticles for controlled delivery of anticancer drugs. Biomaterials 2004; 25(14): 2843-9. [http://dx.doi.org/10.1016/j.biomaterials.2003.09.055] [PMID: 14962562]

[54] Agarwal P, Rupenthal ID. Modern approaches to the ocular delivery of cyclosporine A. Drug Discov Today 2016; 21(6): 977-88. [http://dx.doi.org/10.1016/j.drudis.2016.04.002] [PMID: 27080149]

[55] Barratt G. Colloidal drug carriers: achievements and perspectives. Cell Mol Life Sci 2003; 60(1): 21-37. [http://dx.doi.org/10.1007/s000180300002] [PMID: 12613656]

[56] Saravolatz LD, Pawlak J, Johnson L, *et al.* *In vitro* activities of LTX-109, a synthetic antimicrobial peptide, against methicillin-resistant, vancomycin-intermediate, vancomycin-resistant, daptomycin-

nonsusceptible, and linezolid-nonsusceptible *Staphylococcus aureus*. Antimicrob Agents Chemother 2012; 56(8): 4478-82.
[http://dx.doi.org/10.1128/AAC.00194-12] [PMID: 22585222]

[57] Roberts KD, Azad MAK, Wang J, *et al.* Antimicrobial activity and toxicity of the major lipopeptide components of polymyxin B and colistin: last-line antibiotics against multidrug-resistant Gram-negative bacteria. ACS Infect Dis 2015; 1(11): 568-75.
[http://dx.doi.org/10.1021/acsinfecdis.5b00085] [PMID: 27525307]

[58] Lee C, Choi JS, Kim I, *et al.* Long-acting inhalable chitosan-coated poly(lactic-co-glycolic acid) nanoparticles containing hydrophobically modified exendin-4 for treating type 2 diabetes. Int J Nanomedicine 2013; 8(8): 2975-83.
[PMID: 23976850]

[59] Bergmann-Leitner E, Leitner W. Adjuvants in the driver's seat: how magnitude, type, fine specificity and longevity of immune responses are driven by distinct classes of immune potentiators. Vaccines (Basel) 2014; 2(2): 252-96.
[http://dx.doi.org/10.3390/vaccines2020252] [PMID: 26344620]

[60] Zeng W, Jin L, Zhang F, Zhang C, Liang W. Naringenin as a potential immunomodulator in therapeutics. Pharmacol Res 2018; 135(135): 122-6.
[http://dx.doi.org/10.1016/j.phrs.2018.08.002] [PMID: 30081177]

[61] Pirofski L, Casadevall A. Immunomodulators as an antimicrobial tool. Curr Opin Microbiol 2006; 9(5): 489-95.
[http://dx.doi.org/10.1016/j.mib.2006.08.004] [PMID: 16931122]

[62] Cibulski SP, Mourglia-Ettlin G, Teixeira TF, *et al.* Novel ISCOMs from Quillaja brasiliensis saponins induce mucosal and systemic antibody production, T-cell responses and improved antigen uptake. Vaccine 2016; 34(9): 1162-71.
[http://dx.doi.org/10.1016/j.vaccine.2016.01.029] [PMID: 26826546]

[63] Daubenberger CA. TLR9 agonists as adjuvants for prophylactic and therapeutic vaccines. Curr Opin Mol Ther 2007; 9(1): 45-52.
[PMID: 17330401]

[64] Ryan EJ, Daly LM, Mills KHG. Immunomodulators and delivery systems for vaccination by mucosal routes. Trends Biotechnol 2001; 19(8): 293-304.
[http://dx.doi.org/10.1016/S0167-7799(01)01670-5] [PMID: 11451471]

[65] Agren LC, Ekman L, Löwenadler B, Lycke NY. Genetically engineered nontoxic vaccine adjuvant that combines B cell targeting with immunomodulation by cholera toxin A1 subunit. J Immunol 1997; 158(8): 3936-46.
[PMID: 9103464]

[66] Tarr PE. Granulocyte-macrophage colony-stimulating factor and the immune system. Med Oncol 1996; 13(3): 133-40.
[http://dx.doi.org/10.1007/BF02990841] [PMID: 9106171]

[67] Kanzler H, Barrat FJ, Hessel EM, Coffman RL. Therapeutic targeting of innate immunity with Toll-like receptor agonists and antagonists. Nat Med 2007; 13(5): 552-9.
[http://dx.doi.org/10.1038/nm1589] [PMID: 17479101]

[68] Halbus AF, Horozov TS, Paunov VN. Colloid particle formulations for antimicrobial applications. Adv Colloid Interface Sci 2017; 249(249): 134-48.
[http://dx.doi.org/10.1016/j.cis.2017.05.012] [PMID: 28528626]

[69] Shende S, Ingle AP, Gade A, Rai M. Green synthesis of copper nanoparticles by *Citrus medica* Linn. (Idilimbu) juice and its antimicrobial activity. World J Microbiol Biotechnol 2015; 31(6): 865-73.
[http://dx.doi.org/10.1007/s11274-015-1840-3] [PMID: 25761857]

[70] Usman MS, El Zowalaty ME, Shameli K, Zainuddin N, Salama M, Ibrahim NA. Synthesis,

characterization, and antimicrobial properties of copper nanoparticles. Int J Nanomedicine 2013; 8(8): 4467-79.
[PMID: 24293998]

[71] Durán N, Durán M, de Jesus MB, Seabra AB, Fávaro WJ, Nakazato G. Silver nanoparticles: A new view on mechanistic aspects on antimicrobial activity. Nanomedicine 2016; 12(3): 789-99.
[http://dx.doi.org/10.1016/j.nano.2015.11.016] [PMID: 26724539]

[72] Tiwari P, Vig K, Dennis V, Singh S. Functionalized gold nanoparticles and their biomedical applications. Nanomaterials (Basel) 2011; 1(1): 31-63.
[http://dx.doi.org/10.3390/nano1010031] [PMID: 28348279]

[73] Liu Y, Li R, Xiao X, Wang Z. Molecules that inhibit bacterial resistance enzymes. Molecules 2018; 24(1): 43.
[http://dx.doi.org/10.3390/molecules24010043] [PMID: 30583527]

[74] Wilke MS, Lovering AL, Strynadka NCJ. β-Lactam antibiotic resistance: a current structural perspective. Curr Opin Microbiol 2005; 8(5): 525-33.
[http://dx.doi.org/10.1016/j.mib.2005.08.016] [PMID: 16129657]

[75] Bush K, Fisher JF. Epidemiological expansion, structural studies, and clinical challenges of new β-lactamases from gram-negative bacteria. Annu Rev Microbiol 2011; 65(1): 455-78.
[http://dx.doi.org/10.1146/annurev-micro-090110-102911] [PMID: 21740228]

[76] Kataja J, Seppälä H, Skurnik M, Sarkkinen H, Huovinen P. Different erythromycin resistance mechanisms in group C and group G streptococci. Antimicrob Agents Chemother 1998; 42(6): 1493-4.
[http://dx.doi.org/10.1128/AAC.42.6.1493] [PMID: 9624500]

[77] Okuma K, Iwakawa K, Turnidge JD, *et al.* Dissemination of new methicillin-resistant *Staphylococcus aureus* clones in the community. J Clin Microbiol 2002; 40(11): 4289-94.
[http://dx.doi.org/10.1128/JCM.40.11.4289-4294.2002] [PMID: 12409412]

[78] Hiramatsu K. Vancomycin-resistant *Staphylococcus aureus* : a new model of antibiotic resistance. Lancet Infect Dis 2001; 1(3): 147-55.
[http://dx.doi.org/10.1016/S1473-3099(01)00091-3] [PMID: 11871491]

[79] Happi CT, Gbotosho GO, Folarin OA, *et al.* Polymorphisms in Plasmodium falciparum dhfr and dhps genes and age related in vivo sulfadoxine–pyrimethamine resistance in malaria-infected patients from Nigeria. Acta Trop 2005; 95(3): 183-93.
[http://dx.doi.org/10.1016/j.actatropica.2005.06.015] [PMID: 16023986]

[80] Wang G, Taylor DE. Site-specific mutations in the 23S rRNA gene of Helicobacter pylori confer two types of resistance to macrolide-lincosamide-streptogramin B antibiotics. Antimicrob Agents Chemother 1998; 42(8): 1952-8.
[http://dx.doi.org/10.1128/AAC.42.8.1952] [PMID: 9687389]

[81] Suzuki Y, Katsukawa C, Tamaru A, *et al.* Detection of kanamycin-resistant Mycobacterium tuberculosis by identifying mutations in the 16S rRNA gene. J Clin Microbiol 1998; 36(5): 1220-5.
[http://dx.doi.org/10.1128/JCM.36.5.1220-1225.1998] [PMID: 9574680]

[82] Kumar A, Schweizer H. Bacterial resistance to antibiotics: Active efflux and reduced uptake. Adv Drug Deliv Rev 2005; 57(10): 1486-513.
[http://dx.doi.org/10.1016/j.addr.2005.04.004] [PMID: 15939505]

[83] Nikaido H. Molecular basis of bacterial outer membrane permeability revisited. Microbiol Mol Biol Rev 2003; 67(4): 593-656.
[http://dx.doi.org/10.1128/MMBR.67.4.593-656.2003] [PMID: 14665678]

[84] Wardal E, Sadowy E, Hryniewicz W. Complex nature of enterococcal pheromone-responsive plasmids. Pol J Microbiol 2010; 59(2): 79-87.
[http://dx.doi.org/10.33073/pjm-2010-012] [PMID: 20734751]

[85] Martínez-Martínez L, Pascual A, Jacoby GA. Quinolone resistance from a transferable plasmid.

Lancet 1998; 351(9105): 797-9.
[http://dx.doi.org/10.1016/S0140-6736(97)07322-4] [PMID: 9519952]

[86] Singh R, Smitha MS, Singh SP. The role of nanotechnology in combating multi-drug resistant bacteria. J Nanosci Nanotechnol 2014; 14(7): 4745-56.
[http://dx.doi.org/10.1166/jnn.2014.9527] [PMID: 24757944]

[87] Smillie C, Garcillán-Barcia MP, Francia MV, Rocha EPC, de la Cruz F. Mobility of Plasmids. Microbiol Mol Biol Rev 2010; 74(3): 434-52.
[http://dx.doi.org/10.1128/MMBR.00020-10] [PMID: 20805406]

[88] Robicsek A, Jacoby GA, Hooper DC. The worldwide emergence of plasmid-mediated quinolone resistance. Lancet Infect Dis 2006; 6(10): 629-40.
[http://dx.doi.org/10.1016/S1473-3099(06)70599-0] [PMID: 17008172]

[89] Sharma M, Chauhan PMS. Dihydrofolate reductase as a therapeutic target for infectious diseases: opportunities and challenges. Future Med Chem 2012; 4(10): 1335-65.
[http://dx.doi.org/10.4155/fmc.12.68] [PMID: 22800373]

[90] Su H, Han L, Huang X. Potential targets for the development of new antifungal drugs. J Antibiot (Tokyo) 2018; 71(12): 978-91.
[http://dx.doi.org/10.1038/s41429-018-0100-9] [PMID: 30242283]

[91] Khaw M, Panosian CB. Human antiprotozoal therapy: past, present, and future. Clin Microbiol Rev 1995; 8(3): 427-39.
[http://dx.doi.org/10.1128/CMR.8.3.427] [PMID: 7553575]

[92] Mitscher LA, Pillai SP, Gentry EJ, Shankel DM. Multiple drug resistance. Med Res Rev 1999; 19(6): 477-96.
[http://dx.doi.org/10.1002/(SICI)1098-1128(199911)19:6<477::AID-MED2>3.0.CO;2-W] [PMID: 10557366]

[93] Cowen LE, Sanglard D, Howard SJ, Rogers PD, Perlin DS. Mechanisms of anti-fungal drug resistance. Cold Spring Harb Perspect Med 2015; 5(7): a019752.
[http://dx.doi.org/10.1101/cshperspect.a019752] [PMID: 25384768]

[94] Wernsdorfer WH. Epidemiology of drug resistance in malaria. Acta Trop 1994; 56(2-3): 143-56.
[http://dx.doi.org/10.1016/0001-706X(94)90060-4] [PMID: 8203301]

[95] Ouellette M, Drummelsmith J, Papadopoulou B. Leishmaniasis: drugs in the clinic, resistance and new developments. Drug Resist Updat 2004; 7(4-5): 257-66.
[http://dx.doi.org/10.1016/j.drup.2004.07.002] [PMID: 15533763]

[96] Septimus EJ, Kuper KM. Clinical challenges in addressing resistance to antimicrobial drugs in the twenty-first century. Clin Pharmacol Ther 2009; 86(3): 336-9.
[http://dx.doi.org/10.1038/clpt.2009.122] [PMID: 19571803]

[97] Faria NR, Rambaut A, Suchard MA, *et al.* The early spread and epidemic ignition of HIV-1 in human populations. Science 2014; 346(6205): 56-61.
[http://dx.doi.org/10.1126/science.1256739] [PMID: 25278604]

[98] Bassetti M, Merelli M, Temperoni C, Astilean A. New antibiotics for bad bugs: where are we? Ann Clin Microbiol Antimicrob 2013; 12(1): 22.
[http://dx.doi.org/10.1186/1476-0711-12-22] [PMID: 23984642]

[99] Ribeiro da Cunha B, Fonseca LP, Calado CRC. Antibiotic discovery: where have we come from, where do we go? Antibiotics (Basel) 2019; 8(2): 45.
[http://dx.doi.org/10.3390/antibiotics8020045] [PMID: 31022923]

[100] Caron WP, Mousa SA. Prevention strategies for antimicrobial resistance: a systematic review of the literature. Infect Drug Resist 2010; 3(3): 25-33.
[PMID: 21694891]

[101] Struelens MJ. Multidisciplinary antimicrobial management teams: the way forward to control antimicrobial resistance in hospitals. Curr Opin Infect Dis 2003; 16(4): 305-7. [http://dx.doi.org/10.1097/00001432-200308000-00001] [PMID: 12861082]

[102] Payne DJ, Gwynn MN, Holmes DJ, Pompliano DL. Drugs for bad bugs: confronting the challenges of antibacterial discovery. Nat Rev Drug Discov 2007; 6(1): 29-40. [http://dx.doi.org/10.1038/nrd2201] [PMID: 17159923]

[103] Krivorotova T, Cirkovas A, Maciulyte S, *et al.* Nisin-loaded pectin nanoparticles for food preservation 2016. [http://dx.doi.org/10.1016/j.foodhyd.2015.09.015]

[104] Scott RW, DeGrado WF, Tew GN. De novo designed synthetic mimics of antimicrobial peptides. Curr Opin Biotechnol 2008; 19(6): 620-7. [http://dx.doi.org/10.1016/j.copbio.2008.10.013] [PMID: 18996193]

[105] Giuliani A, Rinaldi AC. Beyond natural antimicrobial peptides: multimeric peptides and other peptidomimetic approaches. Cell Mol Life Sci 2011; 68(13): 2255-66. [http://dx.doi.org/10.1007/s00018-011-0717-3] [PMID: 21598022]

[106] Choi S, Isaacs A, Clements D, *et al.* De novo design and in vivo activity of conformationally restrained antimicrobial arylamide foldamers. Proc Natl Acad Sci USA 2009; 106(17): 6968-73. [http://dx.doi.org/10.1073/pnas.0811818106] [PMID: 19359494]

[107] Sakr A, Brégeon F, Rolain JM, Blin O. Staphylococcus aureus nasal decolonization strategies: a review. Expert Rev Anti Infect Ther 2019; 17(5): 327-40. [http://dx.doi.org/10.1080/14787210.2019.1604220] [PMID: 31012332]

[108] Fernebro J. Fighting bacterial infections—Future treatment options. Drug Resist Updat 2011; 14(2): 125-39. [http://dx.doi.org/10.1016/j.drup.2011.02.001] [PMID: 21367651]

[109] Zhang J, Yang Y, Teng D, Tian Z, Wang S, Wang J. Expression of plectasin in Pichia pastoris and its characterization as a new antimicrobial peptide against *Staphyloccocus* and *Streptococcus.* Protein Expr Purif 2011; 78(2): 189-96. [http://dx.doi.org/10.1016/j.pep.2011.04.014] [PMID: 21558006]

[110] Lang LH. FDA approves use of bacteriophages to be added to meat and poultry products. Gastroenterology 2006; 131(5): 1370. [http://dx.doi.org/10.1053/j.gastro.2006.10.012] [PMID: 17067600]

[111] Wright A, Hawkins CH, Änggård EE, Harper DR. A controlled clinical trial of a therapeutic bacteriophage preparation in chronic otitis due to antibiotic-resistant *Pseudomonas aeruginosa*; a preliminary report of efficacy. Clin Otolaryngol 2009; 34(4): 349-57. [http://dx.doi.org/10.1111/j.1749-4486.2009.01973.x] [PMID: 19673983]

[112] Kutter E, De Vos D, Gvasalia G, *et al.* Phage therapy in clinical practice: treatment of human infections. Curr Pharm Biotechnol 2010; 11(1): 69-86. [http://dx.doi.org/10.2174/138920110790725401] [PMID: 20214609]

[113] Kortright KE, Chan BK, Koff JL, Turner PE. Phage therapy: a renewed approach to combat antibiotic-resistant bacteria. Cell Host Microbe 2019; 25(2): 219-32. [http://dx.doi.org/10.1016/j.chom.2019.01.014] [PMID: 30763536]

[114] O'Flaherty S, Ross RP, Coffey A. Bacteriophage and their lysins for elimination of infectious bacteria. FEMS Microbiol Rev 2009; 33(4): 801-19. [http://dx.doi.org/10.1111/j.1574-6976.2009.00176.x] [PMID: 19416364]

[115] Meinke A, Henics T, Hanner M, Minh D, Nagy E. Antigenome technology: a novel approach for the selection of bacterial vaccine candidate antigens. Vaccine 2005; 23(17-18): 2035-41. [http://dx.doi.org/10.1016/j.vaccine.2005.01.005] [PMID: 15755567]

[116] Heinson A, Gunawardana Y, Moesker B, *et al.* Enhancing the biological relevance of machine learning classifiers for reverse vaccinology. Int J Mol Sci 2017; 18(2): 312. [http://dx.doi.org/10.3390/ijms18020312] [PMID: 28157153]

[117] Merakou C, Schaefers MM, Priebe GP. Progress toward the elusive Pseudomonas aeruginosa vaccine. Surg Infect (Larchmt) 2018; 19(8): 757-68. [http://dx.doi.org/10.1089/sur.2018.233] [PMID: 30388058]

[118] Westritschnig K, Hochreiter R, Wallner G, Firbas C, Schwameis M, Jilma B. A randomized, placebo-controlled phase I study assessing the safety and immunogenicity of a *Pseudomonas aeruginosa* hybrid outer membrane protein OprF/I vaccine (IC43) in healthy volunteers. Hum Vaccin Immunother 2014; 10(1): 170-83. [http://dx.doi.org/10.4161/hv.26565] [PMID: 24064511]

[119] Payne DJ, Cramp R, Winstanley DJ, Knowles DJ. Comparative activities of clavulanic acid, sulbactam, and tazobactam against clinically important beta-lactamases. Antimicrob Agents Chemother 1994; 38(4): 767-72. [http://dx.doi.org/10.1128/AAC.38.4.767] [PMID: 8031044]

[120] Bax HI, de Steenwinkel JEM, ten Kate MT, van der Meijden A, Verbon A, Bakker-Woudenberg IAJM. Colistin as a potentiator of anti-TB drug activity against *Mycobacterium tuberculosis.* J Antimicrob Chemother 2015; 70(10): 2828-37. [http://dx.doi.org/10.1093/jac/dkv194] [PMID: 26183185]

[121] Livermore DM, Mushtaq S, Warner M, *et al.* Activities of NXL104 combinations with ceftazidime and aztreonam against carbapenemase-Producing Enterobacteriaceae. Antimicrob Agents Chemother 2011; 55(1): 390-4. [http://dx.doi.org/10.1128/AAC.00756-10] [PMID: 21041502]

[122] Gin A, Dilay L, Karlowsky JA, Walkty A, Rubinstein E, Zhanel GG. Piperacillin–tazobactam: a β-lactam/β-lactamase inhibitor combination. Expert Rev Anti Infect Ther 2007; 5(3): 365-83. [http://dx.doi.org/10.1586/14787210.5.3.365] [PMID: 17547502]

[123] Lomovskaya O, Bostian KA. Practical applications and feasibility of efflux pump inhibitors in the clinic—A vision for applied use. Biochem Pharmacol 2006; 71(7): 910-8. [http://dx.doi.org/10.1016/j.bcp.2005.12.008] [PMID: 16427026]

[124] Dayyoub E, Frant M, Pinnapireddy SR, Liefeith K, Bakowsky U. Antibacterial and anti-encrustation biodegradable polymer coating for urinary catheter. Int J Pharm 2017; 531(1): 205-14. [http://dx.doi.org/10.1016/j.ijpharm.2017.08.072] [PMID: 28830785]

[125] Dai T, Wang C, Wang Y, Xu W, Hu J, Cheng Y. A nanocomposite hydrogel with potent and broad-spectrum antibacterial activity. ACS Appl Mater Interfaces 2018; 10(17): 15163-73. [http://dx.doi.org/10.1021/acsami.8b02527] [PMID: 29648438]

[126] Pawar V, Bulbake U, Khan W, Srivastava R. Chitosan sponges as a sustained release carrier system for the prophylaxis of orthopedic implant-associated infections. Int J Biol Macromol 2019; 134: 100-12. [http://dx.doi.org/10.1016/j.ijbiomac.2019.04.190] [PMID: 31055114]

[127] Zou P, Chen WT, Sun T, Gao Y, Li LL, Wang H. Recent advances: peptides and self-assembled peptide-nanosystems for antimicrobial therapy and diagnosis. Biomater Sci 2020; 8(18): 4975-96. [http://dx.doi.org/10.1039/D0BM00789G] [PMID: 32931527]

[128] Kingwell K. Vaccines take a shot at antimicrobial resistance. 2018; 229. [http://dx.doi.org/10.1038/nrd.2018.8]

CHAPTER 12

Immunoinformatics - Role in the Identification of New Classes of Antimicrobials

Sunil Krishnan G.[1], **Amit Joshi**[1] and **Vikas Kaushik**[1,*]

[1] *Department of Bioinformatics, School of Bio-Engineering and Bio-Sciences, Lovely Professional University, Punjab, India*

Abstract: Antimicrobials help to restrain or fix the arising irresistible infection in a superior manner anyway, and living creatures require another class of antimicrobials. The new classes of antimicrobial development for the emerging and reemerging pathogenic microbes, the evolution of multidrug-resistant microbes, and the threat of bioterrorism or bioweapons are a global necessity. Integrative genomics, proteomics, and immunoinformatics are powerful tool approaches to design and develop antimicrobials promptly and economically. Natural and artificial antimicrobials for humans, animals, and avians are designed and developed using various immunoinformatics databases, tools, and algorithms. Immunoinformatics plays a great role in dissecting and deciphering genomics, proteomics, and clinical enormous information effectively. The artificial neural network, quantitative matrices and support vector machine algorithms based on immunoinformatics tools would be strong for the planning of adequate customized antimicrobials. The immunoinformatics strategies for antimicrobial improvement are staggeringly utilized for improving living creatures' well-being. The usage of artificial intelligence and machine learning tools is also an asset for immunoinformatics way of antimicrobial design and development. In the new time of pandemic illnesses, progressed immunoinformatics devices play a great role in improving antimicrobials.

Keywords: Anthelmintics, Anti microbial peptides, Anti microbial resistance, Antibacterial, Antifungal, Antimicrobials, Antiprotozoal, Antiviral, Artificial intelligence, Big Data, Bioterrorism and bioweapons, Disease, Drug discovery, Drugs, Genomics, Host, Immunoinformatics, Machine learning, Multidrug resistance, Personalized medicine, Proteomics, Vaccines.

* **Corresponding author Vikas Kaushik:** Department of Bioinformatics, School of Bio-Engineering and Bio-Sciences, Lovely Professional University, Punjab, India; E-mail: vikas.14664@lpu.co.in

M. Amin-ul Mannan & Gaurav Kumar (Eds.)

INTRODUCTION

The agents used to prevent or cure the spreading of infection due to different microorganisms (bacteria, fungus, viruses, and parasites) could be called antimicrobials (antibacterial, antifungal, antiviral, antiprotozoal, and anthelmintics). These include pharmaceutical drugs, vaccines, and therapies that have been prescribed successfully for the treatment of many emerging and reemerging infectious diseases (ERID). However, the evolution of new microbial serotypes/genotypes and antimicrobial resistance (AMR) becomes a new challenge to the medical field for public health management. These challenges allow the innovation and discovery of new antimicrobial agents to overcome resistance, and effectively treat infectious diseases. AMR has been a public health risk [1]. However, in this 21 st century, advanced science and technology could be able to tackle 'big data' bases (disease, host, and pathogen), genomics, proteomics, and immunoinformatics tools [2]. Immunoinformatics is a genomic, proteomics, and bioinformatics approach to immunology, playing a big role in dealing with novel antimicrobial developments. In the battle with infectious diseases, two cost-effective antimicrobial agents such as antibiotics and vaccines, have revolutionized the fight [3]. The impact of these interventions has been phenomenal in reducing mortality and morbidity worldwide. These two greatest accomplishments of modern medicines have saved the most lives globally. Artificial intelligence (AI) and machine learning (ML) algorithms would be helpful for the 'big data' analysis to visualize and interpret the results. These developments may lead to faster and cheaper innovation of novel antimicrobial agents for society. Complex lifecycle and antigenic variability [4] in newly emerged microbes have been giving a further confront in vaccine development [5]. There are reported researches on antigenic variation for immune evasion in *Plasmodium falciparum* [6, 7], Schistosoma haematobium [8], bacterial pathogens [9], Infectious Bursal disease virus (IBDV) [10] and influenza virus [11]. Numerous diseases that affect humans and other species of animal kingdom have been successfully treated with the use of immunoinformatics tools created for vaccines and immunotherapeutics. Several of the vaccines candidates were successfully created for various pathogens and diseases namely schistosomiasis [12], *Helicobacter pylori* [13], Avian Infectious laryngotracheitis virus [14], Retrovirus [15], human papillomavirus [16] neurocysticercosis [17], influenza [18], human immunodeficiency virus (HIV) [19], hepatitis B [20] and fungi [21]. A better understanding of the correlation between immune pathways, host-pathogen protein interaction, antigenic variation in the pathogen, and other genomic variations in the host through immunoinformatics tools have been less time-consuming and allow deep learning for personalized medicine, therapeutic, and vaccines. This immunoinformatics study's success depends on the accuracy of raw data produced from wet laboratory experiments and the superiority of web tool algorithms. The *in vitro* and *in vivo* clinical validation studies were introduced for safety, efficacy and stability validation, followed by immunoinformatics prediction and analysis of antimicrobial agents. In this chapter, we discuss immunoinformatics role in the identification of novel antimicrobials. The progress in the immunoinformatics sector made the complicated vaccine, drug, and therapeutics designing and development easy [22].

IMMUNE SYSTEM AND IMMUNOINFORMATICS

In humans, the innate and adaptive immune system continually interacts with each other to provide a successful immune response. B cell and T cell-mediated adaptive immune systems provide humoral and cell-mediated immunity. Natural killer cells are involved in innate immune responses [23]. Natural and artificial immunization can be referred to as passive and active, respectively. Either passive or active immunization can produce immunity. While activated immunity refers to human intervention, such as vaccination, passive immunity refers to immunity gained by natural causes through a prior encounter with a disease. Vaccines are the best antimicrobial for the prevention of diseases. These vaccines should activate the innate immune system and the adaptive immune system [24]. In this millennium, emerging and reemerging diseases challenge vaccine design and development research in different ways. Immunoinformatics-based vaccine design tools helped to face these new challenges. This new era of vaccine design was made happen through *insilico* high-throughput technologies. Immunoinformatics has developed databases and tools to predict, retrieve, accumulate, restructure, and classify data for antimicrobials and vaccine designing.

ANTIMICROBIALS AND ITS NEW CLASSES

Microorganisms or microbes are too small and undetectable to the human-naked eye. These are found in the environment, human or animal habitats, and even in the body of other living organisms. Most microbes are useful for humans and other living organisms, however, some microbes are pathogenic and treat other living organisms. Synthetic and naturally derived compounds were used to destroy and inhibit the spread or growth of these pathogenic microbes as antimicrobial agents. Disinfectants and antiseptics are the main class of antimicrobials, but these are nonselective agents. The primary action of antimicrobials grouped into antibacterial, antifungal, antiviral, antiprotozoal, and anthelmintics. The details are illustrated in Fig.(**1**).

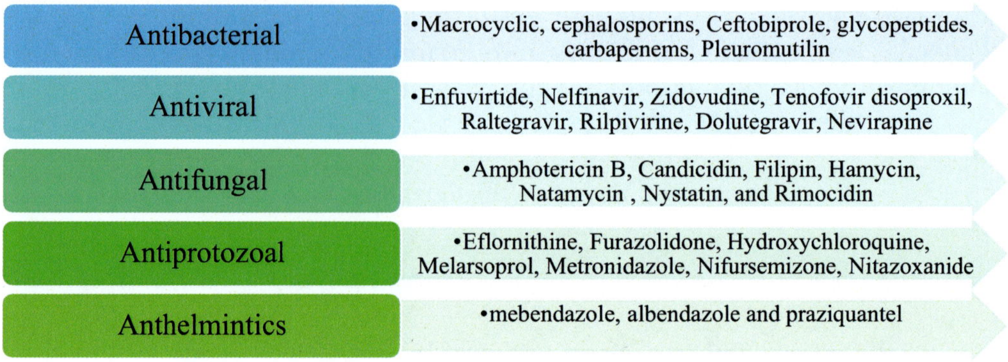

Fig. (1). Classification of antimicrobials and examples based on their activity on the microbes.

Antimicrobials are functionally classified as biocides (destroy microbes) and biostatic agents (inhibit the spread or growth of the microbes). In antimicrobial chemotherapy and prophylaxis, these agents are used to treat or prevent the disease effectively. The ERID, antibiotic resistance, antigenic variation, mutation, and bioterrorism have challenged the identification of new classes of antimicrobials for the global healthcare system. To find a better solution, we need to understand the mechanism of microbes, disease spread, antimicrobial action, function, *etc.* This knowledge could lead to the innovation of novel antimicrobial molecules and therapeutics, as explained in Fig. (**2**).

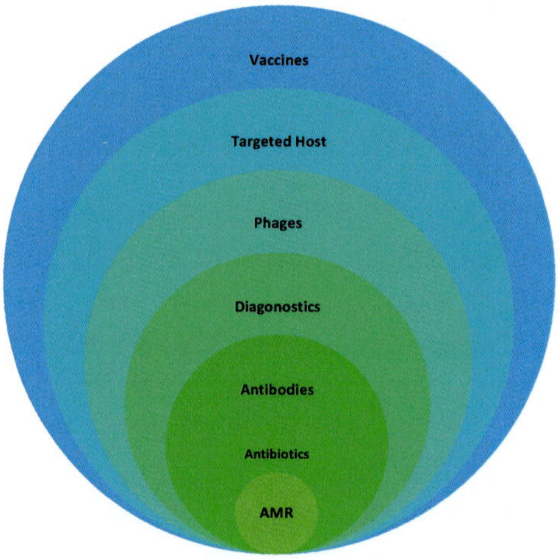

Fig. (2). This figure explained an effective way to fight against AMR with an integrated strategy (vaccines, target host, bacteriophages, diagnostics, antibodies, and antibiotics).

Antiviral

Antiviral drugs are used to treat many viral diseases. These drugs may target specific or multi viruses as an inhibitor for virus multiplication. Many antiviral drugs are USFDA approved for human diseases caused by Human immunodeficiency virus (HIV), Varicella, herpes zoster, Hepatitis B, Influenza, Hepatitis C, Human Parainfluenza, Respiratory Syncytial, and herpes simplex viruses. There are numerous viral disease-specific and broad-spectrum antiviral drugs available for the treatment. The antiviral data were retrieved from the ViPR (https://www.viprbrc.org/brc/antiviralDrug_search) database and listed in Table **1**.

Table 1. The details of USFDA-approved specific and broad-spectrum antiviral drugs.

Viral Disease	Antiviral Drug	Antimicrobial Activity
HIV	Enfuvirtide,Nelfinavir,Zidovudine,Tenofovirdisoproxil,Raltegravir,Rilpivirine,Dolutegravir,Nevirapine, Ritonavir,Efavirenz,Stavudine,Emtricitabine,Elvitegravir,Delavirdine, Didanosine,Tipranavir,Indinavir,Zalcitabine,Abacavir, Atazanavir,Saquinavir,Darunavir,Fosamprenavir,Lopinavir, Maraviroc, and Amprenavir	Specific
Human Hepatitis B	Clevudine, Entecavir, Adefovir, Dipivoxil, and Telbivudine	Specific
Human Herpes	Idoxuridine,Penciclovir,Trifluridine,Famciclovir,Foscarnet,Valaciclovir,Aciclovir, Ganciclovir, and Valganciclovir	Specific
Enterovirus (D and J) Rhinovirus (A and B)	Pleconaril	Broad-Spectrum
human influenza A and B	Oseltamivir(Tamiflu),Rimantadine,Zanamivir,and, Amantadine	Broad-Spectrum
Hepatitis C Virus	Asunaprevir	Specific
HIV and Hepatitis B	Lamivudine	Broad-Spectrum
Hepatitis C, Parainfluenza 2, Influenza A and B, and Respiratory Syncytial	Ribavirin	Broad-Spectrum
Hepatitis C (genotype1) and Respiratory Syncytial	Telaprevir, Simeprevir, Boceprevir	Broad-Spectrum
Hepatitis C and HIV	Ledipasvir	Broad-Spectrum

Antibacterial

Antibacterial drugs are used for the treatment of bacterial elimination, reduction, and multiplication inhibition. The antibacterial drugs could be classified as bactericidal, eliminating the bacteria; the other class is Bacteriostatic, which inhibits bacterial growth. The antibacterial drugs produced from bacteria, plants, or fungi are natural however, artificially synthesized antimicrobials are also available. These drugs act on bacteria by different modes like cell wall synthesis inhibition, disruption in cell membrane permeability, protein, and nucleic acid metabolism. This drug also interferes with other metabolic processes to eliminate the bacteria. The new class of antimicrobial drugs effectively works on replication, spreading the infection machinery of the microbes. Examples of new class antibacterials are macrocyclic antibiotics, newer cephalosporins, ceftobiprole, newer glycopeptides, newer carbapenems, pleuromutilin, glycylcyclines, tigecycline, ketolides, telithromycin, *etc.* The next-generation antimicrobial targets bacterial proteins and virulence factors, modulating the host response pathways. Bacterial segments can be a significant part of bacterial vaccines. Live, attenuated, toxoids, and polysaccharide-bacterial vaccines could be used for the prevention of bacterial infections [13]. Live bacterial expression vectors have been widely used to produce vaccine antigens for hepatitis B and different infectious agents [20].To find a new class of antimicrobial agents for diseases, the role of immunoinformatics is important.

Antifungal

An antifungal or antimycotic is maybe a fungicide that kills the fungus or fungistatic used for the local and systemic treatment or prevention of fungal diseases (athlete's foot, ringworm, candidiasis, cryptococcal meningitis, *etc.*). The antifungal agents classified as Polyenes (Amphotericin B, Candicidin, Filipin, Hamycin, Natamycin, Nystatin, and Rimocidin), Azoles(Bifonazole, Butoconazole, Clotrimazole, Econazole, Fenticonazole, Ravuconazole, Terconazole, Isoconazole, Ketoconazole, and Luliconazole) and Allylamines (amorolfin, butenafine, naftifine, and terbinafine). Fungal RV webserver encourages the prediction of a fungal adhesin-based epitope vaccine [21].

Antiprotozoals

The Antiprotozoals were used to eliminate or inhibit protozoal infections (amebiasis, giardiasis, cryptosporidiosis, microsporidiosis, malaria, babesiosis, trypanosomiasis, Chagas disease, leishmaniasis, and toxoplasmosis). The antiprotozoal molecules which are used for the protozoal disease treatments include Eflornithine, Furazolidone, Hydroxychloroquine, Melarsoprol, Metronidazole, Nifursemizone, Nitazoxanide, Ornidazole, Paromomycin sulfate, Pentamidine, Pyrimethamine, Quinapyramine, Ronidazole, and Tinidazole.

Anthelmintics or Antihelminthics

Helminth diseases are the most widely recognized burdens of mankind, influencing pretty much everyone living in significant destitution. Taenia solium disease tremendously affects human well-being and the economy. Antiparasitic drugs (vermifuges or vermicides) are used to eliminate or inactivate helminths (parasitic worms) from humans or animals without any side effects. The derivatives of Benzimidazole drugs (mebendazole and albendazole) are used for the treatment of threadworms, roundworms, whipworms, tapeworms, and hookworms. The praziquantel drug is effective against schistosomiasis and tapeworm infections. Two inventive pig well-being items (Paranthic®, an antiparasitics from the Morroco and Cysvax® vaccine [17] from India) are currently accessible to handle a significant parasitic (Taenia solium) illness that causes around 30% of human epilepsy or Neurocysticercosis in the impoverished region of the world. Urinary schistosomiasis has been recognized as the primary reason for urogenital infection, and the major rationale behind bladder malignant growth [8]. A few immunoinformatics methods were reported for the urinary schistosomiasis epitope vaccine prediction that might be successful [12].

Plant-sourced Natural and Small Antimicrobials

Natural antimicrobials are derived from natural sources like medicinal and aromatic plants. These antimicrobial compounds could be used as a food preservatives [25]. It has a much wider appreciation in the food industry and consumers. These antimicrobials may reduce the misuse of antibiotics and the use of synthetic preservatives with additional health benefits to consumers. It gives a boost to the research and development of new natural antimicrobials as alternatives in food preservation. These natural compounds have a broad range of antimicrobial effects against many food pathogens [26]. The antimicrobial compounds originated from plant phytochemicals were found effective against bacterial, fungal, and viral infectious diseases [27, 28]. Table **2** lists the antimicrobial derived from natural/plant origin.

Table 2. Antimicrobials derived from natural/plant origin.

Antimicrobial Compound	Plant Name (Scientific Name)	Antimicrobial Action against	References
Berberine	Barberry *Berberis vulgaris,* Goldenseal *Hydrastis Canadensis* Oregon grape *Mahonia aquifolia*	Bacteria, protozoa	[29, 30]
Piperine	Black pepper *Piper nigrum*	Fungi, *Lactobacillus, Micrococcus, E. coli, E. faecalis*	[31]
Carvone	Caraway *(Carum carvi),* spearmint (*Mentha spicata*)	Bacteria, fungi, viruses	[32]
Isothiocyanates (ITCs)	Cauliflower*(Brassica olerace)*	Bacteria, *Helicobacter pylori*	[33, 34]
Tannins	Cascara sagrada *(Rhamnus purshiana)*Eucalyptus *(Eucalyptus globules)* Oak *(Quercus rubra Allium cepa)*	Bacteria, fungi, viruses	[35]
Coumarins	tonka bean Dipteryx odorata, vanilla grass (*Anthoxanthum odoratum*)	Bacteria, fungi, helminths	[36, 37]
Allicin, ajoene	Garlic *Allium sativum* Onion *Allium cepa*	Bacteria, fungi	[38 - 41]
Curcumin	Turmeric *Curcuma longa*	Bacteria, protozoa	[42, 43]

IMPORTANCE OF VACCINE AS AANTIMICROBIAL AGENT

Using vaccines as an antimicrobial agent improved the healthcare system for both animal and human lives. The vaccine candidate identification has to face many challenges like ERID, efficacy, candidate reproducibility, safety, population coverage, mutations, and affordability. These challenges lead the scientific community towards immunoinformatics. Here is an integrative approach of computational technology and immunology to identify the appropriate vaccine candidate for the disease. In this approach, the genomics and proteomics understanding of the host-pathogen relationship leads to the design of superior vaccine candidates. Since 1980 immunoinformatics tools have been used to predict T and B cell immune epitopes as potential vaccine candidates [44]. Analysis of the pathogenic genome and proteome data allows the prediction of protein sequences suitable for epitope prediction. These reverse vaccinology methods allow the screening and selection of suitable epitopes for the vaccine candidate.

IMMUNOINFORMATICS ROLE IN ANTIMICROBIALS

Antimicrobial Peptides

Antimicrobial peptides (AMPs) are common peptides that play a role in generating innate as well as adaptive immune responses; sort of 'nature's anti-infection agents' is very encouraging for tackling antibiotic resistance. The AMPs are classified as non-ribosomal synthesized (N-RMPs) and ribosomally synthesized peptides (RAMPs). Taking into account this, it is profoundly wanted to build up a compelling computational strategy for precisely foreseeing novel AMPs since it can give us more peptide molecules that are valuable for vaccine designing and drug predictions. Antimicrobial peptides are 12 and 50 amino acids, a sole and assorted group of molecules [45].

Antimicrobial Databases and Immunoinformatics Tools

There are many immunoinformatics database tools available to retrieve, predict, and analysis of antimicrobial peptides data. The database can be grouped based on the nature of the data [46]. The antimicrobial peptide databases are categorized as specific databases and general databases. There are various tools and databases available for the better analysis and prediction of various types of antimicrobials. A few databases, like AMSDb, Peptaibol, APD, BACTIBASE CyBase, DBAASP, Peptaibol, and PhytAMP, are dedicated to antimicrobial peptides. The database 'dbAMP' provides a unique dataset for 12389 AMPs and research evidence for its functional activities [47]. The 'Peptaibol' database is used for searching and identifying sequences, 3D structures, biological origin, and reference of Trichoderma and Emericellopsis, the antifungal microbial peptides dataset [48]. There are databases for synthetic antibacterial and antifungal compounds. This tool provides properties, structure, and bioassay of antimicrobial compounds [49]. The 'PhytAMP' is a dedicated database for antimicrobial plant peptides [50]. The antimicrobial consumption database 'ESAC-Net' make available antimicrobial consumption reference data in the European community and the hospital [51]. The antibiotic resistance genes, information about associated products, and phenotypes were identified from the CARD database [52]. The database 'Ocins' was used for the probiotic antimicrobial polypeptides (Ocins) dataset search [53]. Antimicrobial databases were summarized in Table **3**, which allows the user to easily access the information regarding antimicrobial agents for varied microbial populations.

Table 3. Databases for antimicrobial peptides, antiviral, antibacterial chemical, analysis, and prediction.

Name of Database	Description	Web Access Interface Link
AMSDb	Antimicrobial Sequences Database for the searching of annotated AMP sequence data	http://www.bbcm.univ.trieste.it/~tossi/pag1.html
Peptaibol	This database for searching antibiotic peptides (>300) and structural view of *Trichoderma* and *Emericellopsis* fungus	http://public1.cryst.bbk.ac.uk/peptaibol/home.shtml
APD	The Antimicrobial Peptide Database	http://aps.unmc.edu/AP/main.html.
BACTIBASE	A data repository of bacteriocin natural antimicrobial peptides	http://bactibase.hammamilab.org/main.php
CyBase	A database of cyclic peptides and proteins (or cyclotides)	http://research.imb.uq.edu.au/cybase.
DBAASP	Database of Antimicrobial Activity and Structure of Peptides	https://dbaasp.org/
Peptaibol Database	A database of peptaibols, antimicrobial peptides with non-standard amino acids, mainly from fungi (Trichoderma, Emericellopsis)	http://www.cryst.bbk.ac.uk/peptaibol.
PhytAMP	A database dedicated to antimicrobial plant peptides	http://phytamp.pfba-lab-tun.org/main.php
Antimicrobial drug database (AMDD)	A Database for synthetic antibacterial (□2900) and antifungal (□1200) compounds. This tool provides properties, structure, and bioassay. antimicrobial compounds	http://www.amddatabase.info
Antimicrobial chemotherapeutics database (ACD)	Synthetic antimicrobial compounds (~4100) and biological active antimicrobial peptides(~1030)	http://amdr.amu.ac.in/acd
ViPR	USFDA-approved or experimental Antiviral Drug data	https://www.viprbrc.org/brc/antiviralDrug_search.spg?method=ShowCleanSearch&decorator=arena

(Table 3) cont.....

Name of Database	Description	Web Access Interface Link
SING	Antimicrobial combinations for multidrug-resistant strains	http://www.sing-group.org/antimicrobialCombination/
ANTIMIC	A database of antimicrobial sequences with tools for (□1700) searching new AMPs/AMPs sequence data extraction and analysis.	http://research.i2r.astar.edu.sg/Templar/DB/ANTIMIC/
ESAC-Net	The antimicrobial consumption database provides antimicrobial consumption reference data in the European community and the hospital sector.	https://www.ecdc.europa.eu/en/antimicrobial-consumption/surveillance-and-disease-data/database
DRAMP 2.0	A data repository of general (>5,084), patented (>14,739), and clinical (>76) antimicrobial peptides (AMPs).	http://dramp.cpubioinfor.org/
dbAMP	A database for searching functional activities and physicochemical properties of experimentally verified (n= 4271) and putative antimicrobial peptides (n=8118).	http://csb.cse.yzu.edu.tw/dbAMP/
Collection of Anti-Microbial Peptides ($CAMP_{R3}$)	This tool provide information about family-specific (n=45) sequence composition(n=1386) of AMPs	http://www.campsign.bicnirrh.res.in/
The Comprehensive Antibiotic Resistance Database (CARD)	A database of antibiotic resistance genes also provide information about associated products and phenotypes. This database has information about Ontology Terms (n=4550), Reference Sequences(n=3057), SNPs(n=1704), Publications (n=2735), AMR Detection Models (n=3103)	https://card.mcmaster.ca/browse

(Table 3) cont.....

Name of Database	Description	Web Access Interface Link
Antimicrobial Compounds Database	database of all-atom force-field parameters and molecular properties of antibiotics and some beta-lactamase inhibitors	https://www.dsf.unica.it/~gmalloci/abdb/
Ocins database	The probiotic antimicrobial polypeptides (Ocins)	http://ir.cftri.com/id/eprint/14458

IMMUNOINFORMATICS ROLE IN VACCINES DEVELOPMENT

The evolutionary development of drug-resistant pathogenic bacteria has been considered a major threat not only to the pharmaceutical industry but also to the medical fraternity. Nowadays, epitope-based vaccine development is fast and innovative, based on modern mathematical modeling and algorithms. Reverse vaccinology, where genome-to-proteome exploration by ANN (Artificial neural network) approaches are incorporated, screens out bacterial peptides of the desired length. Mostly small nonameric or decameric epitopes were designed in a De-novo fashion by deploying the latest modeling web servers based on HMM (Hidden Markov model) Viterbi algorithm. Various screening servers were used for prediction strategies, including Bepipred for B-Cell epitopes, NetMHC server for T-cell MHCI and MHCII epitopes from bacterial pathogens.

The latest studies on bacterium populations for selecting peptides are based on experimental databases like the virulence factor database, NCBI-Genbank, and IEDB. Database of essential genes (DEG), stores information regarding crucial protein-encoding genomic elements. Latest epitope screening tools include LBTope (B-Cell epitope), ProPred (MHC ClassII), CTLPred (T-cell epitopes), VaxinPAD (Adjuvant prediction), *etc.* In vaccine designing, different epitopes of short length should be combined with better adjuvants for finalization and subjection towards wet lab demonstrations. VacTarBac portal built on Apache HTTP stores all predicted epitopes. IEDB mapping and population coverage assist and guide *insilico* studies for a better selection of epitopes based on immune response generation over a global population [54].

Molecular docking and MD-simulation studies reveal better considerations about interaction in a simulated cellular environment by developing various force-fields at different temperatures, and pressure conditions. For this purpose, modeled epitopes are usually docked with MHC HLA alleles, then binding energies and RMSD values will generate stability criteria for possible interactions. Docking tools like PatchDock, AutoDock Vina, and Dinc server are commonly used to determine the best possible interactions, and MD-simulation tools like Gromacs,

Desmond suite, and MD web are used to determine RMSD as well as RMSF values for docked complexes and to reveal stability under simulated cellular environment. In-silico cloning using the 'Snapgene' tool and 'JCat' server also assisted users to frame the genetic element for cloning the desired gene of interest or peptide-based vaccine. This method also analyzes the codon adaptive index to reveal codon usage in the newly predicted vaccine gene and provide directions for applications of recombinant technology in vaccine designing. Bacterial species like Mycobacterium, Salmonella, Vibrio, Aeromonas, Meningiococcus, and Tropheryma were successfully studied, and epitope-based vaccine works were conducted on them [54]. Such techniques were found to be rapid, and effective in predicting antibacterial agents that can act as putative epitope-based vaccines to elicit immune responses among a variety of patients facing the need for novel treatments for novel pathogens. Computational methods have a significant role in predicting antimicrobial peptides; these assisted the scientific community in fast wet lab analysis and developed rapid therapeutic candidates for novel regimens (Fig. **3**).

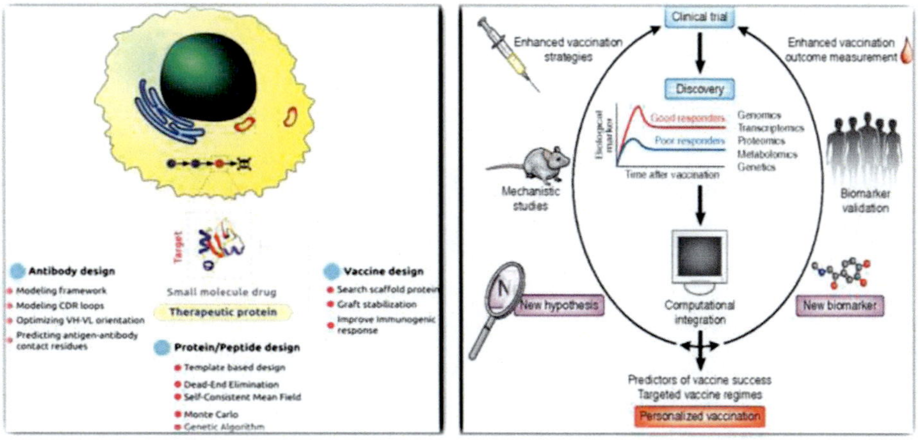

Fig. (3). Significance of the computational method in determining antimicrobial peptides.

Machine Learning in Antimicrobial Peptides

Normally transcribed antimicrobial peptides (AMPs) are a gathering of little, inborn immune molecules, and hold 12–100 amino acids. AMPs have been found in most living things, including bacteriocins, parasitic peptide anti-infection agents, plant thionins and defensins, creepy-crawly arthropods-associated defensins, and cecropins, and amphibian-associated magainins, just as defensins and cathelicidins from higher vertebrates. Attributable to the wide range of antimicrobial action- antibacterial, antifungal, antiviral, also anticancerous, are believed to be more averse to initiating obstruction of these molecules. There were various immunoinformatics approaches have been developed for anticipating new AMPs. The Hidden Markov models (HMMs) were developed to identify and predict AMPs. The BACTIBASE and PhytAMP strategies were unambiguously intended for bacteriocin and plant AMPs. The AntiBP strategy and AntiBP2 technique utilized the Artificial Neural Network (ANN), Quantitative Matrices (QM), and Support Vector Machine (SVM) to predict antibacterial peptides. The Random Forests (RF), SVM, and Discriminant Analysis (DA) are equipped with all classes of AMPs (antibacterial, antifungal, and antiviral). Be that as it may, none of the previously mentioned strategies can distinguish which sorts of highlights are ideal for precisely anticipating and seriously deciphering their organic ramifications [56]. BLASTP was mostly utilized to predict AMPs [57]. Amino acid composition and pseudo-amino acid composition were utilized in many recent studies to code the AMP arrangements [58 - 60].

Antimicrobial Peptide's role in Gastrointestinal Infections

Various AMPs are communicated in the mucosa of the gastrointestinal parcel, where they can adjust intrinsic insusceptible reactions and gut microbiota. Curiously, the antimicrobial elements of AMPs are not the most significant component in the tweak of gastrointestinal maladies. Rather, AMPs may advance certain defensive microbial species and balance intrinsic resistant reactions. The interaction of AMPs with innate immunity and gut microbiota uncovers fascinating medication targets. AMPs may fill in as novel helpful methodologies against gastrointestinal contamination and irritation, yet the use of AMPs and their subsidiaries in treating gastrointestinal infections stays at a beginning phase. AMPs may even legitimately regulate corpulence and metabolic infections. Moreover, AMPs may fill in as biomarkers of gastrointestinal maladies, as the articulation of AMPs is regularly changed during the advancement of gastrointestinal ailments. Protein mimetic or epitope mimetic molecules are 3D epitopes designed and hold therapeutic potentials like drug molecules. Such mimetic molecules show positive ADMET analysis similar to drugs; their principle of interaction relies on protein-protein or DNA-protein association [61, 62].

FUTURE PROSPECTUS

The emerging and reemerging infectious diseases, drug-resistant pathogens, bioterrorism, and bioweapons would pose global health threats and affect the world economy [63]. The advances in immunoinformatics help vaccines and drug discovery in a fast and affordable way when such emergencies. The future development of antimicrobials would be influenced by mainly five situations [64, 65]. The first one would be the multidrug-resistant microbe's emergence and the development of drugs or vaccines against it. The second is the pandemic spread of disease, affecting the global economy and efforts to develop solutions for the global crisis. The third one is the management of zoonotic and complex diseases by effective antimicrobials. The fourth one is the solutions for the epidemics of poor and underdeveloped countries and, finally and not least, the threat of bioterrorism. The era of immunoinformatics and the advancement of science and technology lead to the development of an efficacious, affordable, and safe solution. Our recent research studies on various viruses like SARS-CoV2 [66, 67], Dengue [68, 69], SFTS, and Nipah virus suggested a great interrelationship between both immuno-informatics and chemi-informatics in designing and developing better antimicrobials. Immuno-informatics played various significant roles in designing antimicrobial peptides. In this new reverse-vaccinology era, these antimicrobials have various roles and prospects for developing fast treatment strategies.

CONCLUSION

Antimicrobials inhibit or cure infectious diseases in a better way. The emergence of new and multidrug-resistant microbes challenges the identification of new-generation antimicrobials. Immunoinformatics has abundant scopes in the design and development of antimicrobials in emerging pandemic infectious diseases. In the upcoming era, this would be helpful to reduce the development time and cost of the vaccine, drug, or therapeutics. These immunoinformatics methodologies would be helpful to overcome future challenges in antimicrobial development. There is an abundant number of databases, data analysis software, and artificial neural networks for the identification, design, and development of antimicrobials. The antimicrobials were identified from plants to plasmids. Small and big molecules or peptides are predicted to an efficacious antimicrobials against various diseases. The antimicrobials were used for the treatment of viral, bacterial, helminths, protozoal and fungal diseases. A few viral complex diseases like HIV, Varicella, herpes zoster, Influenza, Human Parainfluenza, Respiratory Syncytial, herpes simplex, Hepatitis B and C are treated with new-generation antiviral drugs. The immunoinformatics methods predicted vaccines for various diseases like Helicobacter pylori, Avian Infectious laryngotracheitis virus, Retrovirus, human papillomavirus Influenza, human immunodeficiency virus, hepatitis B, and Candida albicans. Surface-localized adhesins molecules' ability to elicit an immune response against fungal infections screened through two-layer immunoinformatics methods to predict a vaccine. From big databases to machine learning techniques would be helpful to develop personalized antimicrobials for global pandemic diseases.

CONSENT FOR PUBLICATION

Not applicable.

CONFLICT OF INTEREST

Declared None

ACKNOWLEDGEMENTS

Declared none.

REFERENCES

[1] Poland GA, Whitaker JA, Poland CM, Ovsyannikova IG, Kennedy RB. Vaccinology in the third millennium: scientific and social challenges. Curr Opin Virol 2016; 17: 116-25. [http://dx.doi.org/10.1016/j.coviro.2016.03.003] [PMID: 27039875]

[2] Servín-Blanco R, Zamora-Alvarado R, Gevorkian G, Manoutcharian K. Antigenic variability: Obstacles on the road to vaccines against traditionally difficult targets. Hum Vaccin Immunother 2016; 12(10): 2640-8. [http://dx.doi.org/10.1080/21645515.2016.1191718] [PMID: 27295540]

[3] De Groot AS, Sbai H, Aubin CS, McMurry J, Martin W. Immuno□informatics: Mining genomes for vaccine components. Immunol Cell Biol 2002; 80(3): 255-69. [http://dx.doi.org/10.1046/j.1440-1711.2002.01092.x] [PMID: 12067413]

[4] Ali SA, Almofti YA, Abd-elrahman KA. Immunoinformatics approach for multiepitopes vaccine prediction against glycoprotein B of avian infectious laryngotracheitis virus 2019. [http://dx.doi.org/10.1155/2019/1270485]

[5] Peng M, Mo Y, Wang Y, *et al.* Neoantigen vaccine: an emerging tumor immunotherapy. Mol Cancer 2019; 18(1): 128. [http://dx.doi.org/10.1186/s12943-019-1055-6] [PMID: 31443694]

[6] Recker M, Buckee CO, Serazin A, *et al.* Antigenic variation in Plasmodium falciparum malaria involves a highly structured switching pattern. PLoS Pathog 2011; 7(3): e1001306. [http://dx.doi.org/10.1371/journal.ppat.1001306] [PMID: 21408201]

[7] Sabah SN, Gazi MA, Sthity RA, *et al.* Designing of epitope-focused vaccine by targeting E6 and E7 conserved protein sequences: An immuno-informatics approach in human papillomavirus 58 isolates. Interdiscip Sci 2018; 10(2): 251-60. [http://dx.doi.org/10.1007/s12539-016-0184-5] [PMID: 27640170]

[8] Onile OS, Awobode HO, Oladele VS, Agunloye AM, Anumudu CI. Detection of urinary tract pathology in some Schistosoma haematobium infected Nigerian adults. J Trop Med 2016; 2016: 1-5. [http://dx.doi.org/10.1155/2016/5405207] [PMID: 27635146]

[9] Zheng J, Lin X, Wang X, *et al. In silico* analysis of epitope-based vaccine candidates against hepatitis B virus polymerase protein. Viruses 2017; 9(5): 112. [http://dx.doi.org/10.3390/v9050112] [PMID: 28509875]

[10] Mottram L, Chakraborty S, Cox E, Fleckenstein J. How genomics can be used to understand host

susceptibility to enteric infection, aiding in the development of vaccines and immunotherapeutic interventions. Vaccine 2019; 37(34): 4805-10.
[http://dx.doi.org/10.1016/j.vaccine.2019.01.016] [PMID: 30709726]

[11] Martin W, Sbai H, De Groot AS. Bioinformatics tools for identifying class I-restricted epitopes. Methods 2003; 29(3): 289-98.
[http://dx.doi.org/10.1016/S1046-2023(02)00351-1] [PMID: 12725794]

[12] Onile OS, Fadahunsi AI, Adekunle AA, Oyeyemi BF, Anumudu CI. An immunoinformatics approach for the design of a multi-epitope subunit vaccine for urogenital schistosomiasis. PeerJ 2020; 8: e8795.
[http://dx.doi.org/10.7717/peerj.8795] [PMID: 33062404]

[13] Mhondoro M, Ndlovu N, Bangure D, *et al.* Trends in antimicrobial resistance of bacterial pathogens in Harare, Zimbabwe, 2012–2017: a secondary dataset analysis. BMC Infect Dis 2019; 19(1): 746.
[http://dx.doi.org/10.1186/s12879-019-4295-6] [PMID: 31455256]

[14] Rotman E, Seifert HS. Neisseria gonorrhoeae MutS affects pilin antigenic variation through mismatch correction and not by pilE guanine quartet binding. J Bacteriol 2015; 197(10): 1828-38.
[http://dx.doi.org/10.1128/JB.02594-14] [PMID: 25777677]

[15] Poudel I, Sah K, Subedi S, *et al.* Implementation of a practical and effective pilot intervention against transmission of Taenia solium by pigs in the Banke district of Nepal. PLoS Negl Trop Dis 2019; 13(2): e0006838.
[http://dx.doi.org/10.1371/journal.pntd.0006838] [PMID: 30802248]

[16] Lana DP, Beisel CE, Silva RF. Genetic mechanisms of antigenic variation in infectious bursal disease virus: Analysis of a naturally occurring variant virus. Virus Genes 1992; 6(3): 247-59.
[http://dx.doi.org/10.1007/BF01702563] [PMID: 1329340]

[17] Yewdell JW, Webster RG, Gerhard WU. Antigenic variation in three distinct determinants of an influenza type A haemagglutinin molecule. Nature 1979; 279(5710): 246-8.
[http://dx.doi.org/10.1038/279246a0] [PMID: 86955]

[18] Drabner B, Guzmán CA. Elicitation of predictable immune responses by using live bacterial vectors. Biomol Eng 2001; 17(3): 75-82.
[http://dx.doi.org/10.1016/S1389-0344(00)00072-1] [PMID: 11222981]

[19] Chaudhuri R, Ansari FA, Raghunandanan MV, Ramachandran S. FungalRV: adhesin prediction and immunoinformatics portal for human fungal pathogens. BMC Genomics 2011; 12(1): 192.
[http://dx.doi.org/10.1186/1471-2164-12-192] [PMID: 21496229]

[20] Oaks SC Jr, Shope RE, Lederberg J, Eds. Emerging infections: microbial threats to health in the United States. National Academies Press 1992.

[21] Schijns VE. Activation and programming of adaptive immune responses by vaccine adjuvants. Veterinary Sciences Tomorrow 2001.

[22] Sunita SA, Sajid A, Singh Y, Shukla P. Computational tools for modern vaccine development. Hum Vaccin Immunother 2020; 16(3): 723-35.
[http://dx.doi.org/10.1080/21645515.2019.1670035] [PMID: 31545127]

[23] Arshad MS, Batool SA. Natural antimicrobials, their sources, and food safety 2017.
[http://dx.doi.org/10.5772/intechopen.70197]

[24] Gyawali R, Ibrahim SA. Natural products as antimicrobial agents. Food Control 2014; 46: 412-29.
[http://dx.doi.org/10.1016/j.foodcont.2014.05.047]

[25] Lewis WH, Elvin-Lewis MP. Medicinal plants as sources of new therapeutics. Ann Mo Bot Gard 1995; 82(1): 16-24.
[http://dx.doi.org/10.2307/2399976]

[26] De Clercq E. Antiviral therapy for human immunodeficiency virus infections. Clin Microbiol Rev 1995; 8(2): 200-39.

[http://dx.doi.org/10.1128/CMR.8.2.200] [PMID: 7542558]

[27] Iwasa K, Moriyasu M, Yamori T, Turuo T, Lee DU, Wiegrebe W. *In vitro* cytotoxicity of the protoberberine-type alkaloids. J Nat Prod 2001; 64(7): 896-8.
[http://dx.doi.org/10.1021/np000554f] [PMID: 11473418]

[28] Boberek JM, Stach J, Good L. Genetic evidence for inhibition of bacterial division protein FtsZ by berberine. PLoS One 2010; 5(10): e13745.
[http://dx.doi.org/10.1371/journal.pone.0013745] [PMID: 21060782]

[29] Khan IA, Mirza ZM, Kumar A, Verma V, Qazi GN. Piperine, a phytochemical potentiator of ciprofloxacin against Staphylococcus aureus. Antimicrob Agents Chemother 2006; 50(2): 810-2.
[http://dx.doi.org/10.1128/AAC.50.2.810-812.2006] [PMID: 16436753]

[30] De Martino L, De Feo V, Fratianni F, Nazzaro F. Chemistry, antioxidant, antibacterial and antifungal activities of volatile oils and their components 2009.
[http://dx.doi.org/10.1177/1934578X0900401226]

[31] Park HOWON, Choi KD, Shin IS. Antimicrobial activity of isothiocyanates (ITCs) extracted from horseradish (Armoracia rusticana) root against oral microorganisms. Biocontrol Sci 2013; 18(3): 163-8.
[http://dx.doi.org/10.4265/bio.18.163] [PMID: 24077540]

[32] Fahey JW, Stephenson KK, Wade KL, Talalay P. Urease from Helicobacter pylori is inactivated by sulforaphane and other isothiocyanates. Biochem Biophys Res Commun 2013; 435(1): 1-7.
[http://dx.doi.org/10.1016/j.bbrc.2013.03.126] [PMID: 23583386]

[33] Gradišar H, Pristovšek P, Plaper A, Jerala R. Green tea catechins inhibit bacterial DNA gyrase by interaction with its ATP binding site. J Med Chem 2007; 50(2): 264-71.
[http://dx.doi.org/10.1021/jm060817o] [PMID: 17228868]

[34] Al-Majedy YK, Kadhum AAH, Al-Amiery AA, Mohamad AB. Coumarins: The Antimicrobial agents. Systematic Reviews in Pharmacy 2017; 8(1): 62-70.
[http://dx.doi.org/10.5530/srp.2017.1.11]

[35] Sardari S, Mori Y, Horita K, Micetich RG, Nishibe S, Daneshtalab M. Synthesis and antifungal activity of coumarins and angular furanocoumarins. Bioorg Med Chem 1999; 7(9): 1933-40.
[http://dx.doi.org/10.1016/S0968-0896(99)00138-8] [PMID: 10530942]

[36] Kim S, Kubec R, Musah RA. Antibacterial and antifungal activity of sulfur-containing compounds from Petiveria alliacea L. J Ethnopharmacol 2006; 104(1-2): 188-92.
[http://dx.doi.org/10.1016/j.jep.2005.08.072] [PMID: 16229980]

[37] Iranshahi M, Hassanzadeh-Khayat M, Bazzaz BSF, Sabeti Z, Enayati F. Fruits and antimicrobial activity of the oil. J Essent Oil Res 2008; 20(2): 183-5.
[http://dx.doi.org/10.1080/10412905.2008.9699986]

[38] Reiter J, Levina N, van der Linden M, Gruhlke M, Martin C, Slusarenko A. Diallylthiosulfinate (Allicin), a Volatile Antimicrobial from Garlic (Allium sativum), Kills Human Lung Pathogenic Bacteria, Including MDR Strains, as a Vapor. Molecules 2017; 22(10): 1711.
[http://dx.doi.org/10.3390/molecules22101711] [PMID: 29023413]

[39] Cai Y, Wang R, An MM, Liang BB, Fang Y. *In vitro* bactericidal activity of allicin combined with cefoperazone, tobramycin and ciprofloxacin. Int J Antimicrob Agents 2008; 31(2): 179-80.
[http://dx.doi.org/10.1016/j.ijantimicag.2007.10.009] [PMID: 18083012]

[40] Tyagi P, Singh M, Kumari H, Kumari A, Mukhopadhyay K. Bactericidal activity of curcumin I is associated with damaging of the bacterial membrane. PLoS One. 2015; 10(3): e0121313-e.

[41] Wu D, Kong Y, Han C, *et al.* d-Alanine:d-alanine ligase as a new target for the flavonoids quercetin and apigenin. Int J Antimicrob Agents 2008; 32(5): 421-6.
[http://dx.doi.org/10.1016/j.ijantimicag.2008.06.010] [PMID: 18774266]

[42] Hossain MU, Keya CA, Das KC, *et al.* An immunopharmacoinformatics approach in development of vaccine and drug candidates for West Nile virus. Front Chem 2018; 6: 246. [http://dx.doi.org/10.3389/fchem.2018.00246] [PMID: 30035107]

[43] Yeaman MR, Yount NY. Mechanisms of antimicrobial peptide action and resistance. Pharmacol Rev 2003; 55(1): 27-55. [http://dx.doi.org/10.1124/pr.55.1.2] [PMID: 12615953]

[44] Waghu FH, Gopi L, Barai RS, Ramteke P, Nizami B, Idicula-Thomas S. CAMP: Collection of sequences and structures of antimicrobial peptides. Nucleic Acids Res 2014; 42(D1): D1154-8. [http://dx.doi.org/10.1093/nar/gkt1157]

[45] Jhong JH, Chi YH, Li WC, Lin TH, Huang KY, Lee TY. dbAMP: an integrated resource for exploring antimicrobial peptides with functional activities and physicochemical properties on transcriptome and proteome data. Nucleic Acids Res 2019; 47(D1): D285-97. [http://dx.doi.org/10.1093/nar/gky1030] [PMID: 30380085]

[46] Whitmore L, Wallace BA. The Peptaibol Database: a database for sequences and structures of naturally occurring peptaibols. Nucleic Acids Res 2004; 32(90001) (Suppl. 1): 593D-4. [http://dx.doi.org/10.1093/nar/gkh077] [PMID: 14681489]

[47] Danishuddin M, Kaushal L, Hassan Baig M, Khan AU. AMDD: antimicrobial drug database. Genomics Proteomics Bioinformatics 2012; 10(6): 360-3. [http://dx.doi.org/10.1016/j.gpb.2012.04.002] [PMID: 23317704]

[48] Hammami R, Ben Hamida J, Vergoten G, Fliss I. PhytAMP: a database dedicated to antimicrobial plant peptides. Nucleic Acids Res 2009; 37(Database) (Suppl. 1): D963-8. [http://dx.doi.org/10.1093/nar/gkn655] [PMID: 18836196]

[49] McDonnell L, Armstrong D, Ashworth M, Dregan A, Malik U, White P. National disparities in the relationship between antimicrobial resistance and antimicrobial consumption in Europe: an observational study in 29 countries. J Antimicrob Chemother 2017; 72(11): 3199-204. [http://dx.doi.org/10.1093/jac/dkx248] [PMID: 28961862]

[50] McArthur AG, Waglechner N, Nizam F, *et al.* The comprehensive antibiotic resistance database. Antimicrob Agents Chemother 2013; 57(7): 3348-57. [http://dx.doi.org/10.1128/AAC.00419-13] [PMID: 23650175]

[51] Choyam S, Psn S, Pandey R, Kammara R. Ocins database: a database of bug-busters from *Bifidobacterium*, *Lactobacillus*, and *Enterococcus*. Access Microbiol 2019; 1(4): e000034. [http://dx.doi.org/10.1099/acmi.0.000034] [PMID: 32974523]

[52] Nagpal G, Usmani SS, Raghava GPS. A web resource for designing subunit vaccine against major pathogenic species of bacteria. Front Immunol 2018; 9: 2280. [http://dx.doi.org/10.3389/fimmu.2018.02280] [PMID: 30356876]

[53] Joshi A, Kaushik V. In-Silico Proteomic Exploratory Quest: Crafting T-Cell Epitope Vaccine Against Whipple's Disease. Int J Pept Res Ther 2020; 1. [PMID: 32427224]

[54] Wang P, Hu L, Liu G, *et al.* Prediction of antimicrobial peptides based on sequence alignment and feature selection methods. PLoS One 2011; 6(4): e18476. [http://dx.doi.org/10.1371/journal.pone.0018476] [PMID: 21533231]

[55] Altschul SF, Gish W, Miller W, Myers EW, Lipman DJ. Basic local alignment search tool. J Mol Biol 1990; 215(3): 403-10. [http://dx.doi.org/10.1016/S0022-2836(05)80360-2] [PMID: 2231712]

[56] Eddy SR. Profile hidden Markov models. Bioinformatics 1998; 14(9): 755-63. [http://dx.doi.org/10.1093/bioinformatics/14.9.755] [PMID: 9918945]

[57] Xiao X, Wu ZC, Chou KC. iLoc-Virus: A multi-label learning classifier for identifying the subcellular

localization of virus proteins with both single and multiple sites. J Theor Biol 2011; 284(1): 42-51. [http://dx.doi.org/10.1016/j.jtbi.2011.06.005] [PMID: 21684290]

[58] Robinson JA, DeMarco S, Gombert F, Moehle K, Obrecht D. The design, structures and therapeutic potential of protein epitope mimetics. Drug Discov Today 2008; 13(21-22): 944-51. [http://dx.doi.org/10.1016/j.drudis.2008.07.008] [PMID: 18725320]

[59] Robinson JA. β-hairpin peptidomimetics: design, structures and biological activities. Acc Chem Res 2008; 41(10): 1278-88. [http://dx.doi.org/10.1021/ar700259k] [PMID: 18412373]

[60] Pinner RW, Teutsch SM, Simonsen L, *et al.* Trends in infectious diseases mortality in the United States. JAMA 1996; 275(3): 189-93. [http://dx.doi.org/10.1001/jama.1996.03530270029027] [PMID: 8604170]

[61] Cassell G. Infectious causes of chronic inflammatory diseases and cancer. Emerg Infect Dis 1998; 4(3): 475-87. [http://dx.doi.org/10.3201/eid0403.980339] [PMID: 9716980]

[62] Pisani P, Parkin DM, Muñoz N, Ferlay J. Cancer and infection: estimates of the attributable fraction in 1990. Cancer Epidemiol Biomarkers Prev 1997; 6(6): 387-400. [PMID: 9184771]

[63] Akhtar N, Joshi A, Singh B, Kaushik V. Immuno-Informatics Quest against COVID-19/SARS-CV-2: Determining Putative T-Cell Epitopes for Vaccine Prediction. Infect Disord Drug Targets 2020. [http://dx.doi.org/10.2174/1871526520666200921154149] [PMID: 32957905]

[64] Joshi A, Sharma V, Singh J, Kaushik V. Chemi-Informatic approach to investigate putative pharmacoactive agents of plant origin to eradicate COVID-19 2020. [http://dx.doi.org/10.2174/2666796701999201203210036]

[65] Krishnan S, Joshi A, Kaushik V. T cell epitope designing for dengue peptide vaccine using docking and molecular simulation studies. Molecular Simulation 2020; 46(10): 787-95. [http://dx.doi.org/10.1080/08927022.2020.1772970]

[66] Krishnan S, Joshi A, Akhtar N, Kaushik V. Immunoinformatics designed T cell multi epitope dengue peptide vaccine derived from nonstructural proteome. Microbial Pathogenesis 2021; 150: 104728. [http://dx.doi.org/10.1016/j.micpath.2020.104728]

[67] Joshi A, Sunil Krishnan G, Kaushik V. Molecular docking and simulation investigation: effect of beta-sesquiphellandrene with ionic integration on SARS-CoV2 and SFTS viruses. J Genet Eng Biotechnol 2020; 18(1): 78. [http://dx.doi.org/10.1186/s43141-020-00095-x] [PMID: 33245459]

[68] Joshi A, Joshi BC, Mannan MA, Kaushik V. Epitope based vaccine prediction for SARS-COV-2 by deploying immuno-informatics approach. Informatics in Medicine Unlocked 2020; 19: 100338. [http://dx.doi.org/10.1016/j.imu.2020.100338] [PMID: 32352026]

[69] Kaushik V. *In silico* identification of epitope-based peptide vaccine for Nipah virus. Int J Pept Res Ther 2019; 1-7. [http://dx.doi.org/10.1007/s10989-019-09917-0]

SUBJECT INDEX

A

E

F

G

H

I

K

L

M

N

Made in the USA
Las Vegas, NV
15 November 2026

48bc6dbe-7b34-45aa-8742-4ee4b72f09b5R01